Friction Stir Spot Welding

Friction Stir Spot Welding offers an introduction to friction stir spot welding (FSSW) between both similar and dissimilar metals and materials. It explains the impact of the interlayer in FSSW of different metals with regard to mechanical, metallurgical, wear, thermo-mechanical, and chemical characteristics.

Emphasizing the impact of interlayer on FSSW of different metals, this book discusses the influence of the interlayer in the process as a new technique. Using aerospace and automotive structures as examples, this book explains how their components successfully employ materials like dissimilar aluminum alloys, yielding increased electrical, thermal, and mechanical characteristics. It also considers the reinforcement, effect of tool geometry, wettability, and corrosion behavior of joints.

This book is intended for mechanical, materials, and manufacturing professionals, researchers, and engineers working in the field of FSSW.

Emerging Materials and Technologies

Series Editor: Boris I. Kharissov

The *Emerging Materials and Technologies* series is devoted to highlighting publications centered on emerging advanced materials and novel technologies. Attention is paid to those newly discovered or applied materials with potential to solve pressing societal problems and improve quality of life, corresponding to environmental protection, medicine, communications, energy, transportation, advanced manufacturing, and related areas.

The series takes into account that, under present strong demands for energy, material, and cost savings, as well as heavy contamination problems and worldwide pandemic conditions, the area of emerging materials and related scalable technologies is a highly interdisciplinary field, with the need for researchers, professionals, and academics across the spectrum of engineering and technological disciplines. The main objective of this book series is to attract more attention to these materials and technologies and invite conversation among the international R&D community.

Nanomaterials for Sustainable Hydrogen Production and Storage
Edited by Jude A. Okolie, Emmanuel I. Epelle, Alivia Mukherjee, and
Alaa El Din Mahmoud

Calcium-Based Materials: Processing, Characterization, and Applications
Edited by S.S. Nanda, Jitendra Pal Singh, Sanjeev Gautam, and Dong Kee Yi

Advanced Synthesis and Medical Applications of Calcium Phosphates
Edited by S.S. Nanda, Jitendra Pal Singh, Sanjeev Gautam, and Dong Kee Yi

Non-Metallic Technical Textiles: Materials and Technologies
Mukesh Kumar Sinha and Ritu Pandey

Smart Micro- and Nanomaterials for Drug Delivery: Two-Volume Set
Edited by Ajit Behera, Arpan Kumar Nayak, Ranjan K. Mohapatra, and
Ali Ahmed Rabaan

Smart Micro- and Nanomaterials for Drug Delivery: Volume One
Edited by Ajit Behera, Arpan Kumar Nayak, Ranjan K. Mohapatra, and
Ali Ahmed Rabaan

Smart Micro- and Nanomaterials for Pharmaceutical Applications
Edited by Ajit Behera, Arpan Kumar Nayak, Ranjan K. Mohapatra, and
Ali Ahmed Rabaan

Friction Stir Spot Welding
Edited by Jeyaprakash Natarajan and K. Anton Savio Lewise

For more information about this series, please visit: www.routledge.com/
Emerging-Materials-and-Technologies/book-series/CRCEMT

Friction Stir Spot Welding
Metallurgical, Mechanical and Tribological Properties

Edited by
Jeyaprakash Natarajan and
K. Anton Savio Lewise

CRC Press
Taylor & Francis Group
Boca Raton London New York

CRC Press is an imprint of the
Taylor & Francis Group, an **informa** business

Cover image: Max Khoo/Shutterstock

First edition published 2025
by CRC Press
2385 NW Executive Center Drive, Suite 320, Boca Raton FL 33431

and by CRC Press
4 Park Square, Milton Park, Abingdon, Oxon, OX14 4RN

CRC Press is an imprint of Taylor & Francis Group, LLC

ISBN: 978-1-032-55800-4 (hbk)
ISBN: 978-1-032-55802-8 (pbk)
ISBN: 978-1-003-43228-9 (ebk)

DOI: 10.1201/9781003432289

Typeset in Times
by codeMantra

Contents

List of Contributors

V. Anandakrishnan
Department of Production Engineering
National Institute of Technology,
   Tiruchirappalli
Tiruchirappalli, India

K. Ananthakumar
Department of Mechanical Engineering
Karpagam College of Engineering
Coimbatore, India

T. Arunnellaiappan
Department of Mechanical Engineering
SRM TRP Engineering College
Irungalur, India

M. Baskaran
Department of Mechatronics
   Engineering
K.S. Rangasamy College of Technology
Tiruchengode, India

Lokasani Bhanuprakash
Department of Mechanical Engineering
MLR Institute of Technology
Hyderabad, India

Sampath Boopathi
Department of Mechanical Engineering
Muthayammal Engineering College
Namakkal, India

Balakrishnan Deepanraj
Department of Mechanical Engineering,
   College of Engineering
Prince Mohammad Bin Fahd University
Al Khobar, Saudi Arabia

Arindam Dhar
Department of Metallurgy and
   Materials Engineering
Indian Institute of Engineering Science
   and Technology
Shibpur, India

J. Edwin Raja Dhas
Department of Automobile Engineering
Noorul Islam Centre for Higher
   Education
Kumaracoil, India

M. Satyanarayana Gupta
Department of Aeronautical
   Engineering
MLR Institute of Technology
Hyderabad, India

S. Hari Prasadh
Department of Mechatronics
   Engineering
K.S. Rangasamy College of Technology
Tiruchengode, India

S. Jayalakshmi
School of Mechanical and Electrical
   Engineering
Wenzhou University
Wenzhou, China

N. Jayanth
Centre of Excellence in Product Design
   and Smart Manufacturing
Maulana Azad National Institute of
   Technology
Bhopal, India-462003

N. Jeyaprakash
School of Mechanical and Electrical
Engineering
China University of Mining and
Technology
Xuzhou, China

Sundara Subramanian Karuppasamy
Graduate Institute of Manufacturing
Technology
National Taipei University of
Technology
Taipei, Taiwan

S. Kirubanidhi Jebabalan
Department of Industrial Engineering
University of Trento
Trento, Italy

Ishita Koley
Department of Metallurgy and
Materials Engineering
Indian Institute of Engineering Science
and Technology
Shibpur, India

Avinash Kumar
Department of Metallurgy and
Materials Engineering
Indian Institute of Engineering Science
and Technology
Shibpur, India

Harikishor Kumar
Department of Mechanical Engineering
MLR Institute of Technology
Hyderabad, India

Parshant Kumar
School of Mechanical Engineering
Dr. Vishwanath Karad MIT World
Peace University
Pune, India

V. Lakshmanan
Department of Mechanical Engineering
SRM TRP Engineering College
Tiruchirappalli, India

K. Anton Savio Lewise
Division of Aerospace Engineering
Karunya Institute of Technology &
Sciences
Coimbatore, India

Karuppiah Madhan Muthu Ganesh
Department of Aeronautical
Engineering
MAMSE
Trichy, India

M. Neelavannan
Department of Mechanical Engineering
Dr. Mahalingam College of Engineering
and Technology
Pollachi, India

R. Pandiyarajan
Department of Mechanical Engineering
Agni College of Technology
Chennai, India

P. Parthiban
Department of Production Engineering
National Institute of Technology,
Tiruchirappalli
Tiruchirappalli, India

G. S. Paul Raj
Department of Mechanical Engineering
SRM TRP Engineering College
Tiruchirappalli, India

S. Ponsuriyaprakash
Department of Mechanical Engineering
SSM Institute of Engineering and
Technology
Dindigul, Tamil Nadu

G. Prabu
Graduate Institute of Manufacturing
  Technology
National Taipei University
  of Technology
Taipei, Taiwan.

M. Prabu
Department of Mechanical Engineering
K.S. Rangasamy College of Engineering
Tiruchengode, India

Rabindra Prasad
Department of Mechanical Engineering,
  Amity University,
Gwalior,
Madhya Pradesh, India

Durairaj Raj Kumar
Department of Mechanical Engineering
Sri Ramakrishna College Of Engineering
Perambalur, India-621113

S. Sabarish
Department of Mechanical Engineering
K.L.N. College of Engineering
Madurai, India

J. B. Sajin
Department of Mechanical Engineering
Sree Buddha College of Engineering
Alappuzha, Kerala

Sajivarghese
Department of Mechanical Engineering
Sree Buddha College of Engineering
Alappuzha, Kerala

M. Sanjay
Department of Mechatronics
  Engineering
K.S. Rangasamy College of Technology
Tiruchengode, India

M. Satheesh
Department of Mechanical Engineering
Bahrain Training Institute
Bahrain, Bahrain

S. Sathish
Department of Production Technology,
  Madras Institute of Technology
Anna University, Chromepet,
  Chennai 600044
Tamil Nadu, India

M. Satthiyaraju
Department of Mechanical Engineering
Kathir College of Engineering
Coimbatore, India

Milon D. Selvam
Department of Mechanical Engineering
Kampala International University
Kampala, Uganda

M. S. Senthil Saravanan
Department of Mechanical Engineering
Sree Buddha College of Engineering
Alappuzha, Kerala

S. Senthilraja
Department of Mechatronics
  Engineering
SRM Institute of Science and
  Technology
Chennai, India

R. Arvind Singh
School of Mechanical and Electrical
  Engineering
Wenzhou University
Wenzhou, China

P. Srinivasan
Department of Production Engineering
National Institute of Technology,
  Tiruchirappalli
Tiruchirappalli, India

Paramasivam Thamizhvalavan
Department of Mechanical Engineering,
    Saveetha school of Engineering,
Saveetha Institute of Medical and
    Technical Sciences,
Saveetha University
Chennai, India

Trijo Tharayil
Department of Mechanical Engineering
Sree Buddha College of Engineering
Kerala, Chennai

N. Tiruvenkadam
Department of Mechatronics
    Engineering
K.S. Rangasamy College of Technology
Tiruchengode, India

Kannaian Vijayan
Department of Mechanical Engineering
MAM School of Engineering,
Trichy, India

A. Vivek Anand
Department of Aeronautical
    Engineering
MLR Institute of Technology
Hyderabad, India

Che-Hua Yang
Graduate Institute of Manufacturing
    Technology
National Taipei University of
    Technology
Taipei, Taiwan

J. Yoganandh
Department of Mechanical Engineering
Sri Ramakrishna Engineering College
Coimbatore, India

Jaligama Yogesh
Department of Aeronautical
    Engineering
MLR Institute of Technology
Hyderabad, India

About the Editors

Jeyaprakash Natarajan currently working as an Associate Professor in School of Mechanical and Electrical Engineering, China University of Mining and Technology, Xuzhou City, China. He received his bachelor's degree in Mechanical Engineering from RVS College of Engineering and Technology, Dindigul, India, in 2006. Dr. Jeyaprakash Natarajan received his master's degree in Manufacturing Engineering in 2009 from Government College of Technology, Coimbatore, India, and his PhD degree in Production Engineering from National Institute of Technology, Tiruchirappalli, India in 2018. In 2007, he was appointed to the position of Lecturer at the Department of Automobile Engineering, Thanthai Roever Institute of Polytechnic College, Perambalur, India. In 2009, he was appointed as an Assistant Professor in the Department of Mechanical Engineering at Periyar Maniammai University, Tanjore, India. In 2011, he was appointed to the position of Assistant Professor in the Department of Engineering, Hamelmalo Agricultural College, Keren, Eritrea, North East Africa. In 2013, he was appointed as an Assistant Professor in the Department of Production Engineering, Defence University, Debre Zeit, Ethiopia, Africa. From September 2018 to December 2018, he was a Research Assistant in Additive Manufacturing Center for Mass Customization Production, National Taipei University of Technology, Taiwan. From 2019 to 2020, he was a Postdoctoral Research Associate in Additive Manufacturing Center for Mass Customization Production, Graduate Institute of Manufacturing Technology, National Taipei University of Technology, Taiwan. From January 2021 to July 2022, he was promoted as a Research Assistant Professor in Graduate Institute of Manufacturing Technology, National Taipei University of Technology, Taiwan. Dr. Jeyaprakash Natarajan has published more than 75 peer-reviewed research articles, book chapters, and two books titled *Laser Surface Treatments for Tribological Application* and *Advances in Additive Manufacturing Processes* in 2021. He teaches various postgraduate courses in automotive, mechanical, manufacturing, and laser ultrasound and conducts academic research and consultancy. He is serving as a reviewer for many reputed international journals. Also, he has completed four research projects in the fields of additive manufacturing, surface engineering, tribology, laser cladding, corrosion, etc.

Dr. K. Anton Savio Lewise received his bachelor's degree in Aeronautical Engineering from Noorul Islam College of Engineering, Kumaracoil, Tamilnadu, India, in 2009. He received his master's degree in Aeronautical Engineering in 2012 from Anna University Regional Campus, Tirunelveli, India, and received his PhD degree in Aeronautical Engineering in 2022 from Noorul Islam Centre for Higher Education, Kumaracoil, Tamilnadu, India. In 2009, he was appointed to the position of Lecturer at the Department of Aeronautical Engineering, Tamizhan College of Engineering & Technology, Nagercoil, India. In 2012, he was appointed as an Assistant Professor in the Department of Aeronautical Engineering, Tamizhan College of Engineering & Technology, Nagercoil, India. In 2013, he was appointed as an Assistant Professor in the Department of Aeronautical Engineering, Defence University, Debre Zeit,

Ethiopia, Africa. In 2015, he was appointed as an Assistant Professor in Noorul Islam Centre for Higher Education. In 2022, he was appointed as an Assistant Professor in the Division of Aerospace Engineering, Karunya Institute of Technology and Sciences, Coimbatore, India. He has published more than 25 peer-reviewed research articles, book chapters, patents, and conference proceedings. He teaches various undergraduate and postgraduate courses in aeronautical, aerospace, mechanical, and manufacturing and conducts academic research and consultancy. Dr. K. Anton Savio Lewise is serving as a reviewer for many reputed international journals.

Preface

Friction stir spot welding (FSSW) for similar and different materials was developed as a result of the changing application, which called for the creation of new methods of combining similar and dissimilar materials. Numerous publications have been written about FSSW of comparable and different materials, but the influence of the interlayer in the process has not yet received the same attention as a new technique. As a result, the emphasis of this work is on reporting the impact of interlayer on FSSW of different metals. An introduction to FSSW between different metals is given in Chapter 1.

Many interlayer types connected to the FSSW are described in Chapter 2. Chapter 3 presents the FSSW process parameter optimization. Chapters 4 and 5 explore the types of tool geometries and interlayers being used in the FSSW process. Manufacturing different structures and pieces for the industrial sector requires joining materials that are distinct from one another. Aerospace and automotive structures and components often employ materials like dissimilar aluminum alloys. This is due to their exceptional performance, which includes increased electrical, thermal, and mechanical characteristics. Chapters 6–12 discuss in detail the impact of the interlayer in FSSW of different metals with regard to mechanical, metallurgical, wear, thermo-mechanical, residual stress, and chemical characteristics. Additionally, Chapter 13 discusses the impact of nanopowder on the FSSW of different metals.

Additionally, Chapters 14 and 15 elaborate on the challenges and advanced techniques associated with the FSSW of different metals to produce high-quality welds for a variety of applications. Furthermore, Chapters 16–18 deal with the effect of FSSW tool geometry and how it influences the overall performance of the welding process. The goal of this book is to demonstrate the impact of interlayers on FSSW so that engineers and researchers may examine how these two approaches can be developed further.

Acknowledgments

The editors would like to thank the China University of Mining and Technology (CUMT), Xuzhou City - 221116, China and Karunya Institute of Technology and Sciences (KITS), Coimbatore - 641114, India, for the valuable support and extended facility to carry out this book project in a successful manner. In addition to that, the editors are grateful to Ms. Sonia Tam, Editorial Manager, Taylor and Francis, for her valuable guidance and facilitation. The enthusiasm and thoughtfulness of CUMT/ KITS faculties, students, staffs, and community partners over the years have inspired us to make this book resourceful for other academic colleagues. At last, the editors extend the heartfelt gratitude to family members who were the real inspiration and motivation behind this work.

1 Introduction

*N. Jeyaprakash, Sundara Subramanian
Karuppasamy, and Che-Hua Yang*

1.1 INTRODUCTION

The demands of the industrial revolution have uplifted the usage of metals and alloys in every sector for their technological development. The raw metals and alloys undergo certain fabrication strategies where they can be converted into desired components which can be utilized in aerospace, automobiles, civil, nuclear, electronics, and other industrial sectors. One of the strategies that has evolved with the mankind is the welding technology [1–3]. The welding technique has been reported as one of the oldest fabrication processes that have existed for over a millennium for manufacturing the components. This technique fabricates the components by means of depositing the metals or alloys over other metals. Over the evolution era, welding is defined as one of the fabrication processes where metal or metallic alloys are joined together in order to facilitate mechanical behaviors. Other than the fabrication of new components, the welding technology is used as the treatment method for repairing the worn-out surfaces of complex components, thereby improving their serviceability and eliminating the need for new components [4–6].

In welding, the weld bead is formed and serves as the contact layer between the metals. The terminologies used in welding are shown in Figure 1.1. Here, the faying surface is reported as the contact surface of the two base metals where the weld bead is formed to provide sufficient joint strength [7,8]. The two major categories of the

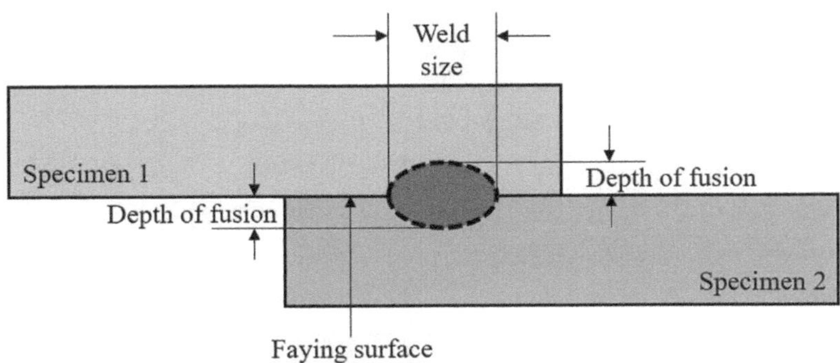

FIGURE 1.1 Terminologies used in the welding process, where the dark grey color represents the weld bead formed at the faying site.

DOI: 10.1201/9781003432289-1

"""

welding process are (i) fusion welding and (ii) solid-state welding (SSW). The former process demands the fusion of the parent materials at the faying site, whereas the latter process does not require the fusion of the faying surface. In both categories, a weld bead is formed, which provides a permanent joint between the parent materials.

In fusion welding, the faying sites are melted which are reported to form a weld bead at the faying site. The melting (fusion) process takes place in the presence of an external heat source, and the filler material is optional in this process if, in this case, the filler material is also fused to form the weld bead. This type needs a heat source to melt the parent/filler material, and a wide range of heat sources, such as electric arcs, high energy beams, and gases, can be adapted for the melting process [9–11]. The major limitations of fusion welding are as follows: (i) the influence of heat-affected zone is unavoidable; (ii) very high level of distortion; (iii) the mechanical characteristics offered by the parent materials are greatly affected due to the intense heating; and (iv) joining dissimilar metals is not suitable because they have various melting points which will affect the weld bead formation, etc. [12,13].

The above-mentioned drawbacks could be replaced by implementing the SSW process. This process will create a permanent joint either by the combined effect of heat and pressure or by means of external pressure. If heat is required, the parent materials are heated below their melting point, thereby avoiding fusion, as in the fusion welding process. This kind has huge advantages over fusion welding, such as (i) minimal influence of heat-affected zone, (ii) no need for filler material, (iii) low level of distortion, (iv) the mechanical behaviors of the parent materials remain unaltered, and (v) it holds good for joining dissimilar materials [14–16]. Depending on the requirements, service environments, cost, and service life, the worker can decide which type of welding process could be suitable to meet the demands. Figure 1.2 shows the classification of the welding process.

Welding	
Fusion	Solid state
1. Arc welding (SMAW, GMAW, TIG, SAW, FCAW, ESW)	1. Cold welding
2. Gas welding (AAW, OAW, OHW, PGW)	2. Roll welding
3. Resistance welding (RSW, RSEW, PW, PEW, FW)	3. Pressure welding
4. Intense energy beam welding (PAW, EBW, LBW)	4. Diffusion welding
	5. Friction welding
	6. Friction stir welding
	7. Forge welding

FIGURE 1.2 Welding classification.

In recent years, components made up of heavy metals in most industrial sectors have been replaced by lightweight metals/alloys, and the demand for new welding methods keeps on rising for creating a permanent joint between light metals such as aluminum. The traditional (fusion) welding techniques lack in joining light/dissimilar metals due to intense heating and high level of distortion. Hence, for lightweight applications, the SSW process is more suitable to create a permanent joint between the materials. A variety of techniques, such as laser spot welding (LSW), ultrasonic welding (UW), self-piercing riveting, and friction stir spot welding (FSSW), can be adapted for the joining process [17–20]. Mostly spot welds are preferred where strength-to-weight ratio is the major key aspect considering the aerospace, automobile, and marine sectors. On comparing the pros and cons of these techniques, most researchers preferred the FSSW technique for producing spot welds in lightweight materials. This chapter gives an introduction (overview) about the FSSW process along with its terminologies.

1.2   FRICTION STIR SPOT WELDING

As discussed earlier, the FSSW process is a type of friction stir welding (FSW) process which can create permanent weld beads in the parent materials. This process does not require any external heat, and the weld bead is formed with the help of external pressure. It is reported as a type of FSW with minimal influence of the heat-affected zone [21–23]. This technique has numerous advantages considering the environmental, metallurgical, and energy aspects as follows:

i. Distortion level: This method does not require any intense heating, which results in a low level of distortion compared to the other welding processes.
ii. Fine microstructure: It can create a fine recrystallized microstructure with the absence of solidification cracking and fewer or no pores.
iii. Dimensional stability: High-quality weld beads were formed with better dimensional stability and repeatability.
iv. No harmful emissions: Since this process does not involve the use of any electric arcs or gases, it is considered as an eco-friendly technique to produce weld beads.
v. Excellent mechanical behavior of the weld beads.
vi. Eliminates the need for degreasing solvents.

The working principle behind the formation of spot welds is schematically depicted in Figure 1.3. Weld bead is formed via four steps, namely, rotation, plunging, dwelling (stirring), and unloading. This process starts with the rotation of the FSSW tool with a threaded pin. The parent materials which undergo welding are arranged in an overlapping fashion. The FSSW tool starts to rotate at a specific speed which is directed onto the parent materials for plunging the tool. While plunging, the FSSW tool heats the overlapping material via friction. This friction softens the overlapping surface, and the threaded pin promotes the plastic flow of materials. Under high compressive pressure, the tool is moved further into the overlapping sheets. After reaching a desired level, the tool is reported to rotate for a predefined time called dwell time, where the parent material expands plastically and forms a weld bead.

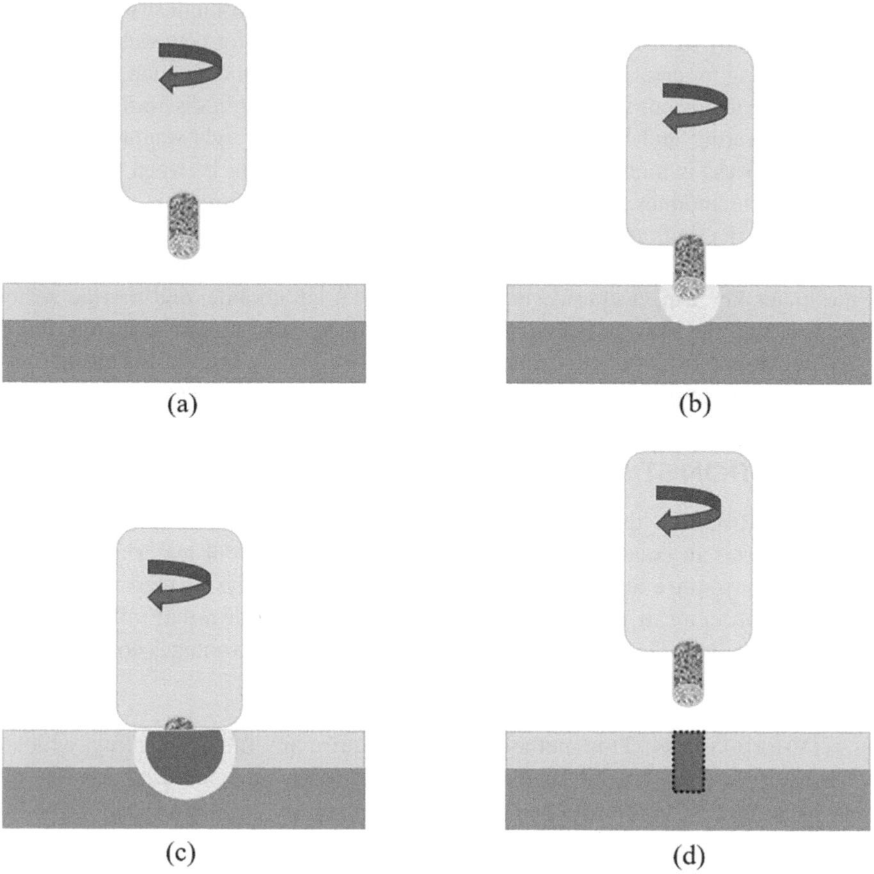

FIGURE 1.3   FSSW working: (a) rotation, (b) plunging, (c) stirring, and (d) unloading.

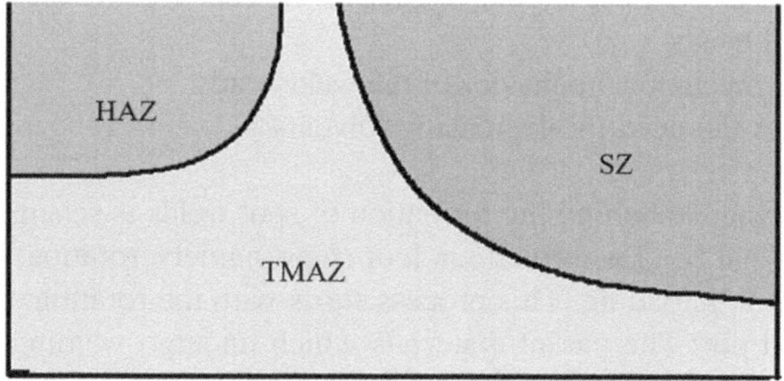

FIGURE 1.4   Zones in the spot welds of FSSW. HAZ, heat-affected zone; TMAZ, thermo-mechanically affected zone; and SZ, stir zone.

In the end, the FSSW tool is drawn out, leaving the plastic deformation. This concludes the FSSW process [24,25]. Figure 1.4 represents the schematic view of the zones in the FSSW process.

1.3 PROCESS CONTROLS IN FSSW

The spot welds can be made via two modes: (i) force control mode and (ii) displacement control mode.

1.3.1 FORCE CONTROL MODE

Here, the factors that are kept constant (kept under control) are the rpm of the rotating tool, the force exerted normal to the faying surface, and the dwell time [26–28]. The schematic representation of the force control mode is provided in Figure 1.5. The operation of the FSSW process under this mode is as follows:

i. The rotating tool is plunged against the faying surface, where the tool's pin is plunged into the test specimen at a constant rpm.
ii. After attaining a predefined level, the force is said to be constant, whereas the FSSW tool keeps on plunging the sample, which will pave the way for the formation of the weld bead at the faying site.
iii. The FSSW tool is retracted when the predetermined dwell time is reached.

1.3.2 DISPLACEMENT CONTROL MODE

The working mechanism of the FSSW process under this mode depends on the effective controlling of the rotating tool's rpm and displacement [29–31]. Figure 1.6 shows the FSSW process operating under this mode. Here, the faying surface experiences a maximum insertion of the rotating tool at a predefined rpm. Then, the tool is retracted, and the force (N) exerted against the faying surface (output) is recorded. The normal force value is less at the initial stage (the pin starts to plunge the sample) and reaches its maximum value when the shoulder of the FSSW tool contacts

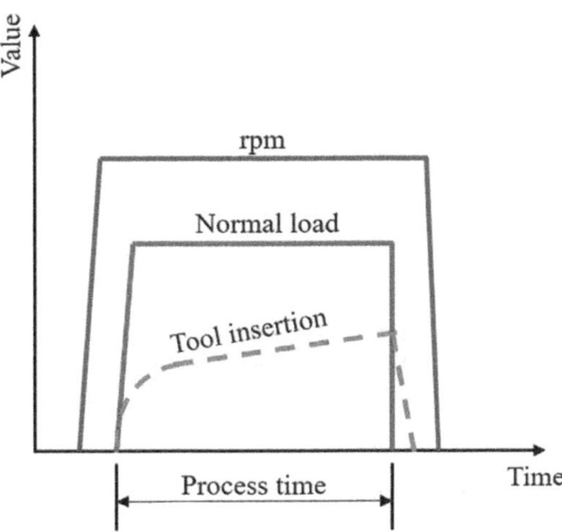

FIGURE 1.5 Force control mode.

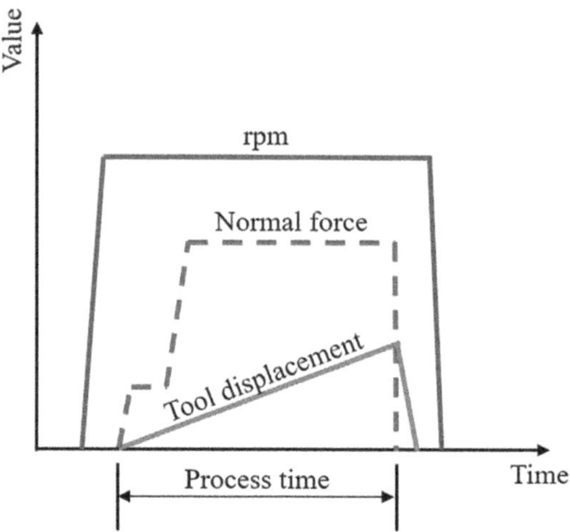

FIGURE 1.6   Displacement control mode.

the faying surface. Both control modes are reported to provide efficient bond (joint) strength between the samples.

In order to produce the FSSW joint via the above-mentioned modes, the samples are arranged in two different configurations, namely, the lap joint and butt joint configurations. In most cases, the lap joint configuration is preferred, where the test specimens are arranged one above the other. The butt joint holds good for the same thickness test specimens, and the weld bead is formed at the seam zone.

1.4   PROCESS PARAMETERS IN FSSW PROCESS

The process parameters are said to be optimized for efficient weld bead formation at the faying site. The major process parameters that are reported to influence the weld bead formation are as follows:

i. **FSSW Rotating Tool's Speed**

   The rpm of the rotating tool is considered as an important criterion that determines the strength of the joint via frictional heat in the spot welding process. This heat determines both the macro- and microstructures of the weld bead, thereby enhancing the mechanical properties such as the joint (bond) strength [32]. The optimization of the tool's speed depends upon the tool's geometry and the type of alloy/metal used in the FSSW process. In general, the mechanical characteristics offered by the weld bead formed at a lower speed are reported to be better than those of the weld bead formed at a greater speed [33,34].

ii. **Dwell Time**

   Another important factor that influences the weld bead is the dwell time. This dwell time determines the intensity of heat formation on the faying surface. During this time, the weld bead is said to be formed as the faying site experiences plasticity [35,36]. Higher dwell time results in a greater amount of weld bead formation, which improves the bond strength. On the other hand, FSSW joints with higher dwell times have yielded a huge amount of intermetallic compounds at the weld bead, which will pave the way for a reduction in the mechanical behaviors of the weld due to the wear and corrosion processes [37,38].

iii. **Plunge Rate**

The plunge rate is defined as the rate of displacement of the plunging tool into the workpiece. On comparing the above-mentioned parameters, the influence of the plunge rate in the FSSW process is comparatively less. In the FSSW process, the higher plunge rate results in greater frictional heat [39]. A greater plunge rate results in the maximum amount of heat generation, which will pave the way for an uneven microstructure with a greater amount of intermetallic compound formation. On the other hand, a low plunge rate will result in improper weld bead formation, which affects the structural integrity of the weld bead [40].

iv. **Tool Pin Height**

As mentioned earlier, the pin is an important criterion in the FSSW process. It generates the required amount of heat, which will promote the melting process at the faying site. During the dwell time, this pin allows the plastic material to flow between the workpiece and forms the weld bead at the faying site [41]. Hence, there is a need for an appropriate height of the pin which promotes material flow. The weld bead formed at a higher pin height is said to provide enhanced tensile and shear strength when compared with the small pin height [42].

v. **Tool Shoulder Diameter**

In the FSSW process, the tool's shoulder diameter plays a vital role in monitoring the quality of the weld bead (spot weld) formed on the test specimen. The tool's shoulder diameter determines the following criteria:

    i. Amount of heat generated and distributed during the plunging and stirring process.

    ii. Required plunging rate (load) to achieve the desired plunging depth.

    iii. Adequate amount of plasticized material flows in the stir zone.

    iv. Microstructural evolution in the spot welds.

    v. Intermetallic compound formation.

    vi. Mechanical characteristics offered by the spot welds.

Moreover, the tool's diameter will serve as an important factor in the case of the pinless FSSW process [43]. In this process, the adequate diameter determines the structural integrity of the weld bead (spot welds). In general, the increased tool diameter will pave the way for enhanced heat generation and distribution in the stir zone, followed by a tremendous improvement in the mechanical characteristics of the spot welds [44].

The above-mentioned parameters need to be optimized in order to have better structural integrity in the spot weld joints. Several literature discussions based on the process parameters are as follows. Tozaki et al. [45] analyzed the rpm of the tool and the dwell time for producing spot weld joints in aluminum alloys. Their study revealed that better spot welds can be obtained at low tool speed, with increased dwell time. At a lower rotational speed, the generated heat tends to be uniform for melting the specimen, and this speed, along with the higher dwell time, results in a better microstructure of the weld bead, followed by improved mechanical behaviors. On the other hand, shorter dwell times at low speed resulted in uneven material flow because of the improper plasticization of the material, and the weld bead formed from this type shows poor structural integrity. Zhang et al. [46] investigated the influence of the tool's rotating speed while producing spot welds in 5052 aluminum alloy. They reported that the increased rotational speed has paved the way for the enhanced growth of grains. This growth rate minimized the effect of the nugget zone. From the hardness distribution, the weld bead formed at low speed has a lower hardness, whereas the hardness increases with higher rotational speed for the 5052 alloy. Also, this higher rotational speed accounts for the dynamic recrystallization of grains, which improved their hardness. Karthikeyan et al. [47] proposed a model for optimizing the process parameters while performing spot welds in the AA2024 alloy. The proposed model works on the response surface approach, considering the plunge rate, tool's speed, dwell time, and tool displacement. Based on the obtained results, they concluded that among the mentioned parameters, the plunge rate provides a greater contribution to the mechanical behaviors exhibited by the spot weld than the other parameters. Furthermore, they suggested that the increased tool displacement at a greater plunge rate offered a better blend of materials during plasticized material flow, followed by enhancing the weld bead properties. Song et al. [48] examined the influence of the tool's plunging (both shoulder and pin) rate and analyzed the morphology and mechanical behaviors of the spot weld made on the AA 6061 alloy. They concluded that the shoulder plunging rate has a significant influence on grain formation when compared with the pin plunging rate, and greater shoulder plunging rate values lead to uneven mixing of material during plasticization [49]. Lee et al. [50] and Wu et al. [51] investigated the bonding strength via grain and hardness distribution of the spot weld produced at greater dwell times. Both studies reported that the increased dwell time resulted in the finer grains, followed by the adequate segregation of the secondary precipitates at the weld zone. This fine distribution of grains and precipitates paved the way for improved bonding strength at the spot weld zone. Furthermore, they suggested that better mechanical behaviors were observed after the heat treatment process. Ojo et al. [52] proposed a model based on the Taguchi approach for selecting the suitable parameters to perform the spot weld via the FSSW process. They observed that the tool's speed had a greater influence on the tensile strength, whereas the tool displacement provided a major contribution to the bond size of the spot welds. Thus, from this literature, it is evident that the optimization of process parameters will depend on the alloy type, working conditions, and so on.

1.5 FSSW TOOL

It is reported that the tool's geometry has a major contribution toward the optimization of the process parameters. Depending on the working condition and the material of the specimen, various types of tool geometry have been proposed. Some of the factors that need to be considered regarding the selection of the FSSW tool are as follows:

i. **Pin Geometry**

In the FSSW process, the material flow during plasticization is promoted by the pin. This pin provides the necessary heat for melting and ensures joint strength [53]. Several researchers worked on the different types of pin geometry and proposed six types that are suitable to carry out the FSSW process, as shown in Figure 1.7.

ii. **Pin Length**

The length of the pin used in the FSSW process acts as the deciding factor for the bonding strength (tensile and shear strength) of the spot welds. Moreover, this length is directly proportional to the thickness of the specimen. For greater thickness, a higher pin length would be the optimal solution for creating better spot weld joints [54].

iii. **Shoulder Geometry**

Another key factor that has a greater contribution to the FSSW process is the shoulder geometry. The shoulder performs three major functions,

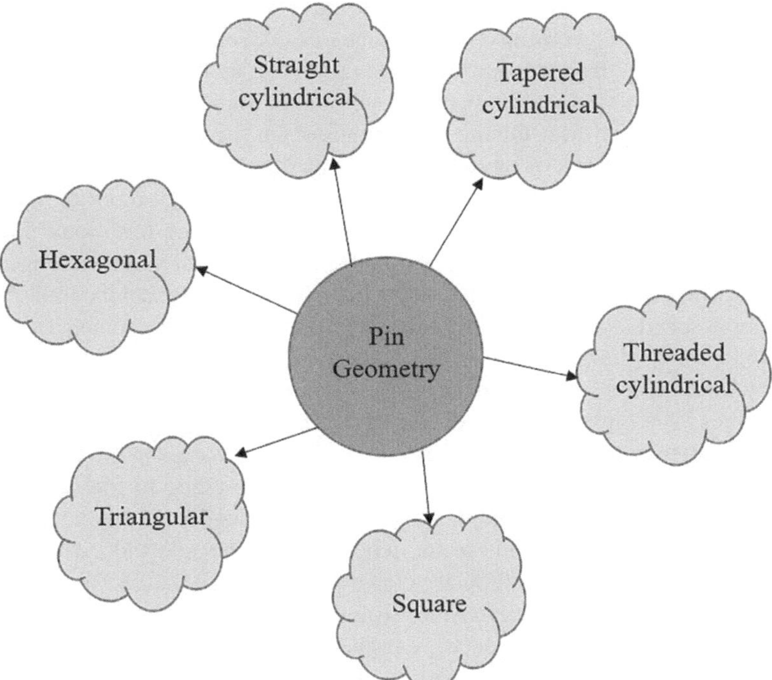

FIGURE 1.7 FSSW – pin geometry types.

as follows: (i) controls the heat generated during the plunging and stirring process; (ii) ensures the necessary load (force) is delivered for creating the spot welds; and (iii) holds the plasticized material flow in the stir zone. The shoulder geometry can be of any shape out of three forms, namely, concave, convex, and flat. Among these three shapes, the concave geometry is reported to have better mechanical strength than the other two shapes [55].

Some of the literature based on the above-mentioned factors is discussed as follows: Badrinarayan et al. [56,57] addressed the hook morphology developed during the FSSW process via triangular and cylindrical pins. While using the triangular pin, they noticed that the obtained hook morphology was discontinuous at the weld zone. The discontinuous hook morphology confirmed the enhanced material flow, which in turn leads to higher plasticization at the weld zone. Moreover, they reported that this hook morphology produced a higher bonding strength during the FSSW process. On the other hand, the morphology obtained from the cylindrical pin tends to be continuous. This pin showed increased material flow in the upward and outward directions, which resulted in a poor spot weld. Furthermore, the weld zone made by the cylindrical pin contained three interzones namely fully bonded, partially bonded, and the zone left unbonded. Only the fully bonded zone has sufficient recrystallization of grains compared to the other zones. The other two interzones, where the material bonding is not stable, would have resulted in poor weld joints. Yuan et al. [58] examined the influence of conventional and off-centered tools along with the tool's rpm while producing spot joints. At lower rotational speeds, both tools offered continuous hook formation due to the uneven material flow during plasticization. On the other hand, the tensile and shear strengths of the weld bead formed by both tools are reported to improve at a higher rotational speed. In particular, they reported that the off-centered tool has better bonding strength than the traditional FSSW tool. Paidar et al. [59] examined the pin geometrical effects along with other process parameters such as shoulder plunging rate and the tool's rotational speed. Among the various combinations of the pin geometries, tool's speed, and plunging rate, better mechanical characteristics were offered by the higher levels of process parameters when compared to their lower values. The higher values resulted in the discontinuous hook morphology, and better grain growth was observed in the weld zone. On the other hand, the low-level values show poor bonding behavior due to the continuous hook morphology at the faying site. Thus, they suggested that a high level of parameter values would result in better strength spot welds, irrespective of the pin geometries [60]. Garg et al. [61] performed the FSSW process in the AA 6061 by optimizing the various process parameters and pin profiles via the Taguchi approach. At higher dwell times and greater rotational speed, the spot welds obtained by the squared pin have better bonding and joint strength when compared with the cylindrical and triangular pin spot welds. The squared pin induces highly intense heat at the faying site when compared with the other two pins at higher process parameter values. Zhou et al. [62] investigated the spot welds obtained using the plain pin and threaded pin tools, along with the parameters. They concluded that the threaded pin promotes the flow of material more than the plain pin tool at the faying site. Also, more intense heat generation was reported while performing the FSSW process with the threaded pin tool. Thus, from the above literature, it is evident that the researchers used various tool geometry in order to produce the spot welds using the FSSW process.

1.6   HEAT GENERATION IN FSSW PROCESS

In the FSW process, the heat generation can be easily modeled since the modeling can be done with steady-state conditions. During modeling, convective heat transfer can be neglected in the governing equations. On the other hand, the FSSW process takes place under non-steady-state conditions, as shown in Figure 1.8. The heat generation analysis can be carried out by solving the governing equations based on the Euler and Langragian approach [63].

Zhang et al. [64] investigated the spot welds made on the AA 6061 alloy by means of FEM simulations. With their boundary conditions, they predicted the interlayer geometries, strain distribution pattern, and load-bearing properties and evaluated their predictions with the experimental results. Saleem et al. [65] proposed a FEM model based on combining the Euler and Lagrangian approaches for investigating the spot weld zone of the AA 6082 alloy. With their model, the process parameter was optimized for its effect on the temperature distribution. Urso et al. [66] examined the mechanical behaviors of the weld zone created in the AA 6060 sheet via a thermo-mechanical simulation. Their study concluded that there is a reduction in the mechanical properties for higher feed rates, whereas better joint strength was reported at the spot welds obtained from lower values. Moreover, the temperature distribution in the weld zone along with the welding forces was modeled, and based on the results, the joint resistance was predicted.

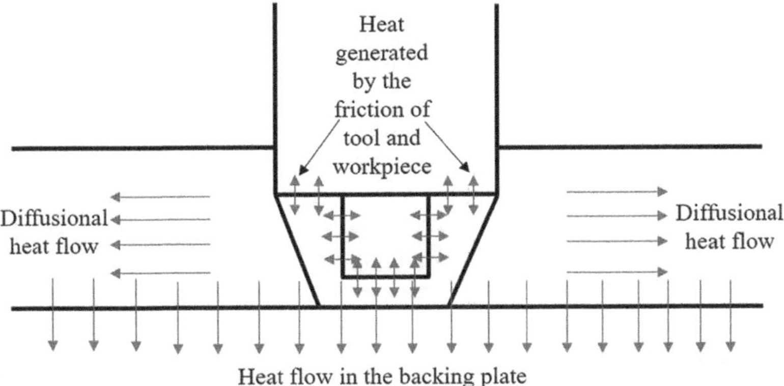

FIGURE 1.8   Schematic representation of heat flow during the FSSW process.

1.7 FSSW IN DISSIMILAR METALS

One of the key characteristics of the FSSW process over conventional techniques is that it can create high-quality weld joints between dissimilar materials. Some of the dissimilar spot joints are discussed below:

i. **FSSW between Al and Mg Alloys**

Suhddin et al. [67] analyzed the spot weld joint made by combining AA 5754 with AZ 31 (Mg alloy). They reported that the weld zone has fine grains that are rich in Al-Mg compounds. The hardness behaviors were improved in the weld zone due to the secondary precipitates, and compared with the similar joints (Al/Al or Mg/Mg), the weld zone made by the Al/Mg dissimilar combination showed notable improvement in the mechanical characteristics such as fatigue life, hardness, and wear resistance. Chowdhury et al. [68,69] combined the AZ31B-H24 (Mg alloy) with the Al alloy. Their studies also confirmed that there is a drastic improvement in the microhardness profile of the weld bead due to the segregation of Al-Mg-rich intermetallic compounds. Choi et al. [70] optimized the process parameters for joining the AZ 31 alloy with the 6K21 Al alloy. By varying the process parameters, a significant weld bead was formed at higher values of the tool's rpm and dwell time. These higher values resulted in the formation of spot welds with better hardness and tensile strength.

ii. **FSSW between Al and Steel**

This is an unusual combination where the formed weld bead exhibits increased structural integrity. Sun et al. [71] combined the 6061 Al alloy and mild steel for producing spot welds and reported that the weld zone has a high tensile failure load. Moreover, while joining, the pin length does not have a significant influence on the formed weld bead. Bozzi et al. [72] examined the spot weld made by combining the AA 6016 alloy with galvanized steel. They reported that the weld zone has greater joint strength with excellent hardness properties.

iii. **FSSW between Al and Cu**

Ozdemir et al. [73] optimized the process parameters to perform spot welding by combining the Al and Cu alloys. At higher rotational speeds, the Cu was reported to melt in an efficient manner, thereby producing finer grains, which are reported to be rich in the Al-Cu matrix. The higher rotational speed with a higher dwell time results in a greater stirring of Al and Cu with dynamic recrystallization, as reported. Heideman et al. [74] combined the AA 6061 alloy with an oxygen-free Cu alloy by varying the process parameters. Discontinuous hook morphology was observed, which confirmed the existence of high-quality spot welds with excellent mechanical behaviors. Moreover, this Al-Cu combination resulted in spot welds with better wear resistance characteristics.

iv. **Polymeric FSSW**

In recent years, polymers have replaced traditional metal parts in the automobile industry due to their high specific strength and low weight.

Because of the tremendous advancements in the polymer industry, the demand for polymers and the joining techniques kept on increasing. Among the various joining techniques, the FSSW process is reported to produce high-quality welds between various polymers and polymer–metal combinations. Lambiase et al. [75,76] analyzed the mechanical characteristics, such as fracture modes and tensile strength, offered by the spot welds created on the polycarbonate sheets. Their study concluded that the plunging depth has a greater influence during plasticization, and a higher plunging depth results in better mechanical characteristics of the spot welds. Saeid et al. [77] investigated the fracture modes while joining PMMA and ABS sheets. By combining a higher tool's rpm, a low plunge rate, and a greater dwell time, efficient spot welds were produced between these sheets, which tend to have greater lap shear strength at the weld zone. Ankit et al. [78] examined the weld bead morphology and the effect of process parameters by performing FSSW on Cu with PMMA. Fine grains were recrystallized at the weld interface at the higher values of the tool's rpm and pin length. By means of elemental mapping, a higher accumulation of polymers was observed at the weld interface, which paved the way for enhancing the tensile-shear strength. Moreover, the polymers can be used as an interface layer while joining metals with non-metals [79].

1.8 DEFECTS AND FRACTURE MODES IN FSSW

The FSSW spot welds do not contain any immunity against defects. Several defects and fracture modes were reported in the FSSW process as follows:

i. **Hook Defect**

Although the FSSW process creates high-quality spot welds when compared to traditional spot welding methods such as resistance spot welding, the spot joints (weld beads) produced by the FSSW process are prone to various structural defects. One of those structural defects that will degrade the mechanical characteristics offered by the spot welds is the hook defect. This type of defect denotes the partial bonding between the base metals. In the lap joint configuration, during plunging and stirring, lack of plunging rate (depth or load) is the main cause of this defect. Moreover, other parameters such as lower dwell time, insufficient rpm of the FSSW's tool, pin's height, etc. also serve as factors that contribute to the hook formation. Usually, the hook may be continuous or discontinuous. A continuous hook results in debonding of the weld joint, whereas a discontinuous hook accounts for higher strength. The hook is prone to several corrosion attacks, which in turn reduce the structural integrity of the weld zone [80].

ii. **Top Sheet Thinning**

This type of defect occurs mostly in the pinless FSSW process and is sometimes reported in the normal FSSW process. When the shoulder is mechanically pressed at a high plunge rate with a selected plunge depth, the top sheet thins [81].

iii. **Weld Flash**

Weld flash denotes the excess material that comes out and settles on the top sheet during the plasticization process. The improper removal of the weld flash results in the weakening of the weld zone. Also, this flash could be eliminated by selecting the appropriate tool under nominal parametric values.

iv. **Root Flaw**

As mentioned, the FSSW process could take place via lap joint and butt joint configurations. Root flaws are associated with the butt joint FSSW process, which induces cracks in the weld zone.

These types of defects could be analyzed via several non-destructive techniques, such as eddy current testing, ultrasonic testing, and radiography images.

Under excessive loading, the FSSW joints experience three fracture modes, such as shear mode, nugget pull-out mode, and mixed mode [82].

1.9 FSSW VARIANTS

After the FSSW process, the spot weld zone contains keyhole-like artifacts. This artifact could serve as a prone zone for many defects, such as wear and corrosion. To overcome this, there are three major variants, namely, pinless FSSW, refill FSSW, and swing FSSW. The pinless FSSW process is similar to conventional FSSW, but it creates spot welds with the pinless tool. The tool's shoulder heats the workpiece, and during the dwell period, plasticization occurs without keyhole formation [83].

1.9.1 Refill FSSW Process

One of the key issues in the conventional FSSW process is the formation of the keyhole. The keyhole is an unavoidable artifact that is created due to the plunging of the tool's pin into the specimen. After prolonged usage, this keyhole will serve as the potential site for promoting defect formation (cracks and wear) in the spot welds. Moreover, the upward movement of the plasticized material flow will leave the weld debris around the keyhole, which acts as the interfacial site and initiates the corrosion, thereby degrading the structural integrity of the spot welds. To overcome this, researchers introduced a new variant of the FSSW process called the refill FSSW process. This process consists of a clamp, a shoulder, and a pin. These three parts work in an independent manner. The clamp will hold the workpiece. During plunging, the shoulder retracts, leaving the pin alone to plunge the workpiece. This retraction will create a reservoir for the weld bead. After plunging, the shoulder moves forward as the pin is retracted. This forward movement will fill the keyhole with excess material [84].

1.9.2 Swing FSSW Process

In the case of swing FSSW, a larger weld area is created due to the slight transverse rotation (swing) of the tool. The spot welds contain a keyhole, but it can be overcome by adapting a pinless tool to the swing FSSW process. On comparing the joint strength, the spot welds made by the swing FSSW process account for higher joint strength than the other variants [85]. In the swing FSSW process, the spot welds are created by the following three steps (procedure), which are explained as follows:

i. To perform the swing FSSW process, first the conventional FSSW process needs to be carried out on the test specimens, which includes rotation, plunging, stirring, and unloading. After the spot weld is made, the tool is moved a certain distance outward from the keyhole, which is termed as offset distance.

ii. Considering this offset distance as the radius, the tool is reported to rotate around the keyhole with the offset radius and form a trajectory path around the original keyhole.

iii. Finally, the tool is retracted, leaving behind the larger weld area (trajectory path), which enhances the bonding strength of the created weld. Thus, the swing FSSW process is reported to produce high-quality spot welds with greater bonding strength when compared with the conventional FSSW process.

1.10 FIXTURES IN FSSW PROCESS

Fixtures play a major role during the formation of weld beads in any type of welding process. Fixtures are defined as structures that are used to hold the specimen rigidly in order to perform the welding. In conventional welding techniques, C-clamps are mostly preferred as the fixtures for producing the weld beads. But in the FSSW process, the spot welds are created on the thin sheets of the base metals, more effort is needed while choosing the type of fixture. In general, the fixture should perform the following functions in order to produce high-quality spot welds [86,87].

i. Anti-vibration: Since the FSSW process involves the rotation of the tool during plunging and stirring, the fixtures should arrest the vibration in order to avoid the distortion phenomenon.

ii. Heat dissipation: A good fixture design should be capable of dissipating the heat from the parent materials to produce spot joints with better structural integrity.

iii. Strength: The fixture should have adequate mechanical strength in order to withstand the rotational and plunging loads without deforming the workpiece.

Based on these factors, the fixtures are chosen for performing the FSSW process.

1.11 CONCLUSION

The FSSW process can create high-quality spot welds compared to other conventional welding techniques. With less energy consumption and zero harmful emissions, spot welds are made via the plasticization of materials. Although FSSW is a variant of the FSW process, it produces highly efficient welds in terms of joint strength, hardness, and wear resistance due to the dynamic recrystallization of the weld bead. The weld bead contains finer grain growth followed by adequate segregation of the secondary precipitates, which paved the way for enhanced mechanical behaviors of the weld bead. This chapter gives an overview of the FSSW process. Both the force control and displacement control modes of operation were discussed. The influence of process parameters, tool geometry, heat generation, FSSW variants, and FSSW of dissimilar metals, was elaborated on with relevant literature.

CONFLICT OF INTEREST

The authors declare that there are no conflicts of interest.

ACKNOWLEDGMENT

The authors would like to acknowledge the editors of this book for providing the opportunity to write this chapter.

REFERENCES

1. Kovacevic R, editor. *Welding Processes.* BoD-Books on Demand; 2012.
2. Davim JP, editor. *Welding Technology.* Springer International Publishing; 2021.
3. Ahmed N, editor. *New Developments in Advanced Welding.* Elsevier; 2005 .
4. Karuppasamy SS, Jeyaprakash N, Yang CH. Microstructure, nanoindentation and corrosion behavior of colmonoy-5 deposition on SS410 substrate using laser cladding process. *Arabian Journal for Science and Engineering.* 2022:1–7. doi:10.1007/s13369-021-06398-6
5. Messler Jr RW. *Principles of Welding: Processes, Physics, Chemistry, and Metallurgy.* John Wiley & Sons; 2008.
6. Olson DL, Dixon RD, Liby AL, editors. *Welding: Theory and Practice.* Elsevier; 2012.
7. Kim JW, Murugan SP, Yoo JH, Ashiri R, Park YD. Enhancing nugget size and weldable current range of ultra-high-strength steel using multi-pulse resistance spot welding. *Science and Technology of Welding and Joining.* 2020;25(3):235–42.
8. Radhakrishnan VM. *Welding Technology and Design.* New Age International; 2005.
9. Jeyaprakash N, Yang CH, Karuppasamy SS, Dhineshkumar SR. Evaluation of microstructure, nanoindentation and corrosion behavior of laser cladded Stellite-6 alloy on Inconel-625 substrate. *Materials Today Communications.* 2022;31:103370.
10. DebRoy T, David SA. Physical processes in fusion welding. *Reviews of Modern Physics.* 1995;67(1):85.
11. Tümer M, Schneider-Bröskamp C, Enzinger N. Fusion welding of ultra-high strength structural steels-A review. *Journal of Manufacturing Processes.* 2022;82:203–29.
12. Natarajan J, Yang CH, Karuppasamy SS. Investigation on microstructure, nanohardness and corrosion response of laser cladded colmonoy-6 particles on 316l steel substrate. *Materials.* 2021;14(20):6183.
13. Deyev GF. *Surface Phenomena in Fusion Welding Processes.* CRc Press; 2005.
14. Sharma Y, Mehta A, Vasudev H, Jeyaprakash N, Prashar G, Prakash C. Analysis of friction stir welds using numerical modelling approach: A comprehensive review. *International Journal on Interactive Design and Manufacturing (IJIDeM).* 2023:1–4.

15. Miranda RM, Gandra JP, Vilaca P, Quintino L, Santos TG. *Surface Modification by Solid State Processing*. Woodhead Publishing; 2013.

16. Jeyaprakash N, Yang CH, Karuppasamy SS, Rajendran DK. Correlation of microstructural with corrosion behaviour of Ti-6Al-4V specimens developed through selective laser melting technique. *Proceedings of the Institution of Mechanical Engineers, Part E: Journal of Process Mechanical Engineering*. 2022;236(5):2240–51.

17. Trivedi, A., S. Bag, and A. De. Three-dimensional transient heat conduction and thermomechanical analysis for laser spot welding using adaptive heat source. Science and Technology of Welding and Joining. 2007;12(1): 24–31.

18. Benatar A, Marcus M. Ultrasonic welding of plastics and polymeric composites. In: *Power Ultrasonics* edited by Juan A. Gallego-Juárez , Karl F. Graff and Margaret Lucas (pp. 205–225). Woodhead Publishing; 2023.

19. Li D, Chrysanthou A, Patel I, Williams G. Self-piercing riveting-a review. *The International Journal of Advanced Manufacturing Technology*. 2017;92:1777–824.

20. Jeyaprakash N, Yang CH, Susila P, Karuppasamy SS. Laser cladding of NiCrMoFeNbTa particles on Inconel 625 alloy: Microstructure and corrosion resistance. *Transactions of the Indian Institute of Metals*. 2023;76(2):599–612.

21. Ikumapayi OM, Akinlabi ET. Recent advances in keyhole defects repairs via refilling friction stir spot welding. *Materials Today: Proceedings*. 2019;18:2201–8.

22. Mehta A, Vasudev H, Jeyaprakash N. Role of sustainable manufacturing approach: Microwave processing of materials. *International Journal on Interactive Design and Manufacturing (IJIDeM)*. 2023:1–7.

23. Feng XS, Li SB, Tang LN, Wang HM. Refill friction stir spot welding of similar and dissimilar alloys: A review. *Acta Metallurgica Sinica (English Letters)*. 2020;33:30–42.

24. Jeyaprakash N, Yang CH, Karuppasamy SS, Duraiselvam M. Stellite 6 cladding on AISI Type 316L stainless steel: Microstructure, nanohardness and corrosion resistance. *Transactions of the Indian Institute of Metals*. 2023;76(2):491–503.

25. Shen Z, Ding Y, Gerlich AP. Advances in friction stir spot welding. *Critical Reviews in Solid State and Materials Sciences*. 2020;45(6):457–534.

26. Sakano R. Development of spot FSW robot system for automobile body members. *In: Proceedings of the 3rd International Symposium of Friction Stir Welding*, Kobe, Japan, 2004.

27. Kumagai M, Tanaka K. Joint properties of friction stir spot welding in a 6000 series aluminum alloy. In: *The 102nd Conference of Japan Institute of Light Metals*, Sapporo; Hokkaido, Japan Paper 2002 (Vol. 124, pp. 17–19).

28. Murugesan P, Satheeshkumar V, Jeyaprakash N, Yang CH, Karuppasamy SS. Effect of α-Al and Si precipitates on microstructural evaluation and corrosion behavior of laser powder bed fusion printed AlSi10Mg plates in seawater environment. *Metals and Materials International*. 2023;29(9) :2515–2532.

29. Fujimoto M. Development of friction spot joining (Report 1)-cross sectional structures of friction spot joints. *Preprints of the National Meeting of Japan Welding Society*. 2004;74:4–5.

30. Fujimoto M. Development of friction spot joining (Report 2)-mechanical properties of friction spot joints. *Preprints of the National Meeting of Japan Welding Society*. 2004;74:6–7.

31. Jeyaprakash N, Yang CH, Karuppasamy SS. Laser cladding of Colmonoy 6 particles on Inconel 625 substrate: Microstructure and corrosion resistance. *Surface Review and Letters*. 2022;29(8):2250102.

32. Tozaki Y, Uematsu Y, Tokaji K. A newly developed tool without probe for friction stir spot welding and its performance. *Journal of Materials Processing Technology*. 2010;210(6–7):844–51.

33. Awang M, Mucino VH. Energy generation during friction stir spot welding (FSSW) of Al 6061-T6 plates. *Materials and Manufacturing Processes.* 2010;25(1–3):167–74.

34. Gerlich A, Avramovic-Cingara G, North TH. Stir zone microstructure and strain rate during Al 7075-T6 friction stir spot welding. *Metallurgical and Materials Transactions A.* 2006;37:2773–86.

35. Farmanbar N, Mousavizade SM, Ezatpour HR. Achieving special mechanical properties with considering dwell time of AA5052 sheets welded by a simple novel friction stir spot welding. *Marine Structures.* 2019;65:197–214.

36. Bagheri B, Alizadeh M, Mirsalehi SE, Shamsipur A, Abdollahzadeh A. The effect of rotational speed and dwell time on Al/SiC/Cu composite made by friction stir spot welding. *Welding in the World.* 2022;66(11):2333–50.

37. Jeyaprakash N, Karuppasamy SS, Yang CH. Application of wear-resistant laser claddings. In: *Handbook of Laser-Based Sustainable Surface Modification and Manufacturing Techniques* edited by Hitesh Vasudev, Chander Prakash (pp. 1–26). CRC Press; 2023.

38. Bilici MK, Yukler AI. Effects of welding parameters on friction stir spot welding of high density polyethylene sheets. *Materials & Design.* 2012;33:545–50.

39. Chu Q, Li WY, Yang XW, Shen JJ, Vairis A, Feng WY, Wang WB. Microstructure and mechanical optimization of probeless friction stir spot welded joint of an Al-Li alloy. *Journal of Materials Science & Technology.* 2018;34(10):1739–46.

40. Karthikeyan R. Establishing relationship between optimised friction stir spot welding process parameters and strength of aluminium alloys. *Advances in Materials and Processing Technologies.* 2022;8(1):1173–95.

41. Tashkandi MA, Al-Jarrah JA, Ibrahim M. Spot welding of 6061 Aluminum alloy by friction stir spot welding process. *Engineering, Technology & Applied Science Research.* 2017;7(3):1629–32.

42. Buffa G, Fanelli P, Fratini L, Vivio F. Influence of joint geometry on micro and macro mechanical properties of friction stir spot welded joints. *Procedia Engineering.* 2014;81:2086–91.

43. Gerlich A, Su P, North TH. Peak temperatures and microstructures in aluminium and magnesium alloy friction stir spot welds. *Science and Technology of Welding and Joining.* 2005;10(6):647–52.

44. Jeyaprakash N, Yang CH, Sivasankaran S. Laser cladding process of cobalt and nickel based hard-micron-layers on 316L-stainless-steel-substrate. *Materials and Manufacturing Processes.* 2020;35(2):142–51.

45. Tozaki Y, Uematsu Y, Tokaji K. Effect of processing parameters on static strength of dissimilar friction stir spot welds between different aluminium alloys. *Fatigue & Fracture of Engineering Materials & Structures.* 2007;30(2):143–8.

46. Zhang Z, Yang X, Zhang J, Zhou G, Xu X, Zou B. Effect of welding parameters on microstructure and mechanical properties of friction stir spot welded 5052 aluminum alloy. *Materials & Design.* 2011;32(8–9):4461–70.

47. Karthikeyan R, Balasubramanian V. Predictions of the optimized friction stir spot welding process parameters for joining AA2024 aluminum alloy using RSM. *The International Journal of Advanced Manufacturing Technology.* 2010;51:173–83.

48. Song X, Ke L, Xing L, Liu F, Huang C. Effect of plunge speeds on hook geometries and mechanical properties in friction stir spot welding of A6061-T6 sheets. *The International Journal of Advanced Manufacturing Technology.* 2014;71:2003–10.

49. Karuppasamy SS, Jeyaprakash N, Yang CH. Application of corrosion-resistant laser claddings. *Handbook of Laser-Based Sustainable Surface Modification and Manufacturing Techniques.* 2023;51(51):27.

50. Lee SH, Lee DM, Lee KS. Process optimisation and microstructural evolution of friction stir spot-welded Al6061 joints. *Materials Science and Technology.* 2017;33(6):719–30.

51. Wu D, Shen J, Lv L, Wen L, Xie X. Effects of nano-SiC particles on the FSSW welded AZ31 magnesium alloy joints. *Materials Science and Technology.* 2017;33(8):998–1003.

52. Ojo OO, Taban E. Hybrid multi-response optimization of friction stir spot welds: Failure load, effective bonded size and flash volume as responses. *Sādhanā.* 2018;43(6):98.

53. Ilman MN. Microstructure and mechanical properties of friction stir spot welded AA5052-H112 aluminum alloy. *Heliyon.* 2021;7(2):e06009.

54. Jeyaprakash N, Yang CH, Ramkumar KR. Microstructure and wear resistance of laser cladded Inconel 625 and colmonoy 6 depositions on inconel 625 substrate. *Applied Physics A.* 2020;126:1–1.

55. Fanelli P, Vivio F, Vullo V. Experimental and numerical characterization of friction stir spot welded joints. *Engineering Fracture Mechanics.* 2012;81:17–25.

56. Badarinarayan H, Yang Q, Zhu S. Effect of tool geometry on static strength of friction stir spot-welded aluminum alloy. *International Journal of Machine Tools and Manufacture.* 2009;49(2):142–8.

57. Badarinarayan H, Shi Y, Li X, Okamoto K. Effect of tool geometry on hook formation and static strength of friction stir spot welded aluminum 5754-O sheets. *International Journal of Machine Tools and Manufacture.* 2009;49(11):814–23.

58. Yuan W, Mishra RS, Webb S, Chen YL, Carlson B, Herling DR, Grant GJ. Effect of tool design and process parameters on properties of Al alloy 6016 friction stir spot welds. *Journal of Materials Processing Technology.* 2011;211(6):972–7.

59. Paidar M, Khodabandeh A, Sarab ML, Taheri M. Effect of welding parameters (plunge depths of shoulder, pin geometry, and tool rotational speed) on the failure mode and stir zone characteristics of friction stir spot welded aluminum 2024-T3 sheets. *Journal of Mechanical Science and Technology.* 2015;29:4639–44.

60. Kim JR, Ahn EY, Das H, Jeong YH, Hong ST, Miles M, Lee KJ. Effect of tool geometry and process parameters on mechanical properties of friction stir spot welded dissimilar aluminum alloys. *International Journal of Precision Engineering and Manufacturing.* 2017;18:445–52.

61. Garg A, Bhattacharya A. On lap shear strength of friction stir spot welded AA6061 alloy. *Journal of Manufacturing Processes.* 2017;26:203–15.

62. Zhou L, Zhang RX, Li GH, Zhou WL, Huang YX, Song XG. Effect of pin profile on microstructure and mechanical properties of friction stir spot welded Al-Cu dissimilar metals. *Journal of Manufacturing Processes.* 2018;36:1–9.

63. Khandkar MZ, Khan JA, Reynolds AP. Prediction of temperature distribution and thermal history during friction stir welding: Input torque based model. *Science and Technology of Welding and Joining.* 2003;8(3):165–74.

64. Zhang B, Chen X, Pan K, Li M, Wang J. Thermo-mechanical simulation using microstructure-based modeling of friction stir spot welded AA 6061-T6. *Journal of Manufacturing Processes.* 2019;37:71–81.

65. Khosa SU, Weinberger T, Enzinger N. Thermo-mechanical investigations during friction stir spot welding (FSSW) of AA6082-T6. *Welding in the World.* 2010;54:R134–46.

66. D'Urso G. Thermo-mechanical characterization of friction stir spot welded AA6060 sheets: Experimental and FEM analysis. *Journal of Manufacturing Processes.* 2015;17:108–19.

67. Suhuddin UF, Fischer V, Dos Santos JF. The thermal cycle during the dissimilar friction spot welding of aluminum and magnesium alloy. *Scripta Materialia.* 2013;68(1):87–90.

68. Chowdhury SH, Chen DL, Bhole SD, Cao X, Wanjara P. Lap shear strength and fatigue life of friction stir spot welded AZ31 magnesium and 5754 aluminum alloys. *Materials Science and Engineering: A.* 2012 ;556:500–9.

69. Chowdhury SH, Chen DL, Bhole SD, Cao X, Wanjara P. Lap shear strength and fatigue behavior of friction stir spot welded dissimilar magnesium-to-aluminum joints with adhesive. *Materials Science and Engineering: A.* 2013;562:53–60.

70. Choi DH, Ahn BW, Lee CY, Yeon YM, Song K, Jung SB. Formation of intermetallic compounds in Al and Mg alloy interface during friction stir spot welding. *Intermetallics.* 2011;19(2):125–30.

71. Sun YF, Fujii H, Takaki N, Okitsu Y. Microstructure and mechanical properties of dissimilar Al alloy/steel joints prepared by a flat spot friction stir welding technique. *Materials & Design.* 2013;47:350–7.

72. Bozzi S, Helbert-Etter AL, Baudin T, Criqui B, Kerbiguet JG. Intermetallic compounds in Al 6016/IF-steel friction stir spot welds. *Materials Science and Engineering: A.* 2010; 527(16-17):4505–9.

73. Özdemir U, Sayer S, Yeni Ç. Effect of pin penetration depth on the mechanical properties of friction stir spot welded aluminum and copper. *Materials Testing.* 2012;54(4):233–9.

74. Heideman R, Johnson C, Kou S. Metallurgical analysis of Al/Cu friction stir spot welding. *Science and Technology of Welding and Joining.* 2010;15(7):597–604.

75. Lambiase F, Paoletti A, Di Ilio A. Effect of tool geometry on mechanical behavior of friction stir spot welds of polycarbonate sheets. *The International Journal of Advanced Manufacturing Technology.* 2017;88:3005–16.

76. Lambiase F, Paoletti A, Di Ilio A. Effect of tool geometry on loads developing in friction stir spot welds of polycarbonate sheets. *The International Journal of Advanced Manufacturing Technology.* 2016;87:2293–303.

77. Dashatan SH, Azdast T, Ahmadi SR, Bagheri A. Friction stir spot welding of dissimilar polymethyl methacrylate and acrylonitrile butadiene styrene sheets. *Materials & Design.* 2013;45:135–41.

78. Pandey AK, Nayak KC, Mahapatra SS. Characterization of friction stir spot welding between copper and poly-methyl-methacrylate (PMMA) sheet. *Materials Today Communications.* 2019;19:131–9.

79. Xie Y, Huang Y, Meng X, Li J, Cao J. Friction stir spot welding of aluminum and wood with polymer intermediate layers. *Construction and Building Materials.* 2020; 240:117952.

80. Xu RZ, Ni DR, Yang Q, Liu CZ, Ma ZY. Influencing mechanism of Zn interlayer addition on hook defects of friction stir spot welded Mg-Al-Zn alloy joints. *Materials & Design.* 2015;69:163–9.

81. Jeyaprakash N, Yang CH, Kumar DR. Machinability study on CFRP composite using Taguchi based grey relational analysis. *Materials Today: Proceedings.* 2020;21:1425–31.

82. Tozaki Y, Uematsu Y, Tokaji K. Effect of tool geometry on microstructure and static strength in friction stir spot welded aluminium alloys. *International Journal of Machine Tools and Manufacture.* 2007;47(15):2230–6.

83. Abdollahzadeh A, Bagheri B, Vaneghi AH, Shamsipur A, Mirsalehi SE. Advances in simulation and experimental study on intermetallic formation and thermomechanical evolution of Al-Cu composite with Zn interlayer: Effect of spot pass and shoulder diameter during the pinless friction stir spot welding process. *Proceedings of the Institution of Mechanical Engineers, Part L: Journal of Materials: Design and Applications.* 2023;237(6):1475–94.

84. Liu Z, Fan Z, Liu L, Miao S, Lin Z, Wang C, Zhao Y, Xin R, Dong C. Refill friction stir spot welding of AZ31 magnesium alloy sheets: Metallurgical features, microstructure, texture and mechanical properties. *Journal of Materials Research and Technology.* 2023;23:3337–50.

85. Wu S, Sun T, Shen Y, Yan Y, Ni R, Liu W. Conventional and swing friction stir spot welding of aluminum alloy to magnesium alloy. *The International Journal of Advanced Manufacturing Technology.* 2021;116:2401–12.

86. Lewise K, Dhas J, Pandiyarajan R. Design, fabrication and analysis of linear clamping fixture for friction stir spot welding. *AIP Conference Proceedings.* 2023;2492:1.

87. Wang DA, Chen CH. Fatigue lives of friction stir spot welds in aluminum 6061-T6 sheets. *Journal of Materials Processing Technology.* 2009;209(1):367–75.

2 Interlayers in FSSW Process
An Overview

N. Jeyaprakash, Sundara Subramanian
Karuppasamy, and Che-Hua Yang

2.1 INTRODUCTION

The need for various joining methods has evolved along with the human era. The joining process is a kind of fabrication process where the components are made by assembling two or more materials. In this process, permanent joints have been made between the parent materials as per the service demand. Based on the working environment and type of materials, the joining process can be carried out in different ways, such as soldering, brazing, fastening, adhesive bonding, and welding [1–3]. The welding technique is one of the joining processes that is extensively used in almost all industrial sectors. Welding is considered as a fabrication process that is used to either manufacture new components or repair existing components. In most cases, other than repairing, the welding technique is preferred as a joining technique where two or more materials are joined together for fabricating a new component [4–6]. This technique creates a highly efficient permanent joint by means of a weld bead formation between the base metals. This weld bead ensures adequate joint strength between the base/parent materials and acts as a support without detaching the base materials. Usually, this weld bead is part of the parent materials, or in some cases, the filler material can be used to form the weld bead in between the materials. The weld bead at the faying site is formed by means of applying heat or external pressure, or in some cases, both heat and pressure are required to form the weld bead. The formed weld bead should be uniform, and the grain growth needs to be fine for better weld bead characteristics [7,8]. Based on the type of heat input and servicing environments, the welding technique is broadly classified into two main categories, namely, fusion welding and solid-state welding [9,10].

Fusion welding is a kind of welding process that involves the application of heat to a significant extent. In this process, the test specimens that need to be welded are heated to their melting point and forcibly welded (fused) together at the melting stage. After subsequent solidification of the faying site, a highly efficient weld bead is formed at the welding site, which provides joint strength between the test specimens.

DOI: 10.1201/9781003432289-2

Thus, a permanent joint is obtained between the base metals via the fusion (melting) process [11–13]. Moreover, this process can be carried out with or without the filler material, but the heating of the base/filler material to its melting point is unavoidable. Since a large amount of heat is involved in the fusion/melting of the materials, fusion welding encounters major limitations as follows [14,15]:

i. Distortion effect: The parts made via fusion welding experience a high level of distortion since a greater amount of heat is involved during the melting of base metals.
ii. Heat-affected zone: The fusion welds contain a greater area than the heat-affected zone. This greater heat-affected zone will pave the way for the formation of defects such as cracks and corrosion, thereby reducing the joint strength offered by the formed weld bead.
iii. Dissimilar metals: This process is good for joining similar metals or metallic alloys. In the case of dissimilar alloys, the fusion welding process is not suitable since dissimilar metallic alloys such as Al-Cu and Al-Mg contain metals with various thermal coefficients and melting points.

These major limitations could be overcome by the solid-state welding process. Solid-state welding does not require a greater amount of heat as in the case of fusion welding. In this process, the weld bead is formed at the faying site in three cases, namely, (i) application of heat, (ii) external pressure, and (iii) both heat and external pressure. Here, permanent joints can be made without the use of filler material [16–18]. Moreover, this technique has the following advantages:

i. More suitable for joining dissimilar metallic alloys and lightweight materials.
ii. Less heat-affected zone as this technique does not involve any melting or fusing of the parent materials.
iii. Low level of distortion with minimal residual stress.
iv. The mechanical characteristics of the parent materials remain unaffected.
v. Enhanced mechanical behaviors offered by the weld bead.
vi. Could be automized.

On the other hand, the usage of lightweight materials such as metallic alloys (Al alloys, Cu alloys, and Mg alloys), ceramics, and polymers has been increasing day by day. It is also predicted that in the future, most of the heavy metal parts made from iron and steel could be replaced by these lightweight materials owing to their outstanding mechanical characteristics [19]. Hence, there is a need for new welding techniques since traditional welding techniques such as fusion welding lack the ability to create high-quality weld joints in lightweight materials [20]. Friction stir spot welding (FSSW) belongs to the category of solid-state welding processes and is reported to produce permanent joints in an efficient manner. This process is said to be more suitable for lightweight materials as it does not demand the melting of lightweight metallic alloys. By means of plunging a rotational FSSW tool against the workpiece, spot welds are created with greater bonding (joint) strength compared to the other welding processes [21–23]. In most cases, the spot welds made by the FSSW process contain the keyhole. The keyhole is formed due to the FSSW tool's pin geometry (pin profile, pin length, and breadth). This keyhole will act as the potential site for defect formation such as hooks, cracks, pores, and corrosion pits, which will minimize the service life and joint strength of the spot welds. Therefore, the keyhole is considered as an unavoidable phenomenon in the FSSW process. Hence, researchers work on the interlayer materials that could strengthen the spot weld and enhance the mechanical characteristics offered by the FSSW joints [24,25]. This chapter is focused on discussing the interlayer materials that could be adapted in the FSSW process, along with their characteristics and relevant literature.

2.2 FSSW PROCESS

Traditionally, resistance spot welding is employed for fabricating spot welds in thin structures. This process creates spot welds by means of two copper electrodes, where the thin sheets are placed between the electrodes. Since copper is a good conductor of electricity, in the presence of electric charges, these two electrodes induce heat at the faying site (a spot where the weld bead or spot weld is formed). This heating locally melts the test specimen at the faying site, and the melt pool formed at the faying site solidifies to form the spot welds [26,27]. This type of spot welding technique has major limitations such as follows:

    i. Not suitable for producing spot welds in polymers and ceramics.
   ii. Limited to thin metallic sheets having a thickness of up to 3 mm.
  iii. More electrical energy is required if the test specimen has higher electrical conductivity than the copper.
   iv. The capital cost is very high.
    v. A limited number of welds can be made.
   vi. The spot welds are prone to more defects such as cracks, pores, and voids, which will affect the fatigue strength offered.
  vii. More skilled operators are needed for producing spot welds.

The above-mentioned drawbacks limit the usage of the resistance spot welding technique in practical applications.

On the other hand, FSSW is a type of solid-state welding technique that is mostly preferred for producing spot welds in both metals (similar as well as dissimilar metals) and non-metals (polymers, ceramics, etc.). This process is reported to be a type (variant) of friction stir welding process. The major difference between friction stir welding and FSSW lies in the direction of the tool movement. In friction stir welding, the tool moves in both linear and transverse directions, whereas in FSSW, the tool does not move in either of the previously mentioned directions. Hence, this FSSW process was reported to produce spot welds more than the linear weld bead, as in the case of the friction stir welding process [28–30].

In the FSSW process, the spot welds are made via four processing steps (Figure 2.1), namely, (i) rotation, (ii) plunging, (iii) stirring, and (iv) unloading [31]. These processing steps are briefly described as follows:

i. Rotation: Here, the FSSW tool starts to rotate at the predefined rotation per minute (rpm) and moves toward the specimens.
ii. Plunging: The rotating tool enters the plunging phase, where the tool is pressed against the workpiece to the required plunge depth.
iii. Stirring: In the plunging phase, the FSSW tool starts to rotate inside the specimen, thereby generating friction at the faying site. This friction induces localized heating, which will pave the way for plasticized (melted) flow of materials at the faying site. This process is reported to hold on for a predefined dwell time.
iv. Unloading: After stirring to the required dwell period, the FSSW tool is retracted back to the original position, leaving the spot welds at the faying site.

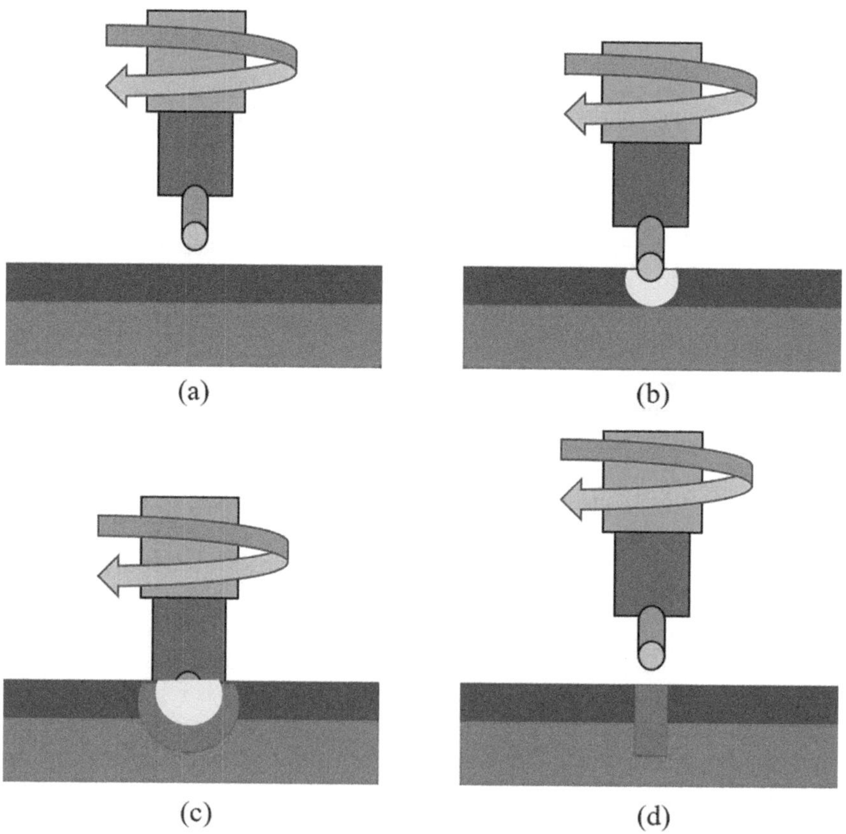

(a)   (b)

(c)   (d)

FIGURE 2.1 Schematic illustration of the FSSW process: (a) rotation, (b) plunging, (c) stirring, and (d) unloading.

Thus, with the aid of these processes, the spot welds are fabricated in similar as well as dissimilar metals and non-metals [32].

Spot welds can be made from material plates that are arranged in any one of the two welding configurations. They are as follows.

2.2.1 LAP JOINT CONFIGURATION

In this configuration, the test specimens are arranged one above the other. In other words, each test specimen is overlapped against the other specimen such that the spot weld is made at the faying site in between the two test specimens. This type of configuration is good for test specimens with different thicknesses. In most cases, this configuration is preferred for producing spot welds via the FSSW process [33,34].

2.2.2 BUTT JOINT CONFIGURATION

The test specimens are arranged in the same plane where the spot welds are produced in the seam zone (contact zone) [35]. This configuration is good for test specimens with different thicknesses. Hence, by arranging the workpiece in one of the two joint configurations, the FSSW spot welds are produced in an efficient manner.

Moreover, the above-mentioned four processes (rotation, plunging, stirring, and unloading) could take place under two major operation modes, such as the force control mode and the tool's displacement control mode [36].

i. Force control mode: As the name implies, the force applied, the rotating tool's rpm, and the withstanding (dwell) time are kept under control. In this mode, the tool's rpm is in direct relationship with the force exerted on the faying site. After plunging the tool, the force is kept constant, whereas the tool keeps on rotating at a constant speed for forming the spot welds [37].

ii. Displacement control mode: This process is a reverse of the previous process, where after plunging the tool's pin to a predefined level, the force has been recorded. Thus, these two modes create permanent spot joints (welds) through the FSSW process [38].

2.3 FSSW PROCESS PARAMETERS

There are some process parameters (Figure 2.2) that need to be controlled for producing spot welds using the FSSW process. They are briefly described as follows:

2.3.1 RPM OF THE TOOL

Since the FSSW process starts with the rotation of the tool, the rpm of the tool is considered as an important factor that determines the generation of heat flux in the FSSW process. The amount of heat produced during the stirring process is directly related to the amount of plasticized material produced [39]. At lower rpm, not sufficient heat is produced at the faying site, and the weld bead formed at low rpm has

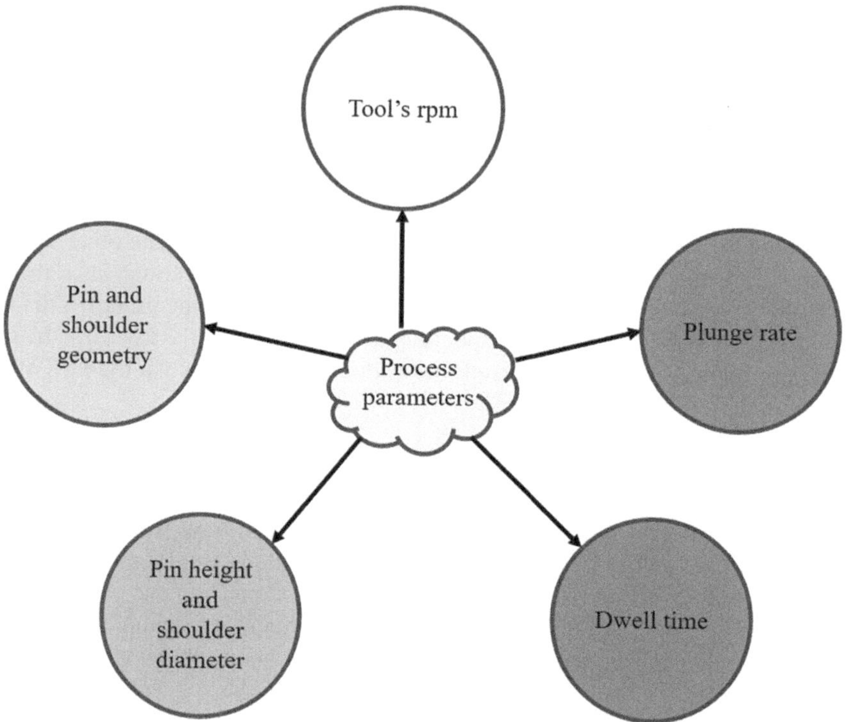

FIGURE 2.2 FSSW process parameters.

indefinite grain distribution and results in poor bonding strength. On the other hand, sufficient heat is generated at higher tool's rpm, which results in a greater amount of plasticized flow of materials at the faying site. Hence, the weld bead formed at a higher rotational speed will have a fine grain distribution [40]. This fine distribution of grains results in enhanced mechanical characteristics offered by the weld bead (spot welds). Moreover, the choice of lower or higher rpm of the rotating tool depends on the type of materials where the spot weld is created. In general perspective, the spot weld formed at a higher tool's rpm tends to be stronger than the spot welds formed at a lower rpm of the FSSW tool [41].

2.3.2 DWELL TIME

The second most important criterion in deciding the joint strength of the weld bead is dwell time. Dwell time can be defined as the time taken to complete the stirring process. During stirring, the frictional heat promotes localized melting, followed by plasticized material flow. In general, the dwell time is directly proportional to the amount of weld bead formation [42]. Higher dwell time results in greater mixing (stirring), thereby enhancing the grain growth rate of the spot welds. Moreover, this higher dwell time results in better mechanical characteristics such as joint strength, hardness, wear, and corrosion resistance offered by the spot welds [43].

2.3.3 PLUNGE RATE

Plunge rate is defined as the amount of force or load exerted against the test surface. In the FSSW process, the tool's pin is pressed (plunged) against the test surface. Hence, the operator needs to ensure the required amount of force or load is delivered by the tool's pin to carry out the plunging process. A lower plunge rate results in uneven plunging and may lead to an improper stir zone [44]. On the other hand, the wear rate of the tool increases with a higher plunging rate. Hence, based on the type of material, the plunge rate is optimized for producing the spot welds in the FSSW process [45].

2.3.4 TOOL'S PIN HEIGHT AND SHOULDER DIAMETER

The two main constituents that constitute the FSSW tool are the shoulder and the pin. The shoulder provides the necessary load or force for plunging the pin inside the test specimens. Moreover, the diameter of the shoulder plays a major role in deciding the amount of weld bead formed in the case of the pinless FSSW process. Higher the diameter, greater the weld bead formation, followed by better mechanical characteristics offered in the pinless FSSW process [46,47]. On the other hand, the pin's height is a key factor in the FSSW process. This height determines the plunging depth, the amount of frictional heat produced, and the area of the stir zone at the faying site. If the stir zone area becomes wider/larger, there is a possibility for better mixing of the plasticized materials during stirring at the faying site. At a lower pin's height, the weld bead does not provide the required bonding strength, whereas at a higher pin's height, the top plate becomes more damaged during the plunging process [48,49]. Hence, the pin's height is optimized based on other factors such as the type of test specimen's materials (Al, Mg, and Cu alloys), rpm of the rotating tool, dwell time, plunge rate, and thickness of the test specimens.

2.3.5 TOOL AND SHOULDER GEOMETRY

Depending on the working condition, the shoulder and tool geometry may vary. As mentioned, the shoulder provides the necessary force to plunge the tool against the workpiece. Researchers proposed three types of shoulder geometries that could be adapted for producing spot joints (welds) in the FSSW process [50–52]. They are as follows:

    i. Concave conical shoulder
   ii. Convex shoulder
  iii. Flat shoulder

Out of these types, the concave conical type is mostly preferred by most of the researchers. In addition to these three shoulder geometries, one can use customized shoulder geometry depending on the service environment. Irrespective of any shoulder geometry, any shoulder should perform the three functions as mentioned below:

i. Ensure the required force or load is delivered during the plunging process.
ii. Monitors or controls the frictional heat generated during the stirring process.
iii. In the case of the refill FSSW process, the shoulder should refill the keyhole with excessive plasticized materials.

Other than the shoulder geometry, six pin geometries (Figure 2.3) have been proposed for producing spot welds in the FSSW process [53–55]. They are as follows:

i. Cylindrical pin
ii. Tapered cylindrical pin
iii. Trapezoidal pin

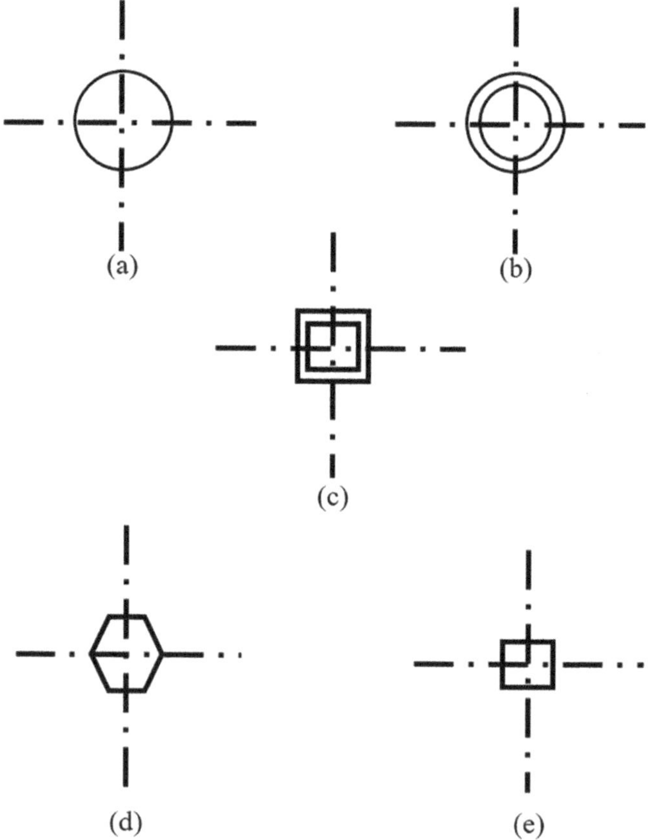

FIGURE 2.3   2D view of the pin profiles used in the FSSW process: (a) cylindrical pin, (b) tapered cylindrical pin, (c) trapezoidal pin, (d) hexagonal pin, and (e) squared pin.

   iv. Hexagonal pin
   v. Squared pin
   vi. Above five pins with threads

2.4 INTERLAYERS IN THE FSSW PROCESS

2.4.1 NEED FOR INTERLAYERS

As mentioned earlier, the FSSW process will produce spot welds by plunging and stirring the tool against the workpiece. The stirring process will create the plasticized flow of materials with which the spot welds (weld beads) are formed at the faying site. If the FSSW process is carried out on similar metallic alloy plates such as Al-Al and Cu-Cu, the obtained spot welds have a fine grain distribution with secondary precipitates, which enhances the mechanical characteristics of the spot welds. On the other hand, if the spot welds are made on dissimilar metallic alloys (Al-Mg alloys, Al-Cu alloys, and Al-Fe alloys), ceramics, and polymers, the spot welds formed on these materials will contain intermetallic compounds (IMCs). These compounds are reported to initiate defects, such as cracks, pores, voids, and hooks, which will minimize the service life of the obtained spot welds [56,57]. The IMCs can be formed under any one of the following conditions:

   i. Not sufficient rpm during stirring: The frictional heat is responsible for creating plasticized material flow during stirring. At lower speeds, not sufficient heat flux is generated at the faying site, which results in the formation of IMCs.
   ii. Shorter dwell time: Lower dwell time results in uneven mixing of plasticized materials at the stir zone, thereby leading to the formation of IMCs.
   iii. Spot welds in dissimilar metallic alloys: While producing spot welds in dissimilar metals, there are more chances for the formation of intermetallic precipitates. These precipitates tend to be brittle and act as interfacial sites, which lead to cracks and hook formation at the spot weld areas, thereby minimizing the joint strength of the weld joints.

2.4.2 CONCEPT OF INTERLAYERS

In order to reduce the influence of these intermetallic precipitates (compounds) and to improve the service life of the spot welds, interlayers are used. As the name implies, the interlayers are materials that are placed between the base metals, which undergo welding. These materials are reported to aid in enhancing the joint strength and other mechanical characteristics offered by the weld bead. Moreover, these materials combine with the parent materials and produce strong phase compounds, thereby decreasing the amount of IMCs on the weld bead. Some of the interlayer materials that are generally used in the FSSW process are copper, zinc, silver, carbon fiber-reinforced polymer (CFRP), adhesives, and so on [58,59]. On the other hand, care should be taken in deciding the accurate interlayer materials for producing the spot welds. Because the unsuitable interlayer material could promote fractures in the spot welds. Moreover, the interlayer material should exhibit mechanical characteristics nearer to the mechanical characteristics of the base metals. Figure 2.4 shows the zones in the formed spot welds.

Spot welds are made with (i) conventional FSSW processes and (ii) FSSW processes with an interlayer (purple). Here, the weld bead (spot weld) contains secondary precipitates, which are formed due to complex alloying reactions between the interlayer material and parent materials (Figure 2.5).

2.4.3 Types of Interlayer

Since interlayers aid in improving the mechanical characteristics of spot welds, a wide range of materials can be preferred as interlayer materials, as shown in Figure 2.6. The choice of interlayer materials depends on the type of base materials used in the

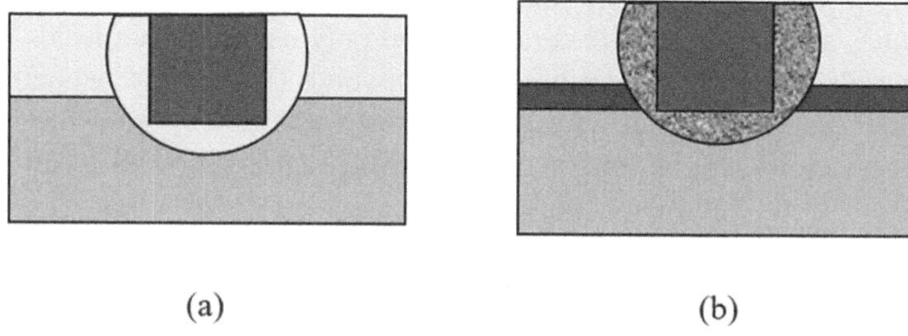

(a)   (b)

FIGURE 2.4   Schematic illustration of both FSSW and FSSW with interlayer.

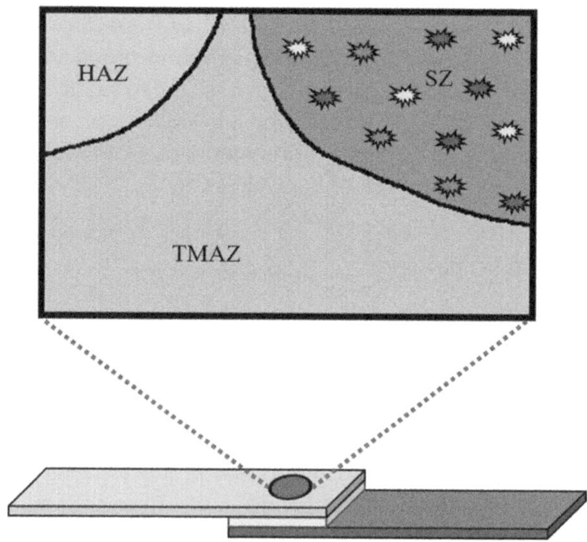

FIGURE 2.5   Zones in the formed spot welds. HAZ, heat-affected zone; TMAZ, thermo-mechanically affected zone; and SZ, stir zone consisting of intermetallic precipitates formed due to the interlayer.

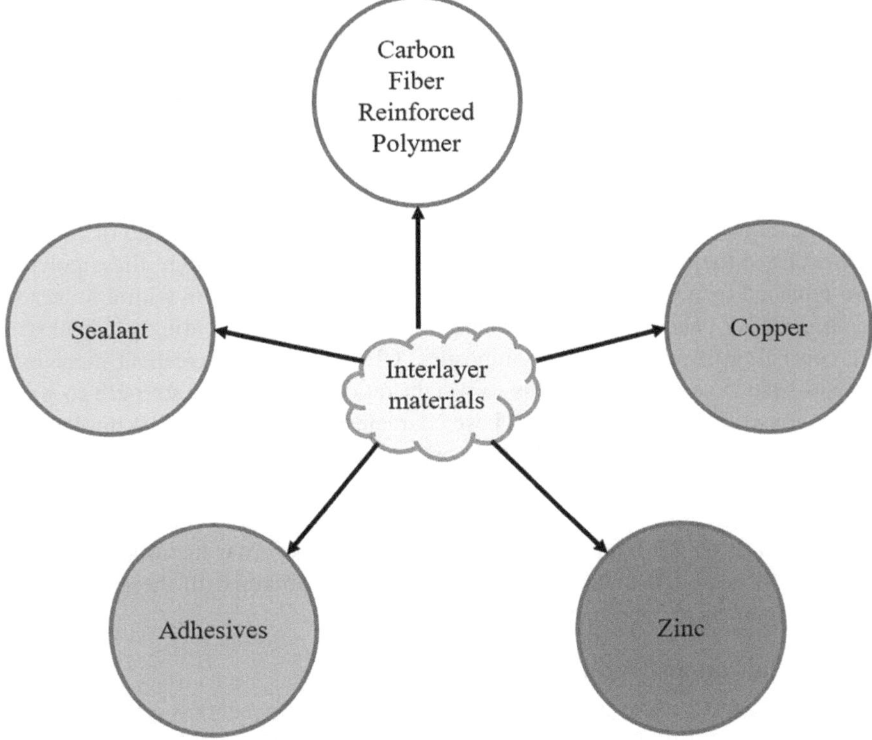

FIGURE 2.6   Types of interlayer materials used in the FSSW process.

FSSW process. This section gives a brief introduction to the types of interlayer materials that could be adapted for producing spot welds via the FSSW process.

2.4.3.1   Zinc Interlayer

For enhancing the properties of spot welds made between dissimilar metals such as Al-Cu alloys and Al-Mg alloys, most researchers preferred zinc as the interlayer material due to its outstanding characteristics, as follows:

a. Solidification rate: The frictional heat induced during the plunging and stirring process effectively melts the thin sheet of Zn between the Al-Cu and Al-Mg alloys and aids in improving the solidification rate of the materials in the stir zone. Moreover, this faster solidification rate avoids the formation of voids, cracks, and other defects, thereby improving the service life of the spot welds.

b. Hardness and corrosion resistance: The spot welds formed with the Zn interlayer were rich in intermetallic precipitates or compounds such as zinc aluminate and Mg_4Zn_7, which are reported to enhance the hardness and wear resistance characteristics. Particularly, the zinc aluminate provides better corrosion resistance properties, which aid in enhancing the durability of the spot welds.

2.4.3.2 Copper Interlayer

Copper has good electrical conductivity behaviors and could be implemented as the interlayer material for producing spot welds in electrical applications. The intermetallic secondary phases formed during the stirring process are rich in copper/cupric phases, which resist corrosion to a greater extent.

2.4.3.3 CFRP Interlayer

In recent years, the usage of polymers has been tremendously increased in almost all sectors due to their outstanding mechanical behaviors. In particular, these polymers have replaced heavy metallic components with lightweight parts with a longer service life. CFRP is one of the most widely used polymers for creating a hybrid structure consisting of metal matrix composites. CFRP possesses excellent mechanical behaviors, such as a high stiffness-to-weight ratio and a greater strength-to-weight ratio, among other characteristics. These characteristics led them to be implemented in aerospace, heat exchangers, sound absorbers, automotive, and other industrial sectors. Moreover, this carbon fiber can be combined with lightweight metallic alloys such as Al, Cu, and Mg alloys to improve the mechanical behaviors offered by them. A thin sheet of CFRP could be placed between the dissimilar metallic alloy sheets (Al-Cu, Al-Mg, Al-Al, etc.), and spot welds can be obtained in these sheets with superior mechanical characteristics.

2.4.3.4 Sealant and Adhesives

In aerospace and nuclear applications, corrosion of components is unavoidable due to the service environment (high temperature and high pressure). Moreover, the spot welds without an interlayer might promote corrosion, which will reduce the bonding strength of the spot welds. Hence, if a material provides combined characteristics such as corrosion resistance and lightweight, such materials can be adapted as interlayer materials in these applications. Sealants and adhesives are polymeric substances that are usually preferred as coating materials to prevent corrosion. A thin layer of sealant or adhesive can be placed in between the dissimilar metallic alloy sheets to improve the corrosion resistance and durability of the spot welds that are made in these sheets. Table 2.1 lists the suitable interlayer material for creating spot welds in various alloys.

2.4.4 Interlayer's Thickness and Its Effects

In the FSSW process, interlayer plays a major role in enhancing the bonding strength of the spot welds. Other mechanical characteristics, such as hardness, corrosion resistance, and anti-wear behaviors, offered by the spot welds are reported to be improved due to the influence of intermetallic precipitates, which are formed with the effect of interlayer materials during the stirring process. The interlayer will increase the solidification rate of the spot welds, thereby reducing the formation of defects (cracks, voids, and hooks) in the spot welds. Hence, the addition of an interlayer could improve the mechanical characteristics offered by the spot welds. The choice of interlayer material solely depends on the base metals or materials where

TABLE 2.1

Interlayer Materials and Their Alloys

Interlayer Materials	Alloys/Materials
Zinc	Al-Cu alloys (6061Al-Cu, AA2024-T3-Cu), Al-Mg alloys (AA7075-AZ31B, AZ31-AZ31, Al-ZEK100, AA6061-T6-AZ61), AA7099-AA6061, AA5083-H112, etc.
Copper	Similar Al and Mg alloys
CFRP	Similar Al alloys (AA5052, AA7075-T6, AA6061, AA1060, AA5182), Dissimilar alloys (AA7075-T6-Ti-6Al-4V)
Adhesives	AA5754-AZ31B, AA6061-AZ31B, and other dissimilar alloys
Sealant	Suitable for similar Al alloys

the spot welds are created. But the thickness of the interlayer used also influences the process parameters in the FSSW process. In general, for creating highly efficient spot welds, interlayers' thickness ranging from several micrometer range to a few millimeter range is preferred by researchers.

The effect of the interlayer's thickness on the process parameters in the FSSW process is discussed as follows:

i. Tool's rpm: As mentioned earlier, the tool's rpm is responsible for the amount of heat generated during the stirring process. This heat determines the size and orientation of the grains formed in the spot welds, followed by the mechanical characteristics offered by the spot weld joints. When an interlayer is introduced, it should start melting at a desired rpm, which would be optimal for the base metals. A thick interlayer will require a higher rpm, which might affect the formation of IMCs. Also, it results in improper mixing, which will pave the way for uneven distribution of IMCs, thereby affecting the mechanical behaviors of the spot welds.

ii. Dwell time: The thickness of the interlayer is directly proportional to the dwell time. Higher thicknesses require longer dwell time to create even distribution of grains and intermetallic precipitate formation in the stir zone. On the other hand, a thin interlayer requires a shorter dwell time for creating efficient spot welds.

iii. Plunge rate: Similar to dwell time, the plunging rate increases with an increase in the thickness of the interlayer. A thick interlayer demands a higher plunging rate for creating efficient plunging depth (stir zone), whereas a lower plunging rate is required for thin interlayers.

iv. Tool's pin height: Pin height determines the size of the stir zone formed during the FSSW process. A thick interlayer requires a higher pin height so that the pin plunges to the required depth for forming the spot welds. On the other hand, very high thickness promotes greater tool wear in the FSSW process. Thus, based on the parent material, the interlayer's thickness is optimized along with its parameters.

2.4.5 Mechanical Perspective of Interlayers

As reported, interlayers aid in improving the joint strength of the spot welds. Other mechanical characteristics such as hardness, creep resistance, and tensile properties are also enhanced by these interlayers.

Xu et al. [60] examined the Zn interlayer effects while producing the spot welds in Mg-Al-Zn alloy sheets. They have compared the microstructure, tensile properties, and hook formation in the spot welds formed with a 0.1-mm thick Zn interlayer and without a Zn interlayer. Their study concluded that the spot welds made with a Zn interlayer are rich in Mg-Zn compounds (α-Mg and Mg_4Zn_7) due to the complex alloying reactions. Moreover, they also reported that the tensile load was increased up to two times due to the influence of these compounds, thereby improving the joint strength offered by the spot welds. Omar Kalaf et al. [61] interrogated the effects of the CFRP interlayer between the similar metallic alloy plates (AA5082 plates) via the FSSW process. They analyzed the microhardness and tensile shear load offered at various rotational speeds (850–2000 rpm) with different dwell times (2–5 seconds) and concluded that the maximum tensile shear load up to 1.8 kN was obtained in the presence of CFRP interlayer at a dwell time of 2 seconds with the tool's rpm of 2000 rpm. Moreover, there is an increase in the microhardness values at higher rotational speeds (2000 rpm) and at higher dwell times (5 seconds). The decreasing trend is observed when the dwell time and rpm of the rotating tool decrease. In addition, the spot welds are rich in aluminum carbide phases due to the effect of the CFRP interlayer. This Al_4C_3 aids in improving the tensile shear load and microhardness of the spot welds at higher values of dwell time and the rotating tool's rpm. Tauqir Nasir et al. [62] used CFRP as the interlayer while joining the AA7075 alloy with Ti-6Al-4V via the FSSW process. The authors have varied the rotational tool's rpm (1000–2000 rpm) and the dwell time (5–10 seconds). Here also, at higher rpm with a higher dwell time (2000 rpm, 10 seconds), in the presence of CFRP as an interlayer, the tensile shear load was enhanced by nearly 17.3% compared to the values obtained at lower rpm with a lower dwell time. This trend was reported while analyzing the microhardness offered. This higher value (microhardness and tensile shear load) was due to the existence of ternary titanium aluminum carbide (Ti-Al-C) phases at the spot welds. Ramzi Gassaa et al. [63] analyzed the use of copper as an interlayer material for producing spot joints in similar metallic sheets made up of AA5754 alloy and concluded that the higher rpm of the tool yields a better microstructure of the weld bead. Moreover, the copper particles formed the strong phases when combined with the Al alloy, thereby enhancing the mechanical characteristics offered by the spot welds. Furthermore, Balsam H. Abed et al. [64] examined the effect of copper interlayer in the spot welds produced on similar metallic alloys (AA 6061-T6 alloy plates). Here, the rpm of the rotating tool is maintained constant at 1800 rpm, whereas the dwell time and plunge depth are varied (dwell time: 10–20 seconds and plunge depth: 0.1–0.3 mm). Their study concluded that at a higher dwell time along with a higher plunge depth, the tensile shear load and the microhardness were higher in the presence of the Cu interlayer. S. H. Chowdhury et al. [65] reported that the spot weld characteristics were improved while using adhesive as an interlayer between Mg-Al alloys. Also, sealant can be used as the interlayer material for producing highly efficient spot welds in the FSSW process [66]. Thus, from these literatures, it can be confirmed that the weld (joint) strength and other mechanical characteristics such as tensile shear load, fatigue and fracture resistance, and microhardness were improved by adding interlayers while producing spot welds via the FSSW process.

2.5 CONCLUSION

Traditional welding techniques lack the ability to join dissimilar metallic alloys and other lightweight materials (polymers, ceramics). FSSW is reported as a variant of the friction stir welding process, which is mostly preferred for fabricating spot welds in both dissimilar and lightweight materials. The spot weld made from the FSSW process is made by mixing the plasticized flow of materials while implementing the FSSW process in dissimilar metallic alloys. The spot welds contain a huge amount of IMCs. These IMCs are brittle and act as interfacial sites that promote cracks, hooks, and other forms of defects, thereby reducing the joint strength of the spot welds. Interlayers are materials that are used in between the parent materials during welding. These interlayers are said to react with the plasticized material flow and form hard phases during the stirring process, which enhances the mechanical characteristics offered by the spot welds. Thus, this chapter gives an introduction to the FSSW process, its process parameters, the need for and concept of interlayers, and the relevant literature.

CONFLICT OF INTEREST

The authors declare that there are no conflicts of interest.

ACKNOWLEDGMENT

The authors would like to acknowledge the editors of this book for providing the opportunity to write this chapter.

REFERENCES

1. LeBacq C, Brechet Y, Shercliff HR, Jeggy T, Salvo L. Selection of joining methods in mechanical design. *Materials & Design*. 2002;23(4):405–16.
2. Campbell FC, editor. *Joining: Understanding the Basics*. ASM International; 2011.
3. Jeyaprakash N, Yang CH, Sivasankaran S. Laser cladding process of Cobalt and Nickel based hard-micron-layers on 316L-stainless-steel-substrate. *Materials and Manufacturing Processes*. 2020;35(2):142–51.
4. Lancaster JF. The physics of welding. *Physics in Technology*. 1984;15(2):73.
5. Phillips DH. *Welding Engineering: An Introduction*. John Wiley & Sons; 2023.
6. Mendez PF, Eagar TW. Welding processes for aeronautics. *Advanced Materials and Processes*. 2001;159(5):39–43.
7. Jeyaprakash N, Yang CH, Duraiselvam M, Prabu G. Microstructure and tribological evolution during laser alloying WC-12% Co and Cr3C2− 25% NiCr powders on nodular iron surface. *Results in Physics*. 2019;12:1610–20.
8. Katayama S, editor. *Handbook of Laser Welding Technologies*. Elsevier; 2013.
9. Chen CM, Kovacevic R. Joining of Al 6061 alloy to AISI 1018 steel by combined effects of fusion and solid state welding. *International Journal of Machine Tools and Manufacture*. 2004;44(11):1205–14.

10. Klochkov IL, Poklaytsky AN, Motrunich SV. Fatigue behavior of high strength Al-Cu-Mg and Al-Cu-Li alloys joints obtained by fusion and solid state welding technologies. *Journal of Theoretical and Applied Mechanics.* 2019;49(2):179–89.

11. Marques ES, Silva FJ, Pereira AB. Comparison of finite element methods in fusion welding processes-a review. *Metals.* 2020;10(1):75.

12. Karuppasamy SS, Jeyaprakash N, Yang CH. Application of Corrosion-Resistant Laser Claddings. *Handbook of Laser-Based Sustainable Surface Modification and Manufacturing Techniques.* 2023;51(51):27.

13. Tümer M, Schneider-Bröskamp C, Enzinger N. Fusion welding of ultra-high strength structural steels-A review. *Journal of Manufacturing Processes.* 2022;82:203–29.

14. Kalpana J, Rao P. A review on techniques for improving the mechanical properties of fusion welded joints. *Engineering Solid Mechanics.* 2017;5(4):213–24.

15. Norman AF, Hyde K, Costello F, Thompson S, Birley S, Prangnell PB. Examination of the effect of Sc on 2000 and 7000 series aluminium alloy castings: For improvements in fusion welding. *Materials Science and Engineering: A.* 2003;354(1–2):188–98.

16. Jeyaprakash N, Karuppasamy SS, Yang CH. Application of wear-resistant laser claddings. In: *Handbook of Laser-Based Sustainable Surface Modification and Manufacturing Techniques* edited by Hitesh Vasudev, Chander Prakash 2023 (pp. 1–26). CRC Press.

17. Nassiri A, Abke T, Daehn G. Investigation of melting phenomena in solid-state welding processes. *Scripta Materialia.* 2019;168:61–6.

18. Cooper DR, Allwood JM. Influence of diffusion mechanisms in aluminium solid-state welding processes. *Procedia Engineering.* 2014;81:2147–52.

19. Natarajan J, Yang CH, Karuppasamy SS. Investigation on microstructure, nanohardness and corrosion response of laser cladded colmonoy-6 particles on 316l steel substrate. *Materials.* 2021;14(20):6183.

20. Kumar Rajak D, Pagar DD, Menezes PL, Eyvazian A. Friction-based welding processes: Friction welding and friction stir welding. *Journal of Adhesion Science and Technology.* 2020;34(24):2613–37.

21. Shen Z, Ding Y, Gerlich AP. Advances in friction stir spot welding. *Critical Reviews in Solid State and Materials Sciences.* 2020;45(6):457–534.

22. Fereiduni E, Movahedi M, Kokabi AH. Aluminum/steel joints made by an alternative friction stir spot welding process. *Journal of Materials Processing Technology.* 2015;224:1.

23. Kao CM, Jeyaprakash N, Yang CH. Modelling and non-destructive measurement of dispersion behaviour of stainless steel and alumina ceramic porous plates. *Ceramics International.* 2021;47(10):14233–43.

24. Çam G, İpekoğlu G, Tarık Serindağ H. Effects of use of higher strength interlayer and external cooling on properties of friction stir welded AA6061-T6 joints. *Science and Technology of Welding and Joining.* 2014;19(8):715–20.

25. Abdollahzadeh A, Shokuhfar A, Cabrera JM, Zhilyaev AP, Omidvar H. The effect of changing chemical composition on dissimilar Mg/Al friction stir welded butt joints using zinc interlayer. *Journal of Manufacturing Processes.* 2018;34:18–30.

26. Podržaj P, Polajnar I, Diaci J, Kariž Z. Overview of resistance spot welding control. *Science and Technology of Welding and Joining.* 2008;13(3):215–24.

27. Zhou K, Yao P. Overview of recent advances of process analysis and quality control in resistance spot welding. *Mechanical Systems and Signal Processing.* 2019;124:170–98.

28. Karuppasamy SS, Jeyaprakash N, Yang CH. Microstructure, nanoindentation and corrosion behavior of colmonoy-5 deposition on SS410 substrate using laser cladding process. *Arabian Journal for Science and Engineering.* 2022:1–7.

29. Tozaki Y, Uematsu Y, Tokaji K. A newly developed tool without probe for friction stir spot welding and its performance. *Journal of Materials Processing Technology.* 2010;210(6–7):844–51.

30. Yang XW, Fu T, Li WY. Friction stir spot welding: A review on joint macro-and microstructure, property, and process modelling. *Advances in Materials Science and Engineering*. Volume 2014, Jan 1, (2014).

31. Gerlich A, Su P, North TH. Tool penetration during friction stir spot welding of Al and Mg alloys. *Journal of Materials Science*. 2005;40:6473–81.

32. Bilici MK, Yükler Aİ, Kurtulmuş M. The optimization of welding parameters for friction stir spot welding of high density polyethylene sheets. *Materials & Design*. 2011;32(7):4074–9.

33. Rashkovets M, Contuzzi N, Casalino G. Modeling of probeless friction stir spot welding of AA2024/AISI304 steel lap joint. *Materials*. 2022;15(22):8205.

34. Jeyaprakash N, Yang CH, Karuppasamy SS, Dhineshkumar SR. Evaluation of microstructure, nanoindentation and corrosion behavior of laser cladded Stellite-6 alloy on Inconel-625 substrate. *Materials Today Communications*. 2022;31:103370.

35. Gao K, Zhang S, Mondal M, Basak S, Hong ST, Shim H. Friction stir spot butt welding of dissimilar S45C steel and 6061-T6 aluminum alloy. *Metals*. 2021;11(8):1252.

36. Jeyaprakash N, Yang CH, Karuppasamy SS, Rajendran DK. Correlation of microstructural with corrosion behaviour of Ti-6Al-4V specimens developed through selective laser melting technique. *Proceedings of the Institution of Mechanical Engineers, Part E: Journal of Process Mechanical Engineering*. 2022;236(5):2240–51.

37. Pan, T-Y. Friction stir spot welding (FSSW): A literature review. SAE Technical Paper 2007-01-1702, 2007*

38. Yuan W, Mishra RS, Webb S, Chen YL, Carlson B, Herling DR, Grant GJ. Effect of tool design and process parameters on properties of Al alloy 6016 friction stir spot welds. *Journal of Materials Processing Technology*. 2011;211(6):972–7.

39. Jeyaprakash N, Yang CH, Karuppasamy SS. Laser cladding of Colmonoy 6 particles on Inconel 625 substrate: Microstructure and corrosion resistance. *Surface Review and Letters*. 2022;29(08):2250102.

40. Patel VV, Sejani DJ, Patel NJ, Vora JJ, Gadhvi BJ, Padodara NR, Vamja CD. Effect of tool rotation speed on friction stir spot welded AA5052-H32 and AA6082-T6 dissimilar aluminum alloys. *Metallography, Microstructure, and Analysis*. 2016;5:142–8.

41. Cox CD, Gibson BT, Strauss AM, Cook GE. Effect of pin length and rotation rate on the tensile strength of a friction stir spot-welded al alloy: A contribution to automated production. *Materials and Manufacturing Processes*. 2012;27(4):472–8.

42. Li G, Zhou L, Zhou W, Song X, Huang Y. Influence of dwell time on microstructure evolution and mechanical properties of dissimilar friction stir spot welded aluminum-copper metals. *Journal of Materials Research and Technology*. 2019;8(3):2613–24.

43. Bagheri B, Alizadeh M, Mirsalehi SE, Shamsipur A, Abdollahzadeh A. The effect of rotational speed and dwell time on Al/SiC/Cu composite made by friction stir spot welding. *Welding in the World*. 2022;66(11):2333–50.

44. Jeyaprakash N, Yang CH, Susila P, Karuppasamy SS. Laser cladding of NiCrMoFeNbTa particles on Inconel 625 alloy: Microstructure and corrosion resistance. *Transactions of the Indian Institute of Metals*. 2023;76(2):599–612.

45. Su P, Gerlich A, North TH, Bendzsak GJ. Energy utilisation and generation during friction stir spot welding. *Science and Technology of Welding and Joining*. 2006;11(2):163–9.

46. Arici A, Mert Ş. Friction stir spot welding of polypropylene. *Journal of Reinforced Plastics and Composites*. 2008;27(18):2001–4.

47. Murugesan P, Satheeshkumar V, Jeyaprakash N, Yang CH, Karuppasamy SS. Effect of α-Al and Si precipitates on microstructural evaluation and corrosion behavior of laser powder bed fusion printed AlSi10Mg plates in seawater environment. *Metals and Materials International*. 2023:1–8.

48. Freeney TA, Sharma SR, Mishra RS. Effect of welding parameters on properties of 5052 Al friction stir spot welds. *SAE Technical Paper*; 2006-01-0969, 2006.

49. Bilici MK, Yükler AI. Influence of tool geometry and process parameters on macrostructure and static strength in friction stir spot welded polyethylene sheets. *Materials & Design*. 2012;33:145–52.

50. Hirasawa S, Badarinarayan H, Okamoto K, Tomimura T, Kawanami T. Analysis of effect of tool geometry on plastic flow during friction stir spot welding using particle method. *Journal of Materials Processing Technology*. 2010;210(11):1455–63.

51. Badarinarayan H, Shi Y, Li X, Okamoto K. Effect of tool geometry on hook formation and static strength of friction stir spot welded aluminum 5754-O sheets. *International Journal of Machine Tools and Manufacture*. 2009;49(11):814–23.

52. Jeyaprakash N, Yang CH, Karuppasamy SS, Duraiselvam M. Stellite 6 cladding on AISI Type 316L stainless steel: Microstructure, nanohardness and corrosion resistance. *Transactions of the Indian Institute of Metals*. 2023;76(2):491–503.

53. Badarinarayan H, Yang Q, Zhu S. Effect of tool geometry on static strength of friction stir spot-welded aluminum alloy. *International Journal of Machine Tools and Manufacture*. 2009;49(2):142–8.

54. Pathak N, Bandyopadhyay K, Sarangi M, Panda SK. Microstructure and mechanical performance of friction stir spot-welded aluminum-5754 sheets. *Journal of Materials Engineering and Performance*. 2013;22:131–44.

55. Lambiase F, Paoletti A, Di Ilio A. Effect of tool geometry on mechanical behavior of friction stir spot welds of polycarbonate sheets. *The International Journal of Advanced Manufacturing Technology*. 2017;88:3005–16.

56. Jeyaprakash N, Duraiselvam M, Aditya SV. Numerical modeling of WC-12% Co laser alloyed cast iron in high temperature sliding wear condition using response surface methodology. *Surface Review and Letters*. 2018;25(7):1950009.

57. Mokabberi SR, Movahedi M, Kokabi AH. Effect of interlayers on softening of aluminum friction stir welds. *Materials Science and Engineering: A*. 2018;727:1.

58. Dong SK, Lin S, Zhu H, Wang CJ, Cao ZL. Effect of Ni interlayer on microstructure and mechanical properties of Al/Mg dissimilar friction stir welding joints. *Science and Technology of Welding and Joining*. 2022;27(2):103–13.

59. Kar A, Kailas SV, Suwas S. Effect of mechanical mixing in dissimilar friction stir welding of aluminum to titanium with zinc interlayer. *Transactions of the Indian Institute of Metals*. 2019;72:1533–6.

60. Xu RZ, Ni DR, Yang Q, Liu CZ, Ma ZY. Influencing mechanism of Zn interlayer addition on hook defects of friction stir spot welded Mg-Al-Zn alloy joints. *Materials & Design*. 2015;69:163–9.

61. Kalaf O, Nasir T, Asmael M, Safaei B, Zeeshan Q, Motallebzadeh A, Hussain G. Friction stir spot welding of AA5052 with additional carbon fiber-reinforced polymer composite interlayer. *Nanotechnology Reviews*. 2021;10(1):201–9.

62. Nasir T, Kalaf O, Asmael M, Zeeshan Q, Safaei B, Hussain G, Motallebzadeh A. The experimental study of CFRP interlayer of dissimilar joint AA7075-T651/Ti-6Al-4V alloys by friction stir spot welding on mechanical and microstructural properties. *Nanotechnology Reviews*. 2021;10(1):401–13.

63. Gassaa R, Hemmouche L, Badji R, Gilson L, Rabet L, Mimouni O. Effect of rotational speed and copper interlayer on the mechanical and fracture behaviour of friction stir spot welds of 5754 aluminium alloy. *Metallurgical Research & Technology*. 2023;120(1):118.

64. Abed BH, Salih OS, Sowoud KM. Pinless friction stir spot welding of aluminium alloy with copper interlayer. *Open Engineering*. 2020;10(1):804–13.

65. Chowdhury SH, Chen DL, Bhole SD, Cao X, Wanjara P. Lap shear strength and fatigue behavior of friction stir spot welded dissimilar magnesium-to-aluminum joints with adhesive. *Materials Science and Engineering: A*. 2013;562:53–60.

66. Kubit A, Wydrzynski D, Trzepiecinski T. Refill friction stir spot welding of 7075-T6 aluminium alloy single-lap joints with polymer sealant interlayer. *Composite Structures*. 2018;201:389–97.

3 Process Parameter Optimization during the Friction Stir Spot Welding Process

R. Pandiyarajan, S. Sabarish, and S. Ponsuriyaprakash

3.1 INTRODUCTION

Thin metal strips are joined using friction stir spot welding (FSSW), wherein the procedure involves using a rotating tool to produce frictional heat and plasticize the aluminum at the joint contact, then forging the material together by applying pressure. Applications of FSSW include the automotive industry for joining aluminum body structures, aerospace for joining aluminum panels and fuel tanks, and electronic packaging for joining aluminum heat sinks and chassis. One way to enhance FSSW connections is by optimizing process variables such as tool rotating speed, welding speed, and axial force. Adjusting the geometry of the instrument, such as the pin's shape and shoulder width, and optimizing joint design, including joint overlap and thickness, and controlling process factors such as workpiece preheating and clamping force are also some methods toward enhancing the weld joint performance. The intense plastic deformation process of friction stir welding (FSW) can cause the layer to fracture as the depth grows. The effect of plunge depth leads to an uneven distribution of particles in the aluminum stir zone. Characterization of FSSW joints can be accomplished through a variety of testing methods, including tensile testing to assess joint strength; microstructural analysis to assess weld quality and any defects, such as voids or cracks; hardness testing to assess weld zone and heat-affected zone hardness; corrosion testing to evaluate the resistance of the joint to corrosion; and fatigue testing to assess joint endurance under cyclic loading. Overall, FSSW is a promising welding technique for joining thin aluminum sheets, with potential applications in various industries. Optimization of the processing parameters but also the joint design can result in superior weld joints with improved mechanical qualities. To ensure the joints' reliability and durability in practical use, it is essential to characterize the joints using the proper testing methods [1].

Due to their excellent mechanical, thermal, and tribological characteristics, aluminum alloy materials are widely used in the production of sports equipment, automobiles, auto components, and aircraft products [2–4]. The reinforcements used are low-cost and well-developed production methods that are used extensively in the above-mentioned industries [5]. FSW, a minimal heat input solid-state welding procedure created by the UK's Welding Institute in 1991 [6], gave rise to FSSW. In comparison to traditional techniques, FSSW is now used to weld aluminum alloy materials because it has superior mechanical characteristics, lower distortion and residual stress levels, and a lower incidence of flaws [7]. The temperature and grain growth in the stir zone are influenced by a number of factors, including welding speed, axial force, tool geometry, tool substance, tool tilt angle, gripping force, and geometry. However, these two variables are most influenced by the tool's spinning speed [8–10]. Stronger connections were observed when welding aluminum alloys with a square-profiled FSSW tool tip made of high-carbon, high-chromium (HCHCr) steel [11]. To achieve the best results, the numbers for optimal conditions should be sought after. Software called Design-Expert is used to identify the optimal conditions. To obtain the highest tensile strength for an aluminum FSSW solder nugget, multi-objective optimization using a surface method has been used. The ideal welding parameters are determined using the desirability method to optimize the mechanical properties [12]. The process variables that affect the FSSW joint performance have been identified using analysis of variance (ANOVA) and main effect graphs [13].

Without using any filler substance, FSSW is capable of joining two or more elements. By inserting a rotating instrument into the materials to be united, frictional heat is produced, softening the materials and causing them to adhere to one another. When fusing materials such as aluminum and other lightweight metals, which are challenging to solder using traditional techniques, FSSW is especially helpful.

3.1.1 Introduction of FSSW Process

FSW and FSSW both generate frictional heat using a spinning implement. However, in FSSW, the tool is designed to create a spot weld instead of a continuous weld. This results in a stronger joint because the heat input is concentrated in a small area, which reduces distortion and produces a more uniform weld [14].

FSSW has several advantages over conventional welding processes, including:

1. It does not require any filler material, which reduces cost and weight.
2. It produces a stronger joint than conventional welding processes.
3. It can be used to connect components that are challenging to weld together using standard techniques.
4. It produces a high-quality weld with minimal distortion.

Multiple companies, including the automobile, aircraft, and marine sectors, can use FSSW. It works especially well to connect lightweight elements such as titanium, magnesium, and aluminum [15].

3.1.2 Influencing Input Parameter of FSSW

The strength and functionality of the completed joint might fluctuate depending on several variables that affect the FSSW process. Some of the main input factors that might impact the FSSW process include the following [16]:

- Tool rotational speed: This refers to the speed at which the FSSW tool rotates. Higher rotational speeds can increase the heat input and soften the material, but they may also increase the risk of defects such as material expulsion.
- Tool plunge depth: This refers to the depth to which the FSSW tool is plunged into the material being welded. Deeper plunges can increase the heat input and improve the quality of the weld, but they may also increase the risk of defects such as voids.
- Tool geometry: The size and form of the FSSW tool are referred to as tool geometry, and they can have an impact on the heat intake and material movement during the welding process. For particular materials and sizes, different tool designs might be preferable.
- Clamping force: Clamping force is the term for the energy used to keep the materials together while the FSSW procedure is being performed. Higher clamping pressures can enhance the weld's quality and lower the possibility of flaws like material ejection, but they also run the risk of causing deformation or deformity.
- Welding speed: This refers to how quickly the FSSW tool goes along the joint that's being welded. Higher welding speeds can increase productivity but may also reduce the quality of the weld and increase the risk of defects such as porosity or incomplete fusion.
- Material properties: The material properties of the materials being welded can also influence the FSSW process, including their melting point, thermal conductivity, and ductility. Materials with high melting points or low ductility may be more difficult to weld using FSSW.

Optimizing these input parameters is critical to achieving a high-quality FSSW weld. The optimal input parameter values may depend on the specific application, material, and joint being welded. Therefore, it is important to carefully evaluate and adjust the input parameters to achieve the desired results [17–19].

3.1.3 FSSW Tool Types and Significance

For FSSW, a variety of tool types are available, each with unique benefits and drawbacks. Some of the most typical varieties are listed as follows [20]:

- Conical tool: One of the most commonly used FSSW tools. It has a conical shape with a flat end and is suitable for joining materials with thicknesses up to 4 mm.
- Threaded tool: This tool has threads on its surface, which helps to remove excess material and reduce the formation of defects. It is suitable for joining materials with thicknesses up to 6 mm.
- Shouldered tool: This tool has a shoulder that contacts the material being welded, which helps to control the heat input and reduce the formation of defects. It is suitable for joining materials with thicknesses up to 8 mm.
- Triflute tool: This tool has three flutes on its surface, which helps in reducing defect formation and enhancing the weld joint quality. It is suitable for joining materials with thicknesses up to 8 mm.

The choice of tool depends on several factors, including the materials being welded, the thickness of the materials, and the desired properties of the weld. For example, a conical tool may be suitable for joining thin materials, while a shouldered tool may be more appropriate for thicker materials. Utilizing a threaded or trifluted tool may aid in enhancing the weld's quality and minimizing flaws. It is impossible to exaggerate the importance of selecting the proper instrument for FSSW. The weld's strength, longevity, and corrosion resistance can all be significantly influenced by the tool's shape and substance. The joint's stability may be jeopardized by flaws such as voids, fractures, and porosity that can be caused by using the incorrect instrument. Therefore, when choosing an FSSW instrument, it is crucial to closely consider the materials being joined as well as the intended weld characteristics [21].

3.1.4 Introduction of Optimization Technique

Finding the ideal welding process parameters for FSSW or FSW [22] procedures may be done using a variety of optimization techniques. Some of the most popular techniques are as follows:

1. Response surface methodology (RSM): A model that illustrates how the welding process's variables interact. The data from several tests created as part of RSM are used to build the output response. The best welding process factors can be discovered using optimization methods such as gradient-based optimization, genetic algorithms, or particle swarm optimization after the model has been constructed [23].
2. Taguchi method: A statistical method called the Taguchi methodology is used to identify the optimal welding process variables by designing several experiments. The method uses an orthogonal array to reduce the number of trials necessary. After that, statistical analysis is employed to determine the effect of each process parameter on how the output responds [24].
3. Genetic algorithms: To discover the ideal welding process parameters, genetic algorithms use a population-based search method that resembles the action of natural selection. The algorithm produces a population of feasible solutions, assesses their fitness based on the output response, and then employs selection, crossover, and mutation operators to produce new solutions that are, ideally, superior to the earlier ones [25].
4. Particle swarm optimization: Finding the ideal welding process settings entails using a population-based search algorithm that mimics the behavior of a flock of birds or a school of fish. The algorithm creates a colony of possible solutions, assesses each one's fitness using the output response, and then modifies each solution's location and velocity with the best ones so far discovered [26].
5. Simulated annealing: Simulated annealing involves using a probabilistic algorithm that simulates the process of annealing metals to find the optimal welding process parameters. The algorithm starts with an initial solution

and then iteratively modifies it by randomly changing one or more process parameters. The modifications are accepted or rejected based on a probability function that depends on the fitness of the solution and the temperature parameter [26].

The choice of optimization method will depend on the specific application, as each method has its strengths and limitations. The optimal welding process parameters obtained through optimization methods can help improve the quality and efficiency of FSSW or FSW processes [27].

3.1.5 MECHANICAL AND METALLURGICAL CHARACTERIZATION OF FSSW JOINTS

FSSW typically produces a weld joint through the plasticization and mixing of the materials in the joint by the use of large amounts of frictional forces without the need for melting. Such a degree of mechanical force on the joint warrants mechanical and metallurgical characterization to evaluate the quality and integrity of the joint, and to ensure the joint is capable of realizing the required performance specifications. Here are some of the commonly used techniques for mechanical and metallurgical characterization of FSSW joints:

- Tensile testing: A typical mechanical testing technique for assessing the flexibility and strength of FSSW joints is tensile testing. Typically, joints are used to create tensile examples that are then evaluated in a universal testing apparatus. The test findings can be used to calculate the final tensile strength, yield strength, elongation, and fracture toughness.
- Microstructure analysis: An optical lens, scanning electron microscopy (SEM), or transmission electron microscopy (TEM) will be used to examine the details of the FSSW joint. The microstructure of the substance can be used to observe the spread of intermetallic compounds, particle size, grain structure, and other elements that may have an impact on the joint's characteristics.
- Hardness testing: Hardness testing is a non-destructive mechanical testing method that can be used to evaluate the hardness and strength of the FSSW joint. Hardness testing is typically performed using a microhardness tester or a Vickers hardness tester.
- Metallography: Metallography is a technique used to prepare and examine metal samples to reveal the microstructure of the material. Metallography involves cutting, grinding, and polishing the FSSW joint, followed by etching with a chemical solution to reveal the microstructure.
- Fractography: Fractography is the examination of a material's split surface to ascertain the reason for failure. Fractography can help to spot flaws or other features that may have led to the failure, as well as the mode of fracture, such as a ductile or brittle fracture.

By using a combination of these techniques, the FSSW joint's mechanical and metallurgical properties can be characterized comprehensively, which is crucial for the evaluation of joint quality and its application suitability [28].

The current study relies on the weld nugget and hardness tensile strength to fabricate and assess the efficacy of FSSW joints in aluminum alloy. Tool rotational speed (rpm), axial load (kN), and welding speed (mm/min) were used in establishing a three-variable-three-level parameter Box–Behnken design matrix using Design-Expert software (v8). The tensile strength (MPa) and hardness (HRC) of the friction stir spot welded piece of aluminum alloy are determined to be the output parameters to be determined using the developed linear regression equation. The desirability technique has been used to enhance regression models to optimize the tensile strength of the welding process parameters.

3.2 METHODOLOGY

3.2.1 Box–Behnken Design

The Box–Behnken technique frequently employs response surface designs, particularly when estimating the F-statistic and using three levels and three components. The data are estimated using a variety of techniques, including sums of squares, mean sums of squares, degrees of freedom, changed determinations of the coefficient, and other types of determination coefficients. The ANOVA's sum of squares was used to calculate the proportional shares of each component. The generated data are put into the model, which generates reaction surfaces and contour plots. The optimal design factors are found using the results of the desirability method.

3.2.2 Desirability Approach

The Design-Expert software (v8) offered flexibility and accessibility using the desirability method. The desirability technique is used to emphasize the value of every unique response. The multiple response optimization issue makes use of the general desirability function method. The goal of this strategy is to transform the many replies into a performance that has no dimensions. The dimensionless individual desire function ranges from 0 to 1, with 0 being the least desirable and 1 being the most desirable. The equation may be used to calculate the overall desirability function from the individual desirability values. Low and high values must both be present in the aim. For each response optimization in this investigation, the same is applied. The following desired function determines the tensile strength and hardness minimum and maximum values. If the answer is less than 0, dn equals zero for $0 \leq dn \leq 1$.

In this paper, we determine each response of the individual desirability function using Equations (3.1–3.4). The weight fill is using the change of shape in each and every desirability function goal. The weights are used to place more importance on the lower and upper boundaries. The target value is important. The range of weight is between 1 and 0.1. While the weights less than and greater than one give less importance and more importance, respectively. If the weight is equal to one, the individual desirability response varies linearly from 0 to 1. Based on the objective function for desirability, a relative rating can be given to each answer in relation to other replies. As indicated by (+) for the least important value 1 to the most important value 5 (+++++), the significance shifts from least to most essential. When different responses are given varying degrees of importance, Equation (3.5) provides the complete objective function.

Maximum desirability will be determined by

$$di = \begin{cases} 0, Yi \leq Li \\ (\dfrac{Yi-Li}{Hi-Li})^{w}, Li < Yi < -Hi \\ 1, Yi \geq Hi \end{cases} \tag{3.1}$$

$$di = \begin{cases} 0, Yi \leq Li \\ (\dfrac{Hi-Yi}{Hi-Li})^{w}, Li < Yi < -Hi \\ 1, Yi \geq Hi \end{cases} \tag{3.2}$$

$$di = \begin{cases} (\dfrac{Yi-Li}{Ti-Li})^{w2}, Li < Yi < Ti \\ (\dfrac{Yi-Hi}{Ti-Hi})^{w2}, Ti < Yi < Hi \\ 0, \text{Otherwise} \end{cases} \tag{3.3}$$

The definition of desirability is given by the equation, which depends on the range of the goal:

$$di = \begin{cases} 1, Li < Yi < Hi \\ 0, \text{Otherwise} \end{cases} \tag{3.4}$$

$$D = \left(\Sigma r_i \sqrt[n]{\left[\prod_{i=1}^{n} d_i^{ri} \right]} \right)^{1/\Sigma r_i} \tag{3.5}$$

where n is the total number of responses, L and H are the low and high values, and Ti is the desired answer value for the ith response [17].

3.2.3 Experimental Design Process Parameter

The Box–Behnken approach is used for the experimental design using Design-Expert software (v8). This approach has three stages and three elements [29]. Because of this, the quality of the joint is affected by a variety of welding variables, including welding speed, axial weight, feed, tool shape, and substance, among others.

Three factors—axial weight (kN), tool rotational speed (rpm), and welding speed (mm/min)—determine the highest possible welding quality and power (kN). This study only takes three parameters into account. The level of the characteristic is displayed in Table 3.1 as a minimum, maximum, and medium level. Table 3.2 displays the results of 17 distinct parameterized Box–Behnken response surface trails. Each path must be conducted a minimum of three or five times in order to ascertain the mechanical and metrological behavior of the welded connections. The collected data are imported into the same software, and RSM is used to build the matrix using a polynomial and a regression equation for each answer. The adequacy metrics are considered for the optimal fit using the same software. Figure 3.1 shows the overall methodology of the present work.

TABLE 3.1

Process Parameter Levels

Range	−1	0	1
Rotational speed (rpm)	$X-1$	$X0$	$X1$
Welding speed (mm/min)	$Y-1$	$Y0$	$Y1$
Axial force (kN)	$Z-1$	$Z0$	$Z1$

TABLE 3.2

Experimental Design Matrix and Observed Results

No. of Trail	Rotational Speed (rpm)	Welding Speed (mm/min)	Axial Load (kN)	Tensile Strength (kN/mm²)	Joining Efficiency (%)	Hardness (HV)
1	−1	−1	0	T1	EF1	H1
2	1	−1	0	T2	EF2	H2
3	−1	1	0	T3	EF3	H3
4	1	1	0	T4	EF4	H4
5	−1	0	−1	T5	EF5	H5
6	1	0	−1	T6	EF6	H6
7	−1	0	1	T7	EF7	H7
8	1	0	1	T8	EF8	H8
9	0	−1	−1	T9	EF9	H9
10	0	1	−1	T10	EF10	H10
11	0	−1	1	T11	EF11	H11
12	0	1	1	T12	EF12	H12
13	0	0	0	T13	EF13	H13
14	0	0	0	T14	EF14	H14
15	0	0	0	T15	EF15	H15
16	0	0	0	T16	EF16	H16
17	0	0	0	T17	EF17	H17

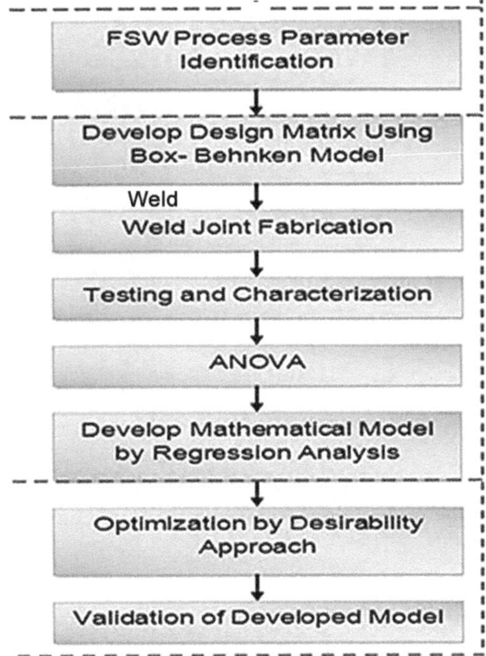

FIGURE 3.1 Overall flow of optimization process.

3.3 EXPERIMENTAL WORK

3.3.1 Tool Fabrication for FSSW

In order to execute the experiment, specialized FSSW equipment was required. At this point, the tool's design is crucial since a good tool may increase both the weld quality and the fastest feasible welding speed. High-carbon, high-chrome D2 steel is required for the fabrication of FSSW instruments. There are high wear and impact resistance qualities. The fabrication of an FSSW tool made of D2 steel with a square pin shape is depicted in detail in Figure 3.2. The temperature of D2 steel should be warmed gently to 650°C, and then it can be raised to 800°C. After being heated, it changes from having 53 to 63 HRC.

Due to its high chromium content, D2 steel has modest corrosion resistance when hardened. Table 3.3 displays the chemical make-up of high-carbon high-chromium D2 steel.

3.3.2 FSSW Process

As a result of the friction created by the revolving tool and the workpiece material during this operation, the area close to the FSSW tool becomes softened. The FSSW machine and the basic operation of the CAD drawing are shown in Figure 3.3a and b. An integrated factorial or fractional factorial design is absent from a response surface

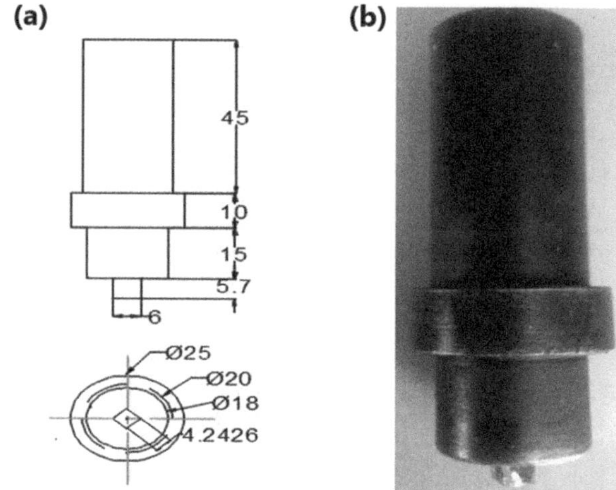

FIGURE 3.2 FSSW tool: (a) tool dimensions and (b) photograph of specialized tool after fabrication (unit: mm).

TABLE 3.3
FSSW Tool Chemical Composition

Element	Mn	Si	Co	Cr	Mo	V	P	Ni	Cu	S	C
%	0.61	0.59	1.1	12	0.9	1.1	0.03	0.31	0.25	0.03	1.45

design known as a Box–Behnken design. It is possible to create three-level three-factor surface designs utilizing Box–Behnken patterns. The three equally spaced values, often coded as −1, 0 and +1, which represent the three-level three-factor surface design parameters, are used to put each factor or independent variable.

These are the crucial settings that must be made before beginning the FSSW merging procedure. To the pneumatic controls, the constructed instrument is attached and changed. The rotational speed is now set in accordance with the specifications, and the welding speed is then set by changing the pressure in the pneumatic controls. Set the most crucial number, the load factor. The instrument now descends, penetrating and moving in the X direction. After the tool has moved the full feed distance, the pressure is removed, and the previous steps are performed again in accordance with the specifications.

3.4 RESULT AND DISCUSSION

3.4.1 DEVELOPING EMPIRICAL RELATION

The Design-Expert program's fit summary tab makes the most polynomial solution recommendations. The tensile strength (T) and Rockwell hardness (R) of the joint

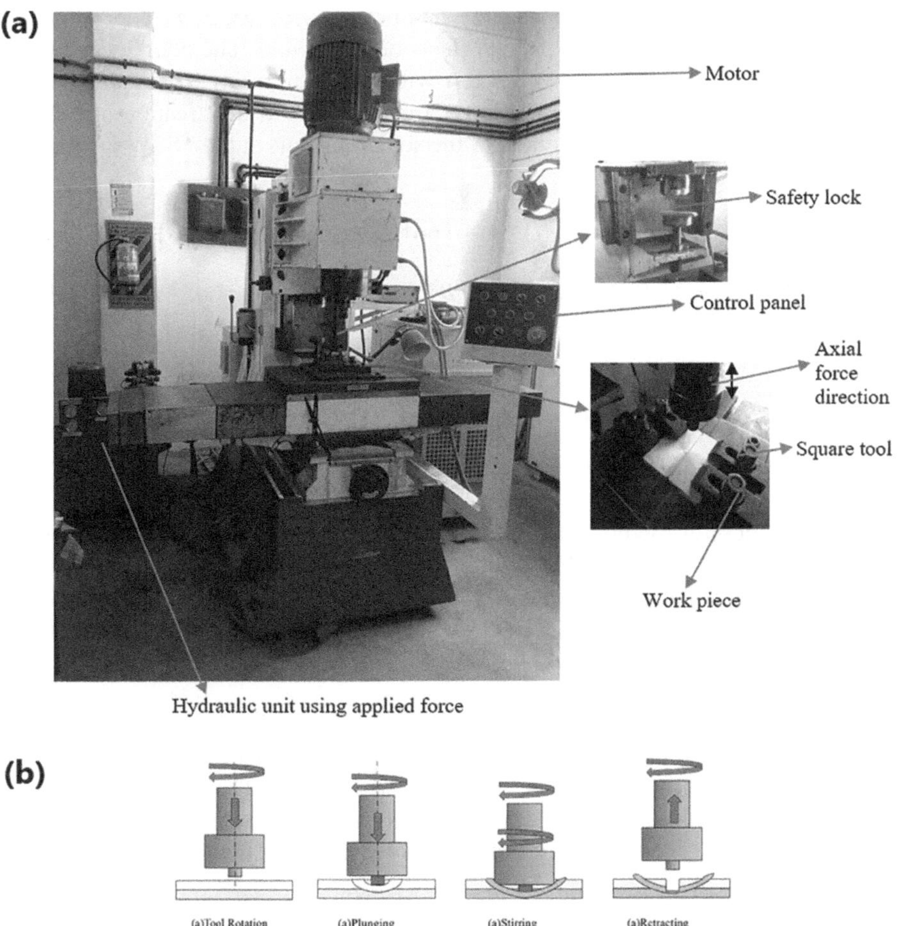

FIGURE 3.3   (a) Friction stir welding machine and (b) working process of FSSW process.

nugget are its mechanical properties (H) in Equations (3.6 and 3.7), while Equation (3.8) shows the polynomial relations.

$$\text{Mechanical properties} = f\,(\text{load, speed, welding speed}) \tag{3.6}$$

$$\sigma_T = f\,(F,\,N,\,S) \tag{3.7}$$

$$(\sigma_b)\ \text{or}\ (\sigma_H) = b_0 + b_1(N) + b_2(S) + b_3(F) + b_{11}(N_2) + b_{22}(S_2) +$$
$$b_{33}(F_2) + b_{12}(NS) + b_{13}(NS) + b_{23}(SF) \tag{3.8}$$

where b_0 is the expected result and the numbers $b1$, $b2$, $b3$,..., bn depend on the individual main and relational special effects of the factors. The confidence level is set at 97% when using the Fisher's F test with the Design-Expert software (v8) to analyze all coefficients. Just these criteria are employed in the creation of the final modes. Tensile strength and hardness of the solder nuggets at the FSSW junctions in Equations (3.9 and 3.10).

$$\text{Tensile strength} = +202.33 + 11.02 * A - 0.66 * B + 12.36 * C + 4.00 * A * B$$

$$-15.55 * A * C - 5.17 * B * C - 28.35 * A^2 - 31.48 * B^2 - 60.03 * C^2$$

$$(3.9)$$

$$\text{Hardness} = +45.50 - 1.86 * A - 1.26 * B - 0.75 * C + 0.80 * A * B + 2.13 *$$

$$A * C - 0.87 * B * C + 0.72 * A2 + 2.73 * B2 + 4.40 * C2 \qquad (3.10)$$

The effectiveness of the novel algorithm is evaluated using the ANOVA methodology. The findings of the ANOVA for joint tensile strength and particle hardness are shown in Tables 3.4 and 3.5, respectively. If the F-ratio value is less than the typical F-ratio value at a 97% confidence level, the model is considered to be sufficient.

3.4.2 WELD NAGGED HARDNESS

The weld nugget hardness was measured at three separate places in the middle of the thickness, and the average number was used for the study. The study clearly shows that the base metal has a lesser hardness than the stir zone, which is independent of the rotational zone. The disparity in hardness between the heat-impacted zone and

TABLE 3.4

ANOVA for Tensile Strength

Source	Sum of Squares	Degree of Freedom	Mean Square	F-Value	p-Value, Prob > F
Model	21099.29	9	2344.37	90.05	<0.0001
Lack of fit	97.50	3	32.50	1.98	0.35
Pure error	32.67	2	16.33		

TABLE 3.5

ANOVA for Weld Nugget Hardness

Source	Sum of Squares	Degree of Freedom	Mean Square	F-Value	p-Value, Prob > F
Model	161.50	9	17.94	19.18	0.0023
Lack of fit	4.18	3	1.40	5.57	0.1560
Pure error	0.5	2	0.25		

the zone during churning is what causes the particulate refinement in the stir zone. Figure 3.4 depicts the joint created using a tool with the lowest RPM hardness in the HAZ area on the receding side. The retreating side of the tool has significantly less toughness than the advancing side, regardless of how rapidly it turns. The HRC hardness is determined in the weld nugget region of the junction, which was created with a tool rotating at 800 rpm. The highest HRC value was found in the area of the joint where the weld nugget was formed using a tool with a 600 rpm rotational speed. An axial force of 5 kN and a welding speed of 600 rpm were used to determine the hardest possible material. This explains why this joint has better mechanical characteristics. Figure 3.4 illustrates how tool rotating speed factors affect the microhardness of an aluminum alloy (a–c).

3.4.3 Responses and the Impact of FSSW Process Parameters (Tensile Strength and Hardness)

The interplay between spinning speed and welding speed is the most important relative to all other components put together. Since it is correlated with heat production, tool rotational speed in FSSW is more sensitive than other factors. Regardless of the welding speed, low turning rates cause little heat to be produced, which in turn causes inadequate material movement and poor plasticization in the stir zone, both of which reduce tensile strength. Both the irregular material flow and the formation of coarse granules in the stir zone are caused by high spinning speed, which also results in high heat generation and heat transmission to the base material.

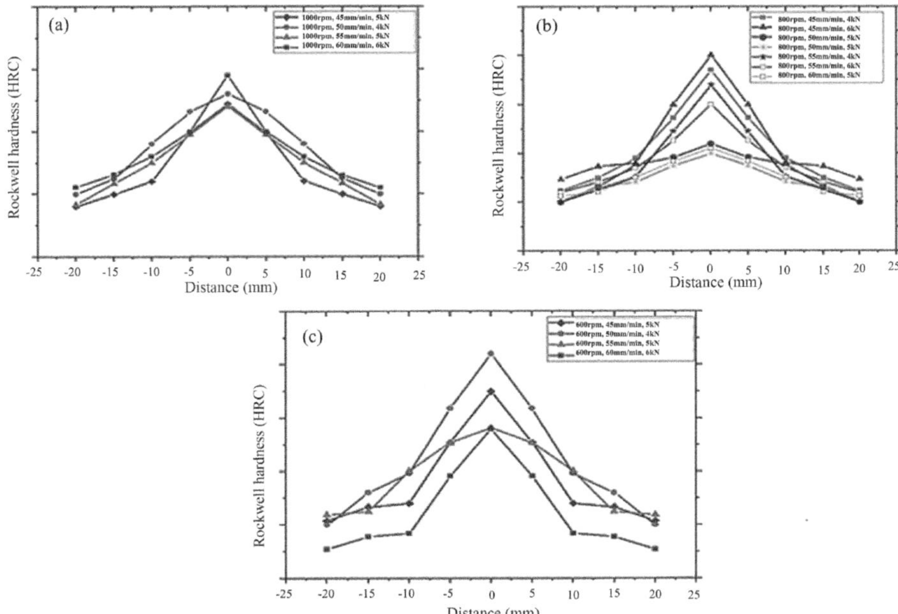

FIGURE 3.4 Microhardness and the impact of tool spinning speed factors.

The FSSW should receive the ideal amount of heat to improve tensile strength. Welding pace has a big impact on how efficiently FSSW of aluminum alloy is produced. Because the softened region is smaller for higher welding speeds compared to lesser welding speeds, with increased welding speed, the weld's tensile strength rises proportionately. Because the heat intake is lower and the cooling rate is greater at faster welding speeds, there is less time needed for the formation of grain structures.

Therefore, these areas have less power. The parts are produced based on the crucial value. If the speed is less than the critical value, the joints produced are defect-free; however, if the speed is greater than the critical value, there is a chance that the joints contain flaws. The perturbation plot in Figure 3.5 illustrates how the FSSW factors affect the improved design's tensile strength. If the reference points travel along with the selected reference point, the differences in reaction can be seen in the graph. When a component has a steep incline or curve, the reaction is sensitive. The graph clearly shows that the tool's rotating speed has the greatest impact on the joint's tensile strength.

The 2D contour and 3D surface graphs of the tensile strength and weld nugget hardness as affected by the tool's rotating speed and welding speed, respectively, are shown in Figures 3.6a,b and 3.7a,b.

Figure 3.8 shows that the turning speed is more responsive than the welding speed (e.g. at constant axial force of 5 kN). When welding, the revolving speed is less susceptible to changes in the hardness of the weld particles than the circle speed (e.g. at a constant 5 kN axial force). Compared to welding speed, rotating speed is more resistant to fluctuations in weld nugget toughness (e.g. at a steady axial force of 5 kN.)

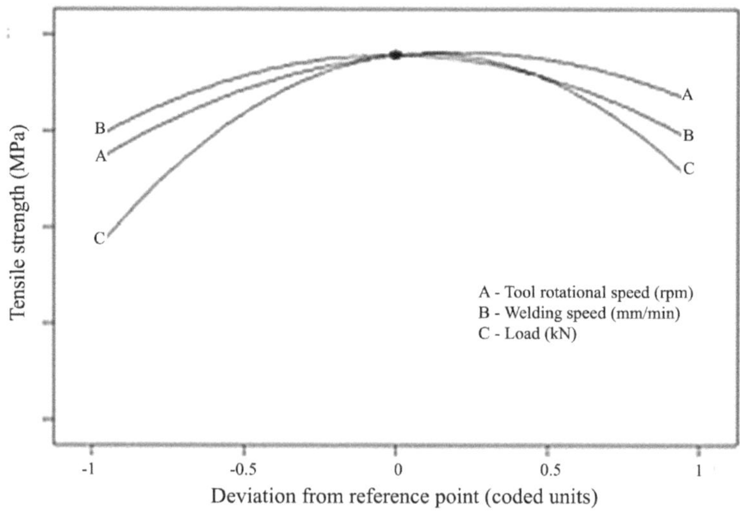

FIGURE 3.5 The perturbation plot of tensile strength as a consequence of FSSW parameters.

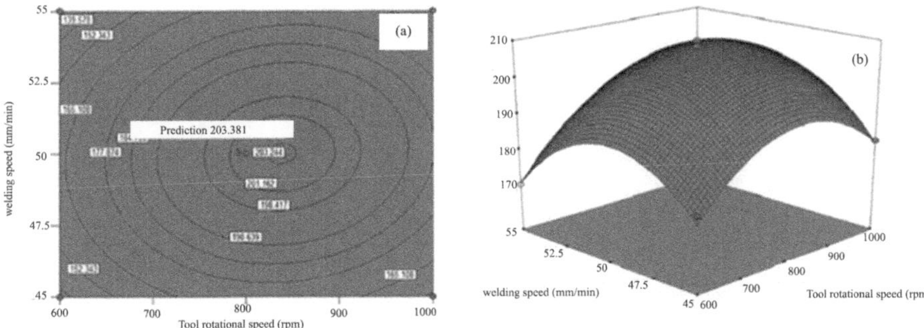

FIGURE 3.6 The contour graphs (a) depicting the impact of rotational speed and (b) depicting the effect.

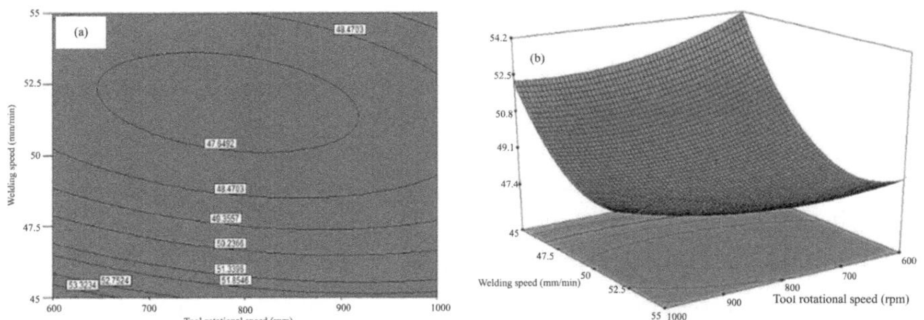

FIGURE 3.7 The contour graphs: (a) demonstrating the impact of rotary speed and (b) demonstrating the impact of welding speed on joint nagged hardness (HRC).

3.4.4 Optimization Method

Weld nugget handling should be tough since it exhibits an increase in strength. Hence, combining the study of strength and toughness makes sense. The best weld state must be determined through an optimization study in order to achieve the intended mechanical characteristics of the welded join. When the models are finished, they should be reviewed for appropriateness. The optimization criteria may then be used to select the optimal welding conditions. Increasing tensile strength and weld nugget hardness without restricting the welding settings is one of the study's primary goals. The second objective is to increase turning and welding speeds while maintaining optimum tensile strength, notch tensile strength, and weld nugget toughness: Table 3.6 contains them.

The findings of the testing and refining plainly indicate that a higher spinning rate is required to enhance tensile strength and weld nugget hardness. These findings demonstrate that rotating speed affects tensile strength and weld nugget hardness.

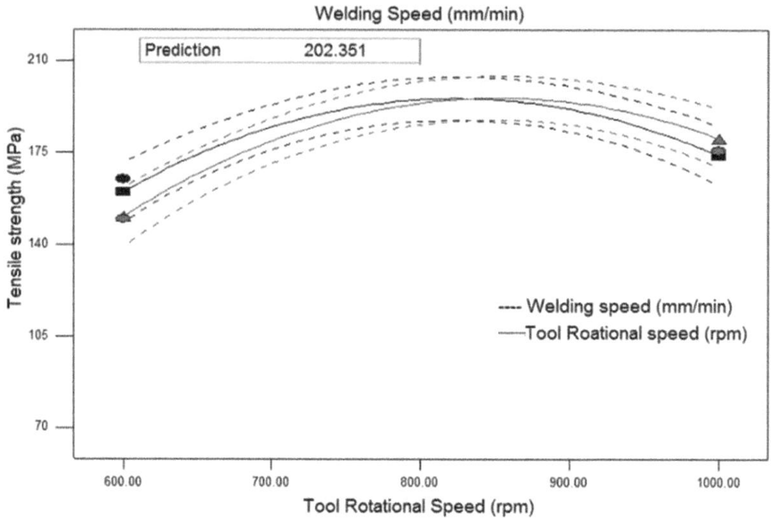

FIGURE 3.8 Interaction effect of tensile strength based on rotational speed and welding speed.

TABLE 3.6

Criteria's Used in Optimization Techniques

Sl. No	Input and Output Parameter	Range		Importance	First Criterion	Second Criterion
		Minimum	Maximum			
1	Tool rotational speed (rpm)	$X-1$	$X1$	3	In range	Maximize
2	Welding speed (mm/min)	$Y-1$	$Y1$	3	In range	Minimize
3	Load (kN)	$Z-1$	$Z1$	3	In range	Minimize
4	Tensile strength (MPa)	Min $R1$	Max $R1$	5	Maximize	Maximize
5	Hardness (HRC)	Min $R2$	Max $R2$	5	Maximize	Maximize

The improvement in the ideal welding situation in terms of maximal tensile strength and weld nugget hardness is shown in Tables 3.7 and 3.8.

It is plainly clear that the graphical representation makes it possible to choose the appropriate welding conditions visually. Overlay graphs are used to present the results, and this type of graphic is highly useful in determining the optimal welding parameter values for different reactions. The above-mentioned elements are represented by the area that is highlighted on the overlay maps in Figure 3.9. Due to the consistent material flow and production of smaller grain particles over the stir zone, the optimal weld settings, as displayed in Table 3.7 and Figure 3.9, generated enhanced mechanical properties. Table 3.8 and Figure 3.10 show the second criteria that gives minimum mechanical properties due to excess heat input and coarse grain size.

TABLE 3.7

First Criterion of Optimal Solution Obtained by Design-Expert (Example)

Experiment No.	Input Parameter 1 (rpm)	Input Parameter 2 (mm/min)	Input Parameter 1 (kN)	Response 1 (MPa)	Response 1 (Hv)	Desirability
1	789.9802	48.51518	5.160389	200	46.23976	1
2	788.3704	48.5357	5.156971	199.9999	46.23934	1
3	786.4673	48.57025	5.159226	200.0001	46.23773	1
4	786.528	48.552	5.141335	199.9998	46.23599	1
5	783.7729	48.60129	5.139114	199.9998	46.23135	1

TABLE 3.8

Second Criterion of Optimal Solution Obtained by Design-Expert (Example)

Experiment no	Input Parameter 1 (rpm)	Input Parameter 2 (mm/min)	Input Parameter 1 (kN)	Response 1 (MPa)	Response 1 (Hv)	Desirability
1	852.9251787	48.33356411	5.112137017	199.9999987	45.77780169	0.88279997
2	850.8841927	48.32145991	5.117257693	200.0000678	45.80607231	0.882783695
3	851.8041507	48.35054034	5.134891761	199.9999221	45.8029799	0.882771934
4	845.1422832	48.27520991	5.106753563	199.9999914	45.86753689	0.882760811
5	841.5460546	48.26308582	5.065130548	200.0000084	45.87243197	0.882752759

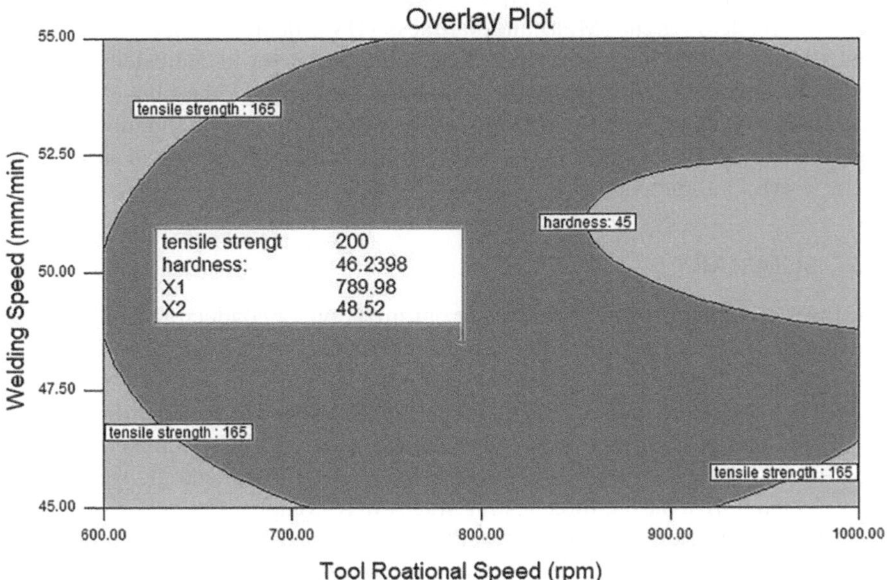

FIGURE 3.9 Plot overlays displaying the First criterion's area of ideal welding conditions.

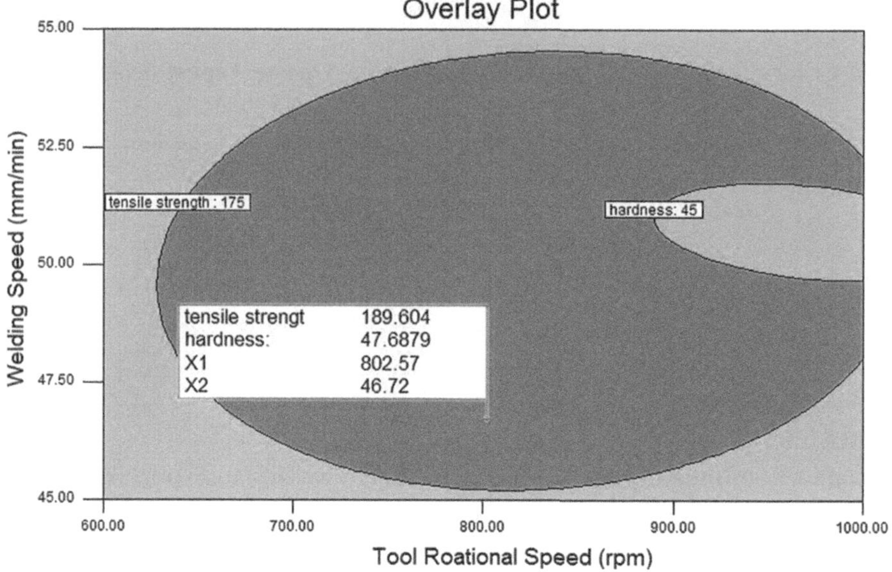

FIGURE 3.10 Plot overlays displaying the second criterion's area of ideal welding conditions.

3.4.5 CONFORMITY ANALYSIS

To assess the model derived from the optimization outcomes aimed at enhancing weld conditions, three tests were undertaken. These tests scrutinized the correlation between tensile strength (MPa) and hardness (HV), detailing their actual values, expected values, and the corresponding percentage of error in dedicated tables. The evaluation encompassed not only the actual and anticipated values but also the testing conditions, along with the findings of the percentage error validation examination. Actual values, expected values, and testing circumstances, as well as the results of the % error validation test.

3.5 SUMMARY

The investigation into process parameter optimization for conducting FSSW on various aluminum alloys led to the following conclusions:

- The connection between the reactions (tensile shear force and hardness) and the input process factors was determined using the Box–Behnken experimental design strategy (tool rotation speed, dwell time, and plunge depth). To maximize the Ts, three testing stages were completed.
- In order to examine how the welding process variables impacted the finished product, perturbation plots and contour plots were employed. The element having the most influence was found to be TRS.

- The developed model is put to the test for validation of Ts by contrasting the outcomes of each of the 17 experiments with those expected under comparable welding circumstances, assessing the rates of errors for accuracy and reliability. The rates of errors

REFERENCES

1. Pandiyarajan, R., Maran, P., Murugan, N., Marimuthu S., and Sornakumar, T. (2019) Friction stir welding of hybrid AA 6061-ZrO2-C composites FSW process optimization using desirability approach, *Materials Research Express*, 6(6). doi:10.1088/2053-1591/ab084e.
2. Pandey, A. K., Nayak, K. C., and Mahapatra, S. (2019) Characterization of friction stir spot welding between copper and poly-methyl-methacrylate (PMMA) sheet, *Materials Today Communications*, 19, 131–139.
3. Suban, A. A. K., Perumal, M., Ayyanar, A., and Subbiah, A. V. (2017) Microstructural analysis of B4C and SiC reinforced Al alloy metal matrix composite joints, *The International Journal of Advanced Manufacturing Technology*, 93, 515–525.
4. Umar, M. J., Palaniappan, P. L. K., Maran, P., and Pandiyarajan, R. (2020) Investigation and optimization of friction stir welding process parameters of stir cast AA6082/ZrO2/B4C composites, *Materials Science-Poland*, 38(4), 715–730. doi:10.2478/msp-2020-0082
5. Aita, C. A. G., Góss, I. C., Rosendo, T. D. S., Tier, M., Wiedenhöft, A., and Reguly, A., (2020) Shear strength optimization for FSSW AA6060-T5 joints by taguchi and full factorial design, *Journal of Materials Research and Technology*, 9(6), 16072–16079.
6. Ghosh, A., Yadav, A., and Kumar, A. (2017) Modelling and experimental validation of moving tilted volumetric heat source in gas metal arc welding process, *Journal of Materials Processing Technology*, 239, 52–65.
7. Abbass, M. K., Hussein, S. K., and Khudhair, A. A. (2016) Optimization of mechanical properties of friction stir spot welded joints for dissimilar aluminum alloys (AA2024-T3 and AA 5754-H114), *Arabian Journal for Science and Engineering*, 41(11), 4563–4572.
8. Kumar, P., Yadav, A., and Kumar, A. (2020) Electron beam processing of sensors relevant Vacoflux-49 alloy: Experimental studies of thermal zones and microstructure, *Archives of Metallurgy and Materials*, 65, 1147–1156.
9. Crushan, R., and Ashoka Varthanan, P. (2023) Optimization of dissimilar AA5052-H32 and AA5083-H111 aluminium FSW joints with scandium interfacial layer, *Materials and Manufacturing Processes* 2023 Aug 18;38(11):1372–84.
10. Anugrah, S., and Vikas, U. (2022) Effect of welding parameters and tool pin features on mechanical properties of dissimilar aluminum alloy friction stir welds, *Advances in Materials and Processing Technologies*. 2022 Oct 20:1-1.
11. Anton Savio, K., Lewise, J., Edwin Raja, D., and Pandiyarajan, R. (2022) Optimising aluminium 2024/7075 friction stir welded joints, *Advances in Materials and Processing Technologies*, 8(4), 4579–4597. doi:10.1080/2374068X.2022.2079226
12. Derazkola, H. A., García, E., Eyvazian, A., and Aberoumand, M. (2021) Effects of rapid cooling on properties of aluminum-steel friction stir welded joint, *Materials*, 14, 908
13. Akinlabi, E. T., Osinubi, A. S., Madushele, N., Akinlabi, S. A., and Ikumapayi, O. M. (2020) Data on microhardness and structural analysis of friction stir spot welded lap joints of AA5083-H116, *Data in Brief*, 33, 106585.
14. Kumar, A., Kumar, S., Mukhopadhyay, N. K., Yadav, A., and Winczek, J. (2020) Effect of SiC reinforcement and its variation on the mechanical characteristics of AZ91 composites, *Materials*, 13, 4913.

15. Gerlich, A., Avramovic-Cingara, G., and North, T. (2006) Stir zone microstructure and strain rate during Al 7075-T6 friction stir spot welding, *Metallurgical and Materials Transactions A*, 37(9), 2773–2786.

16. Ashok Kumar, R., Pandiyarajan, R., Raghav, G. R., Nagarajan, K. J., Rengarajan, S., Satheesh Pandian, D., and Sudhakar, K. (2023) Role of traversing speed and axial load on the properties of friction stir welded dissimilar AA6101-T6 AND AA1350 aluminium alloys, *Journal of the Chinese Institute of Engineers*, 46(1), 81–85, doi:10.1080/02533839.2022.2141338

17. Saravanan, N. et al. (2023) Optimization and characterization of surface treated Lagenaria siceraria fiber and its reinforcement effect on epoxy composites, *Pigment & Resin Technology*, 52(2), 273–284, doi:10.1108/PRT-08-2021-0093

18. Saravanan, N. et al. (2023) Optimization and analysis of dry sliding wear process parameters on cellulose reinforced ABS polymer composite material, *Pigment & Resin Technology*, 52(2), 203–210. doi:10.1108/PRT-06-2021-0067

19. Prabakaran, M. P. et al. (2020) Optimization of Nd: YAG pulsed laser welding process parameters on dissimilar metal joints, *Wutan Huatan Jisuan Jishu*, 16, 25–35, doi:10.37896/whjj16.06/094.

20. Andrade, D. G., Sabari, S., Leitão, C., and Rodrigues, D. M. (2021) Shoulder related temperature thresholds in FSSW of aluminium alloys, *Materials*, 14(16), 4375.

21. Bagheri, B., Rizi, A. A. M., Abbasi, M., and Givi, M. (2019) Friction stir spot vibration welding: Improving the microstructure and mechanical properties of Al5083 joint, *Metallography, Microstructure, and Analysis*, 8(5), 713–725.

22. Baskoro, A., Riyanto, A., Arifardi, M., and Rupajati, P. (2020) Influence of tools diameters and plunge depth on mechanical properties of micro friction stir spot welding materials A1100. In: *International Conference on Materials Science and Manufacturing Engineering*. Volume 727, (MSME 2019) 7–9 November 2019, Guangzhou, China.

23. Ponsuriyaprakash, S., Udhayakumar, P., and Pandiyarajan, R. (2023) Optimization and analysis of dry sliding wear process parameters on cellulose reinforced ABS polymer composite material, Pigment & Resin Technology, Volume 52 (2),203–210.

24. Maria jayaprakash, A., Senthilvelan, T., and Vivekananthan, K. P. (2013) Optimisation of shock absorber process parameters using failure mode and effect analysis and genetic algorithm. *Journal of Industrial Engineering International*, Volume 9 (18), 9–18.

25. Sobh, A. S., Sayed, E. M., Barakat, A. F., and Elshaerr, R. N. (2023) Turning parameters optimization for TC21 Ti-alloy using Taguchi technique. *Beni-Suef University Journal of Basic and Applied Sciences*, Volume 12 (20), 12–20.

26. Zeng, R., and Wang, Y. (2018) A chaotic simulated annealing and particle swarm improved artificial immune algorithm for flexible job shop scheduling problem. *Journal on Wireless Communication and Network*, Volume 2018 (101),101.

27. Chu, Q., Li, W., Yang, X., Shen, J., Vairis, A., Feng, W., and Wang, W. (2018) Microstructure and mechanical optimization of probeless friction stir spot welded joint of an Al-Li alloy, *Journal of Materials Science & Technology*, 34(10), 1739–1746.

28. De Castro, C. C., Plaine, A. H., Dias, G. P., de Alcantara, N. G., and dos Santos, J. F. (2018) Investigation of geometrical features on mechanical properties of AA2198 refill friction stir spot welds, *Journal of Manufacturing Processes*, 36, 330–339.

29. Ebrahimpour, A., Mostafapour, A., and Nakhaei, M. R. (2021) Application of response surface methodology for weld strength prediction in FSSWed TRIP steel joints, *Welding in the World*, 65(2), 183–198.

4 Types of Tool Geometry

*Milon D. Selvam, S. Kirubanidhi Jebabalan,
and M. Neelavannan*

4.1 INTRODUCTION

The welding procedure is one of the inevitable manufacturing methods in the manufacturing industry. In welding, there are many types, namely resistance, fusion, solid-state, and high energy density. The topic of study, friction stir spot welding (FSSW), comes under the broad classification of the solid-state welding method. In this joining method, bonding is achieved by the application of pressure, which results in deformation, leaving us with no requirement for melting or adding any filler material [1,2]. The mechanism underlying the process is that, at an elevated temperature under the melting point without experiencing severe deformation, the atoms of the workpiece diffuse across the interface to fill any leftover gaps and finish the connecting procedure [1].

The weld joint configuration is very important in the welding process, and some of the kinds of weld joint arrangements are lap joints, butt joints, T-butt joints, edge butt joints, fillet joints, multiple lap joints, etc., but when it comes to FSSW, mostly lap joints are adapted [1]. The foundation of FSSW is generally the linear friction stir welding (FSW) concept, and it was invented in 2001 to overcome resistance spot welding (RSW). The technologies RSW and FSSW are suitable to perform spot-welding sequencing on materials that possess low melting temperatures, but RSW possesses certain limitations such as high thermal conductivity, consumption of power, and current, making it quite intricate to join aluminum alloys. The RSW process is a traditional resistance welding process, whereas the FSSW is a more advanced version of FSW [2,3].

The capacity of FSSW to combine materials that are often difficult to fuse using conventional welding processes is one of the technology's key benefits. This includes materials with high melting points, dissimilar materials, and those prone to distortion or degradation during fusion welding. By avoiding the need for melting, FSSW minimizes heat-affected zones, reducing the risk of defects and preserving the material's desirable properties [1,2]. The FSSW process offers numerous benefits, such as improved joint strength, enhanced fatigue resistance, and reduced distortion. It is also a more environmentally friendly process compared to fusion welding, as it eliminates the need for filler materials and reduces the generation of harmful fumes and spatter. Due to its adaptability and efficiency, FSSW has found use in a number of sectors, including shipbuilding, electronics, aircraft, and the automobile industry. Aluminum and magnesium alloys, which are often employed in the production of modern vehicles and airplanes, are particularly well-suited to it [1–3].

DOI: 10.1201/9781003432289-4

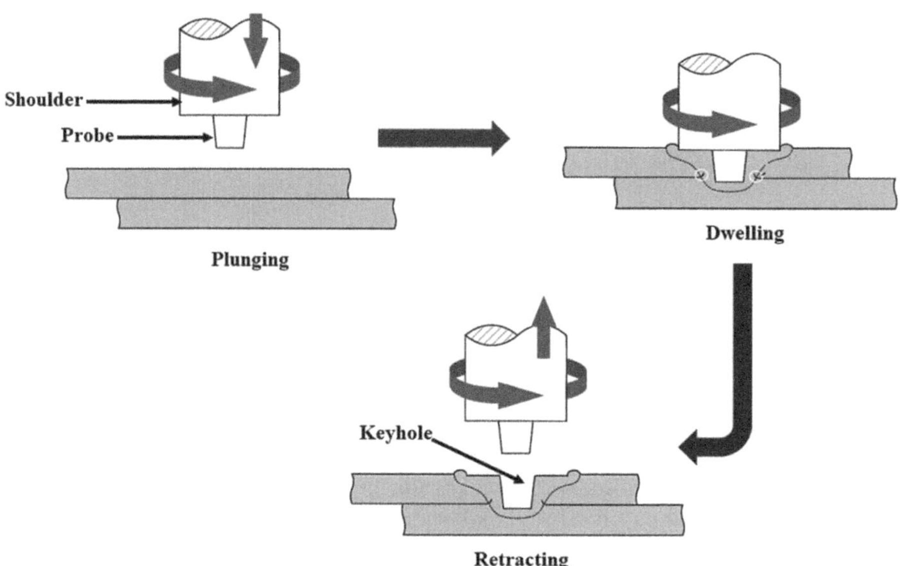

FIGURE 4.1   Schema of conventional FSSW process.

FSSW is a spot-welding method and an off-shoot to FSW on the process side, but with a crucial difference: there is rotation but no translation of the tool during FSSW. In the technology aspect and on the application side, it finds a suitable alternative to riveting. It is a three-stage process where a non-consumable tool plunges, stirs, and then retracts. The plunging operation takes place into sheets of materials, mostly placed in a lap joint configuration [4].

Figure 4.1 shows the FSSW operating schema. Once the spinning non-consumable tool has reached the necessary depth, it is lowered into the materials to be connected. After the third stage, in which the tool retracts from the welded connection and leaves a friction stir spot (FSS) weld, the process proceeds to the following stage, called dwelling. During dwelling, the tool is retained for the necessary amount of time. The heat produced and material plasticization surrounding the pin, which is brought on by tool penetration and dwell duration throughout the operation, dictate the weld's mechanical characteristics [5–7].

4.1.1   CLASSIFICATION OF FSSW

In the FSSW technique, there are various classifications, such as conventional FSSW, pinless FSSW, and refill FSSW. As we know, FSSW as a process was developed in 2001, so in these two decades, it has seen a lot of developments and a variety of processes have been explored. This section gives some insights into the classification of FSSW.

4.1.1.1 Conventional FSSW

Mazda patented conventional FSSW, often referred to as plunge FSS welding (PFSSW), in 2003 [8,9]. The Mazda RX-8 was the first vehicle to adopt this method of attaching the rear door panels. The Mazda MX-5 sports car's aluminum trunk lid was joined to the steel bolt retainers in 2005, creating a unique connection [10].

The weld spot is made using a spinning, non-consumable tool with a tapered shoulder and a threaded probe in the traditional form of FSSW. When the tool spins and is inserted into the workpieces, pressure and heat are generated at a specific location. In the welding process, the interaction of heat and pressure results in the development of a weld [6]. Three phases make up the mechanism of this procedure. In the first step, the tool is spun at the correct revolutions per minute (rpm) before the tool pin is lowered at a certain rate and depth onto the sheets that are held in the necessary joint configuration. The tool pin stays in place for a brief period of time, allowing the workpieces to connect, and then retracts, leaving an unattractive keyhole. The amount of material swept by the tool determines the shape of the hole [1,3]. Although this process has a number of benefits, including a 90% reduction in energy use and a 40% reduction in capital investment, the aesthetically unappealing keyhole that forms at the weld center causes a reduction in weld size and a lack of bonding strength, which reduces joint strength and causes the formation of hard intermetallic at the weldment [3]. This led to an alternative process known as the pinless FSSW process.

4.1.1.2 Pinless FSSW

As mentioned in the item above, the primary issue of keyhole creation in conventional FSSW results in the failure of welded joints and may be remedied by rapidly increasing the bonded area. A novel development in FSSW called pinless FSSW was created in 2009 to solve this problem, and academics have been paying close attention to it recently [11]. Pinless FSSW is a solid-state welding technology that joins two or more metal components without the need for fasteners or any extra materials, as shown in the process diagram in Figure 4.2. The procedure is similar to that of traditional FSSW [3,5], but the mechanism of weld formation is a bit different when compared to the conventional process. The grain refinement caused by the pinless tool, which causes the top sheet to migrate to the center of the weld and causes the bottom sheet material to be displaced upward, has a significant impact on it. The limitation of this process is that pinless FSSW requires a more complex tool design, which will lead to higher tool costs compared to conventional FSSW.

The ability of a pinless tool in the FSSW process to quickly and efficiently connect lightweight and high-strength metallic materials is crucial in application fields including energy, automotive, and aerospace. The procedure is summarized by drawing attention to Garg and Bhattacharya's [12] results that, for both comparable and different FSS-welded joints, the highest shear tensile strength was reached utilizing pinless tools. This result was due to the creation of stronger connections together with the eradication of flaws such as hook formation that manifest themselves during dissimilar welding in the traditional FSSW technique.

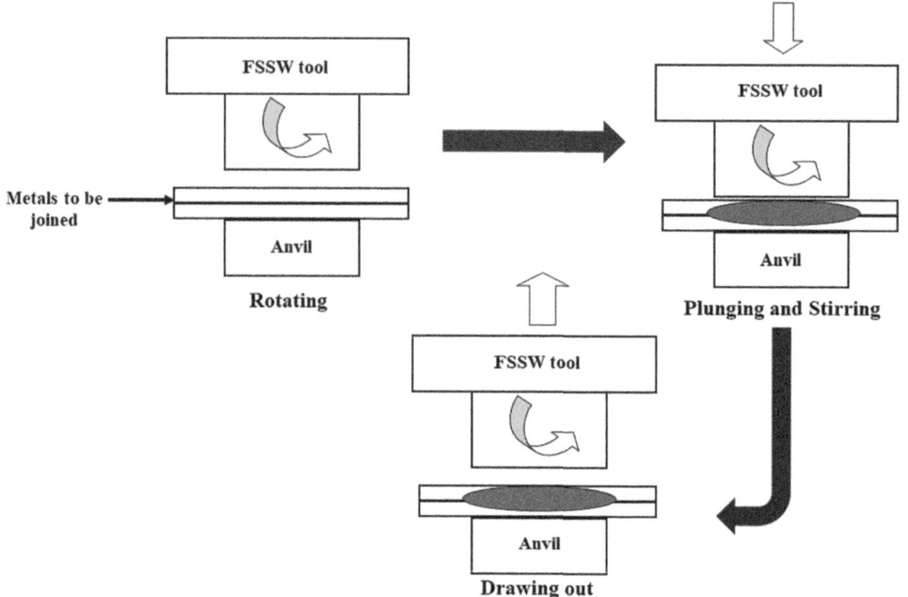

FIGURE 4.2 Pinless FSSW process.

4.1.1.3 Refill FSSW

A research center in Germany, specializing in material processing research, known as GKSS, invented a solid-state joining method called the refill FSSW process. It was invented and patented in 2004. The primary objective of this process was to join lightweight materials with low melting points, such as aluminum, with magnesium. By permitting the joining of various combinations, such as non-ferrous to ferrous metals and polymer to metals, the technique has significantly advanced in recent years [13]. Figure 4.3 depicts the refill FSSW process. The operation window involves parts like a non-consumable tool that comprises two rotating components: the probe and a shoulder, which are brought together with the help of a static clamp ring. Compared to other FSSW processes, the major improvement observed is that the rotating elements have independent vertical movement, allowing the production of a weld that doesn't have an exit hole after the tool has been extracted [14]. The joining mechanism resembles the back extrusion process, which is a plastic deformation process involving the displacement of the material.

In comparison with conventional FSSW, the refill FSSW procedure is considered to be more complex. It can be further subdivided into two types: the pin plunge type and the sleeve plunge method. Welded joints produced by the sleeve plunge type yield superior strength due to the presence of a larger nugget zone when compared to the pin plunge variant. When a comparison is made based on energy consumption, the sleeve type demands higher consumption [3,15].

Double-acting tools, which are different variations of the refill FSSW technique, have been developed much more recently. The two-step FSSW is a challenging welding procedure that increases weld cycle times and expenses [3].

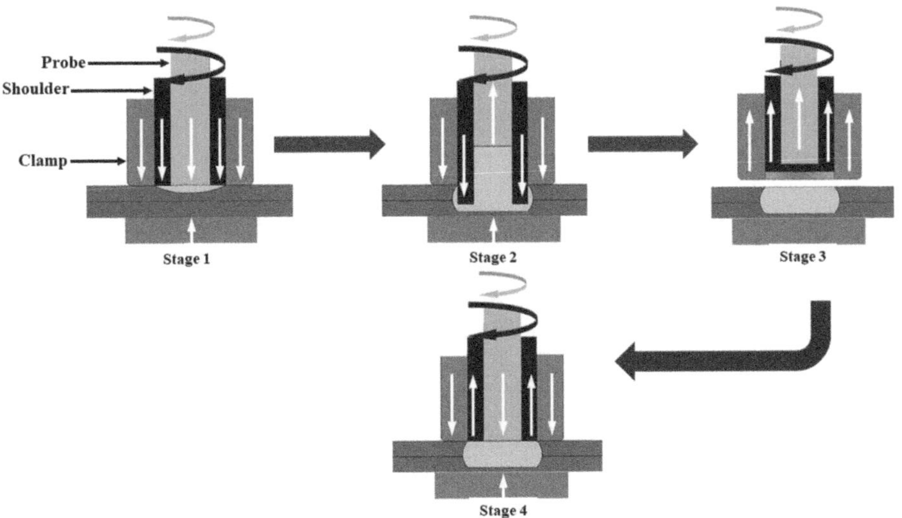

FIGURE 4.3 Refill FSSW process.

There are other welding processes like the one discussed above, such as swept FSSW, stitch FSSW, and swing FSSW, which have not gained popularity in the academic and industrial arenas [3,5]. The most investigated and researched FSSW process variants are the conventional, pinless, and refilled FSSW processes. Literature reviews clearly state that these three variants can be considered as spot-welding processes.

4.2 FSSW TOOL GEOMETRY AND DESIGN

The geometry of the FSSW tool has a major impact on the FSSW's quality and associated productivity [3,15]. The geometrical factors of the FSSW tool, comprising a pin and shoulder, should not only ensure the quality of the weldment but also create the required conditions for thermoplastic deformation and mass transfer. The durability, strength, and minimum penetration force of the tool could be determined when the tool is introduced into the welding zone [3,4].

4.2.1 TOOL GEOMETRY

The welding tool, or FSSW tool, which consists of a pin and shoulder, is a crucial component of the FSSW apparatus. Figure 4.4 shows the schematic for the typical FSSW tool configuration. The tool is vulnerable to the heat produced during the FSSW process and the material flow from the source metal, which affects the microstructure and mechanical characteristics of the weldment. The efficiency and quality of the weldment might be greatly increased by using the proper FSSW tool design [15]. Table 4.1 is a list of the different FSSW tool geometry parameters.

According to the material stated above, the FSSW tool has a pin and a shoulder, with the shoulder playing a crucial part in the FSSW welding process. When the pin

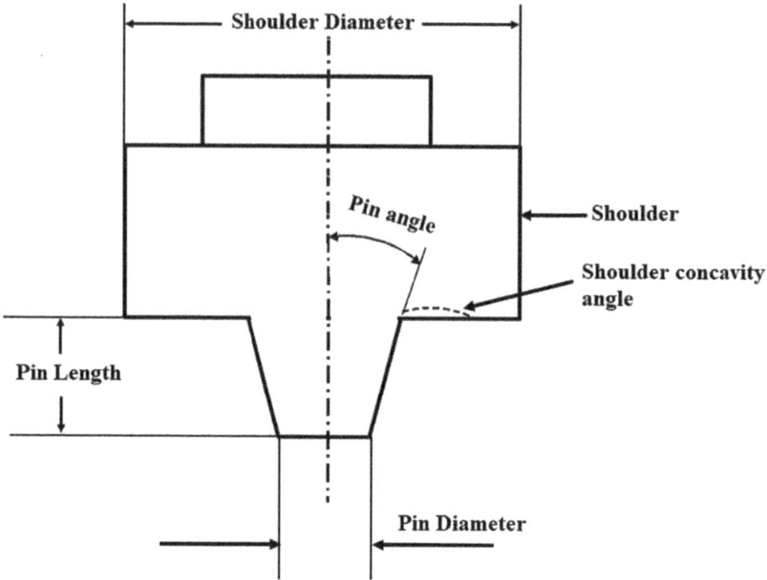

FIGURE 4.4 FSSW pin geometry.

TABLE 4.1
FSSW Tool Geometrical Factors [15]

Geometrical Factors	Range
Pin diameter	5–11.25 mm
Pin length	4–7 mm
Pin angle	0°–25°
Shoulder concavity angle	0°–12°
Shoulder diameter	15–35 mm
Tool profiles	6 types

is inserted into the workpiece, it is supported by it. The friction that develops when the tool shoulder touches the workpiece is what creates the heat needed for the FSSW welding process. It stabilizes the rotation of the tool and the plastic deformation of the material. Finally, the tool shoulder serves as a border, keeping the plastically formed material outside of the welded zone and preventing the mixing of materials with marginally different characteristics.

The pin contributes far less to heat production than the tool shoulder does. Driving the plasticized material into the welded zone is the pin's primary purpose. Heat generation (heat input), heat dispersion, and material flow rate are the parameters that are most affected by the FSSW tool shoulder and pin shape.

According to ASTM specifications, in order to produce high-quality FSSW joints, both the tool itself and the tool's shape must be carefully designed. Table 4.2 is a list of all the different tools utilized in FSSW.

TABLE 4.2

Literature of FSSW Tools

Sl. No.	Welded Material	Tool Material	FSSW Pin			FSSW Shoulder		Joint Strength (Shear Force/Shear Strength)	Ref.
			Diameter (mm)	Length (mm)	Morphology	Diameter (mm)	Morphology		
1	Al5083 and C10100	H13 tool steel	6	1.2	Cylindrical	16	Flat	2.6 kN	16
2	AA5083-O	High-carbon high-chromium steel	6	5.8	Straight square	18	-	288 MPa	17
3	AA1050/C10100	H13 tool steel	5	2.5	Conical pin	14	concave	7.07 kN	18
4	AA2024	High-carbon steel	5.4	0.8	Threaded	16.2	flat cylindrical	9.39 kN	19
5	AA6061	Tool steel	-	-	Tapered threaded	15	-	8163 N	20
6	AA6061-T6 and AZ31B	Tool steel	5	5	Tapered threaded	-	-	3.61 kN	21
7	AA2024-T3	Tool steel AISI D3	5	3	Straight cylindrical	18	flat	-	22
8	AA1060 and C11000	H13 tool steel	5	4	Conical pin	15	concave	4.8 kN	23
9	AA1050 H24	AISI D2	3	4.5		10	Concave	4.8 kN	24
10	AA 1050 and pure copper	Tool steel	5	2.8, 4, 5	Threaded	20	Flat	1.8 kN, 3.9 kN, 3.2 kN	25
11	6061 Al-T6 and pure copper	H13 tool steel	4	1.83, 2.6	Threaded	10	Concave	1.7 kN, 2 kN	26
12	Al and Cu	Tool steel	6	1.5, 2.5, 6	Tapered and threaded	16	flat	2.8 kN, 3.4 kN and 4.6 kN	27
13	AZ31B and AA6061	Tool steel	5	-	-	18	-	3.6 kN	28
14	Al 6022-T4 and Al 7075-T6	H13 tool steel	6.4	1.2	Groove	14.5	-	-	29
15	Aluminum alloy and ST16 steel	Tool steel	5.2		Cylindrical threaded	18	-	3745 N	30
16	AA5052-H32	Tool steel	3.5	3.2	Cylindrical	20	Flat	4330 N	31

The FSSW tool's fundamental profile is made up of a linear-shaped rod with two distinct parts: a shoulder that is somewhat thicker and a pin that extends from the shoulder. The thickness of the workpiece that has to be welded determines the tool's size. The FSSW technique includes bringing a tool that is rotating at a high speed into contact with the surface of the workpieces. The workpiece is next pierced by the pin to a depth that is about equivalent to its thickness. The tool's shoulder automatically hits the surface when the pin enters the workpiece, causing the tool to travel along the joint line at the feed rate. The friction created by the tool pin's shoulder moving back and forth causes the metal to heat up. The metal is pushed out into the empty area behind the moving tool as a result of the metal deforming plastically along the spinning tool. The tool's shoulder restricts the amount of material that may be driven out from above. The spinning tool is removed from the joint when the joining procedure is finished [3,5]. Figure 4.5 shows the various FSSW pin profiles that have been adopted in practice.

High hardness, high heat resistance, wear resistance, and low thermal conductivity should all be characteristics of the FSSW tool material. This is especially true for welding tool materials made of steel, titanium, nickel alloys, etc. Surface hardening and coating methods are occasionally utilized to enhance the tool's quality attributes. To regulate the flow of the plasticized material, unique profile incisions are produced on the pin and shoulder surfaces. The pin might be square, tapered, threaded, triangular, hexagonal, or straight cylindrical in form. Additionally, the shoulder's surface might be conical, concave, or flat [2–5].

The fundamental building blocks for the advancement of FSSW innovation are the FSSW tool configuration features, such as tool material and tool geometry/design, and the FSSW process variables, such as plunge rate, rotational speed,

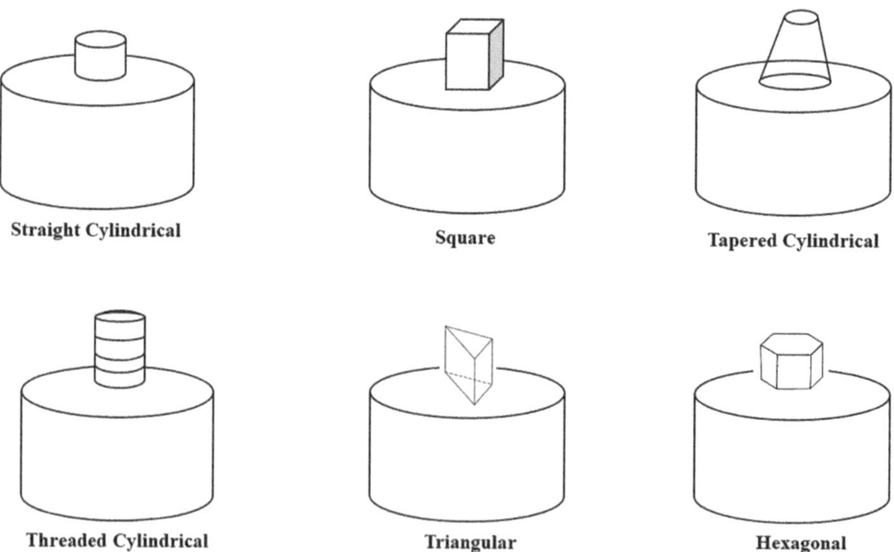

FIGURE 4.5 FSSW pin profiles.

plunge profundity, and dwell duration. These variables fundamentally affect heat production, material stream, force, hub force, weld integrity, and weld tool deprivation mechanisms. Deformities such as holes, absence of holding, fractured top-off, or insufficiently reinforced sections may result from lower heat generation and an inadequate material stream. However, with greater intensity input, rapid disintegration causes grain growth or inappropriate relaxation of the surrounding heat-affected zone (HAZ). It is crucial to get a deeper understanding of the factors affecting these variables in order to study the best FSSW tool design and interaction boundaries. The effect of pre-and post-treatment on the strength of the weld should also be taken into consideration [6,9].

The shape and design of the FSSW tool directly affect the weldment. The mechanical characteristics of the weldment would be affected by the heat produced as a result of friction during the FSSW process and the flow of material. Mubiayi and Akinlabi [23] examined the effects of tool geometry on Al/Cu while using a conical pin and concave shoulder (CCS) and flat pin and flat shoulder (FPS). The FPS could withstand the most shear force.

The geometry and size of the FSSW welding device are also considered key factors influencing the strength of the FSSW weldment. In welding joints without an entrance on the upper plate, joining usually takes place based on metallurgical bonding at the connecting point. Factors such as high strain rates and welding temperatures can also contribute to creating a better weldment. Larger welding devices have a positive advantage, particularly when joining dissimilar materials such as aluminum and copper. Only a shoulder is included on the pinless FSSW tool, which is viewed as a less desirable example of a suitable pin tool. The primary benefit of refilling FSSW, in this example, is that different thicknesses may be welded utilizing a changeable pin or sleeve length. To combine materials with low melting points, such as aluminum and magnesium alloys, tool steels are the best materials currently available. The application of FSSW to industrially applicable components and materials with high melting points, such as steel alloys, is the goal of futuristic development.

4.2.2 FSSW Tool Materials

In the FSSW process, various types of materials are being used in order to ensure effective joining and durable tool life. For the FSSW process, the choice of tool material is crucial since it heavily depends on the materials that will be welded. Table 4.3 lists a few of the common tool materials.

4.2.3 Influence of the FSSW Tool

Aluminum and magnesium alloys are the principal materials that are joined with FSSW because of their comparatively low melting points [13]. The welding of copper, nickel, and titanium alloys, as well as steels, by this method was also successfully performed [14]. With the help of FSSW, aluminum alloys up to 70 mm thick are welded in one pass. FSSW makes it possible to obtain lap joints of aluminum sheets with a thickness of 0.2 mm or more. The welding speed of 6082 alloys with a 5 mm thickness may reach up to 1 mm/s. The main factors that influence the process of

TABLE 4.3
FSSW Tool Materials [5,32,33]

FSSW Tool Materials	Description
High-speed steel (HSS)	High-speed steel is a popular tool material due to its good combination of wear resistance, toughness, and thermal stability. HSS tools can withstand the high rotational speeds and pressure during the FSSW process.
Tool steel	Tool steels, including various grades such as D2, A2, and M2, are chosen for their high hardness, abrasion resistance, and toughness. They can be suitable for FSSW applications where the materials being welded are not excessively challenging.
Carbide-based tools	Carbide tools, such as cemented tungsten carbide, are exceptionally hard and wear-resistant, making them suitable for welding abrasive materials or for extended tool life requirements.
Ceramics	Some ceramic materials, such as aluminum oxide or silicon nitride, can be used as FSSW tools. Ceramics are known for their high hardness and resistance to wear, but they may be brittle and require careful handling.
Composite	Composite tools, combining different materials such as tungsten carbide and steel, can offer improved performance by combining the advantages of each material.

the FSSW are welding speed, tool rotation frequency, tool pressing and movement forces, tool inclination angle, and dimensions [15,16]. The type of material to be welded, its thickness, and welding speed all affect the pushing and moving forces. Aluminum alloy samples from AA7010 to AA7651 were welded at a speed ranging from 1 to 2.5 mm/s with a tool rotating at 180–660 rpm. This welding experiment revealed that as the rotation speed increases, the heat input into the metal increases, and a microstructure with more uniform grains forms in the weldment [14,17]. To a certain extent, this also improves the strength and plasticity of the material. To obtain ideal circumstances, it is important to raise the tool's rotation speed in addition to the welding speed. To avoid welding faults, however, it is required to guarantee the material's characteristics, dependability, and manufacturability and to carefully choose the production methods that are most suited for generating the material's uniform microstructure [15,16,18].

4.3 PARAMETERS OF FSSW

FSSW is a derivative of FSW; this derivate process, when compared with FSW, not been able to gather a wider audience but has been gaining momentum in the last decade among researchers in academia as well as in mechanical factories. The three unique zones created by the FSSW process are the stir zone (SZ), also known as the nugget zone, the thermo-mechanically affected zone (TMAZ), and the heat-affected zone (HAZ) [3]. This topic focuses on how the parameters affect these zones, and in turn, enhance or decrease the strength and mechanical properties of the weldment. The readers must understand these correlations, as each parameter will focus on each zone for each zone formation, i.e., the bonding of these parameters does play

TABLE 4.4
FSSW Parameters [15]

FSSW Parameters	Range
Tool rotation speed	560–1120 rpm
Plunge depth	0.2–1.2 mm
Delay before retracting	0–60 s
Dwell time	8–90 s

an important role. For a welding process to be accepted by the end user/customer, the welding strength plays a significant role in determining the strength of the tool used to weld. Its geometrical profile does play a role, but behind the tool is something known as the parameters. Tool geometry and weld parameters are quite interwound, but efficiently optimized parameters only help in the successful movement of the tool across the weld line [3]. So, this subtopic will be entirely devoted to helping the readers comprehensively understand the parameters accompanying the process and how it affects the weld; this conceptual understanding will be the key takeaway in this subtopic.

Rotational speed, the amount of time before retraction, plunge depth, and dwell time are the process variables that are essential for the success of FSSW technology. In Table 4.4, the standard FSSW settings are presented.

Choosing the proper value for the process parameters is crucial. The trial-and-error method is the most basic and unscientific approach that may be applied. Data fitting to a second-order quadratic mode using artificial neural networks, response surface methodology, design of experiments methodologies, signal/noise (S/N) ratio, and analysis of variance using the Taguchi approach are just a few examples of the scientific tools that have seen a surge in use in the last 15 years in determining process parameters [34,35]. The purpose of optimizing the welding parameters is to reduce cycle time and the impact of tool force on the component. The outputs, such as axial force and torque, are significantly affected by welding parameters. The heating rate, weld temperature, weld integrity, and formation of a welding defect known as hook geometry all depend on the welding parameters [36,37].

The importance of understanding process parameters lies in the fact that controllable inputs due to certain transformational activities give rise to uncontrollable output process parameters and hence should be understood so that users select optimized parameters.

4.3.1 ROTATIONAL SPEED

The rate at which a welding tool spins around its axis is referred to as tool rotation. It is one of the most important factors in the production of heat. Internal parameters such as heat production and material flow, as well as outputs such as torque, axial force, weld integrity, and tool degradation processes, are all impacted by the rotational speed. Tool torque, which is responsible for creating the thermal cycle and energy in the form of heat production, contributes to the energy created during FSSW [37,38].

The rotating speed in FSSW is frequently expressed in rpm. Because of the increased friction caused by the increased rotation rate, the materials are stirred and mixed more vigorously, which has a favorable effect on the bonded or welded area. The weld strength rises as a result of this productive partnership. As the tool shoulder remains in touch with the SZ material during the whole process, the parameter rotating speed mostly affects the SZ of the weldment.

According to the literature, friction stir processing and friction stir welding both call for tool rotation rates between 200 and 1000 rpm [39]; however, for friction stir welding, we need tool rotation speeds of more than 1000 rpm to achieve the necessary mechanical characteristics in the SZ area. In order to achieve proper bonding in the SZ region, efficient energy utilization, and keeping up with another competitive welding process such as RSW, the SZ region requires very little energy. This is because lower tool speeds allow energy to be dissipated into the adjacent heat sinks. Therefore, in order to maximize heat input, shorten the welding process, and ultimately achieve the goal of a strong bond, we must increase tool rotation speed [40,41].

4.3.2 PLUNGE RATE

The rate at which the tool applies downward pressure to the workpiece and moves over the weld joint is referred to as the plunge rate. It is expressed in mm/min. Since the material undergoes substantial modification as a result of the heating rates and stresses involved, which are induced by high temperatures, this step is extremely important in understanding the physics behind the process. Research in this area has been concentrated on studying the effect of plunge rate as it determines the heating rate, efficiency of energy utilized, and role of plastic deformation. These critical factors are investigated as important in determining weld quality [42]. The plunge rate has a substantial impact on energy use. A lower dive rate leads to greater energy efficiency since the welding cycle lasts less time. Higher plunge rates are typically used for metals with low melting points, but for materials with high melting point, the plunge rate needs to be very low to provide enough heat for the metal to plasticize, which then reduces axial stresses. The SZ width reduces as the tool penetration rate rises, limiting the bonding area. Additionally, a faster plunge rate increases the axial force, which shortens the tool's lifespan [43,44].

4.3.3 PLUNGE DEPTH

The term 'plunge depth' refers to how deeply the tool penetrates the workpiece that has to be connected. The depth to which a tool may penetrate is dependent on the length of the pin, as was demonstrated in the preceding subtopic with specific emphasis on tool geometry. The breadth of the weld strength and bond size is significantly influenced by the pin's length [45]. Shoulder strength may be lost owing to excessive plunging if the length of the pin exceeds the zone of material flow formed by the tool [46].

The macrostructure parameters of the weld joint, such as tensile strength, are significantly influenced by the shoulder plunge depth. The literature unequivocally demonstrates that when the plunge depth rises, the SZ's microhardness rises as well. The SZ's grain sizes exhibit elongation at the microstructure level since this increases the joint under study's mechanical characteristics by indicating a grain's resistance to plastic deformation [47,48].

4.3.4 Dwell Time

As heat generated from the tool shoulder in conventional FSSW is relatively small, dwell time is a crucial parameter to account for. A dwell duration is specified, which is the number of seconds the tool shoulder is allowed to rest on the workpiece in order to allow material flow and noticeably enhance heat generation. The energy input necessary for the establishment of the stir zone and the creation of a bonded region between the workpieces is provided by a crucial parameter known as dwell time [49]. Longer dwell times on metals with higher melting points will enhance the heat input and provide the predicted temperature needed for plastic flow [50]. When dwell time is granted, there are certain drawbacks. One of them is an increase in welding cycle time, which might be a serious problem if FSSW and RSW weld cycles are compared. Less HAZ softening of heat-treatable Al alloys results from quick welding cycles of less than one second [51].

4.4 CASE STUDIES

4.4.1 Case Study 1: Effect of FSSW Tool Geometries on the Static Strength of Joined Polyethylene Sheets [15]

In this case study, it was examined how the FSSW tool affected the static strength when attaching sheets of 4 mm thick high-density polyethylene. In this study, six different types of FSSW tool pin profiles, including cylindrical straight (CS), cylindrical threaded (CT), cylindrical tapered (TC), square (SQ), triangular (TR), and hexagonal (HG), with various tool geometrical factors, including pin length, shoulder diameter, and pin angle, were selected to process the FSSW. To determine the static weld strength of the weldment, shear tests for the manufactured lap-joined polyethylene sheets were carried out while the welding procedures were being carried out and monitored at room temperature.

The production of the FSSW tool, which was made from alloy steel grade AISI/SAE1040 and subsequently hardened up to 40 HRC using a heat treatment method, is a crucial part of this work. Each of the six selected FSSW tool profiles contains pins that are around 5.5 and 7.5 mm in length and size, respectively. The pin angle of the tapered cylindrical shape is 15°. By calculating the diameter of the CSA that was connected to the turning pin, the size of the pin in the square, triangular, and hexagonal pin profiles was determined. Additionally, the diameter of the pin's bottom side was measured in order to determine the size of the cylindrical tapered, cylindrical straight, and cylindrical threaded pin profiles. Figure 4.6 shows how the six various FSSW tool pin profile types affected the static shear tensile strength.

As mentioned, the diameter of the cylindrical straight profile pin was 7.5 mm, whereas the cylindrical tapered pin was designed with a pin angle of 15°, and the

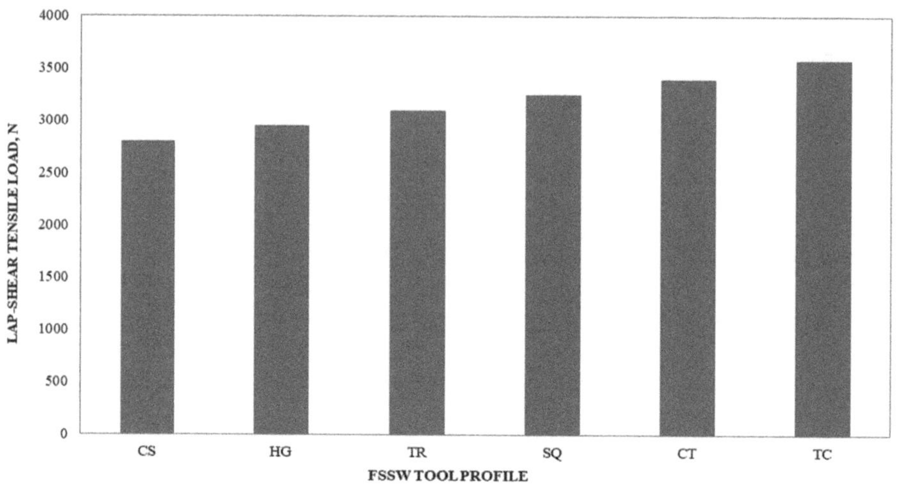

FIGURE 4.6 Lap-shear strength of polyethylene sheets welded with different FSSW tool profiles.

design factor for the bottom diameter was 7.5 mm. The diameter of the cylindrical threaded profile pin was also 7.5 mm. Likewise, the measurement of the other three pin profiles, such as square, triangular, and hexagonal, was also maintained at 7.5 mm. The study elucidates that the cylindrical tapered pin profile gives the ultimate strength, followed by the cylindrical threaded, square, triangular, hexagonal, and finally the straight cylindrical produced the poorest strength. From this study, it is evident that the tool profile geometry has a significant influence on the mechanical property (shear strength) of the FSS-welded polyethylene sheets.

4.4.2 Case Study 2: Effect of FSSW Tool Geometry on the Morphology and Mechanical Properties of Acrylonitrile Butadiene Styrene (ABS) [52]

An FSSW tool with a flat morphological pin was created, manufactured, and used in this case study to join acrylonitrile butadiene styrene (ABS) sheets. The weldment morphology and the mechanical characteristics were found to be influenced by the FSSW pin geometries. ABS sheets were FSS-welded using an FSSW tool that had a pin and shoulder. AISI1045 alloy steel was used to make the FSSW tool in this investigation.

It is important to determine how the angle between the tensile direction and the longitudinal orientation of the keyhole affects the mechanical characteristics of the weldment. As a result, three distinct keyhole orientations, including 0°, 45°, and 90°, were selected for this investigation.

The CS and surface appearances of the ABS samples produced by the FSSW method with the aid of tool pin profiles such as cylindrical (CY), triangular (TA), triflute (TF), and flat (FL) with the factors of 500 rpm tool rotational speed, 10 mm/min

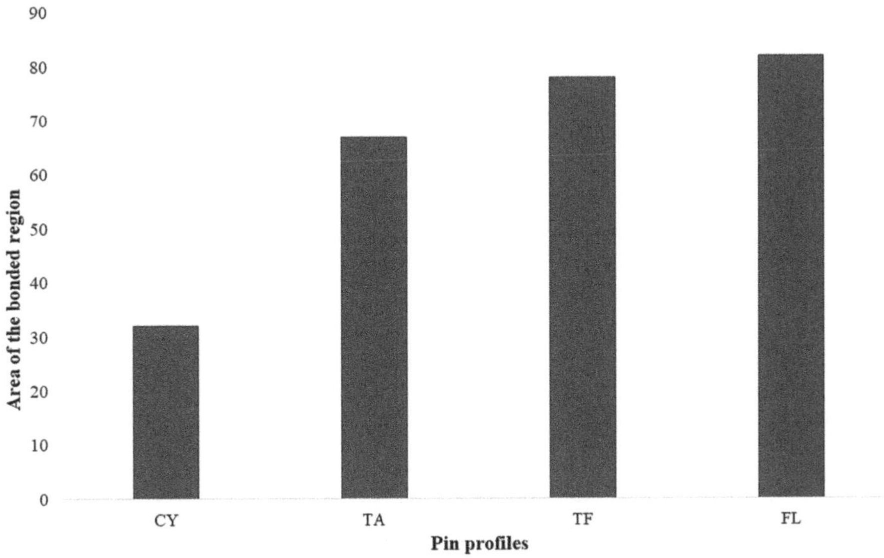

FIGURE 4.7 Area of the bonded region for the four tool pin profiles.

tool plunge rate, 400 rpm tool stirring phase, and 12 s dwell time have been noted. On the spot-welded ABS, it was noted that all the sample surfaces were clean and devoid of CS flaws. The keyhole generated was found to be significantly influenced by the size of the pin profile; larger keyholes were seen when the pin profile was likewise large.

The area and diameter of the bonded regions on the ABS sheets produced by the four-pin profiles are portrayed in Figures 4.7 and 4.8, respectively. The area of the bonded region (A_b) can be calculated from Equation (4.1):

$$A_b = \frac{\pi \times D_b^2}{4} - A_k \tag{4.1}$$

where D_b is the diameter of the bonded region and A_k is the keyhole area. This study explored that the D_b of the weldment that was fabricated by using a triangular pin profile was approximately 11.32 mm, which was bigger than the D_b formed by the other three pin profiles. Additionally, it was found that the material relieved by the triangular pin profile was larger than that by the flat and triflute pin profiles. This is because the triangle pin design generated more frictional heat, which caused the ABS material to soften and the D_b to increase.

It was observed that the A_b formed in the case of the cylindrical pin profile is small due to the bigger keyhole, whereas the flat pin profile yielded the largest A_b. The shear fracture surface morphology of the four samples fabricated with different pin profiles is portrayed in Figure 4.8.

Figure 4.9 illustrates the lap-shear strength of ABS sheets welded using the four tool pin profiles. The sample with a flat pin profile was found to be able to withstand

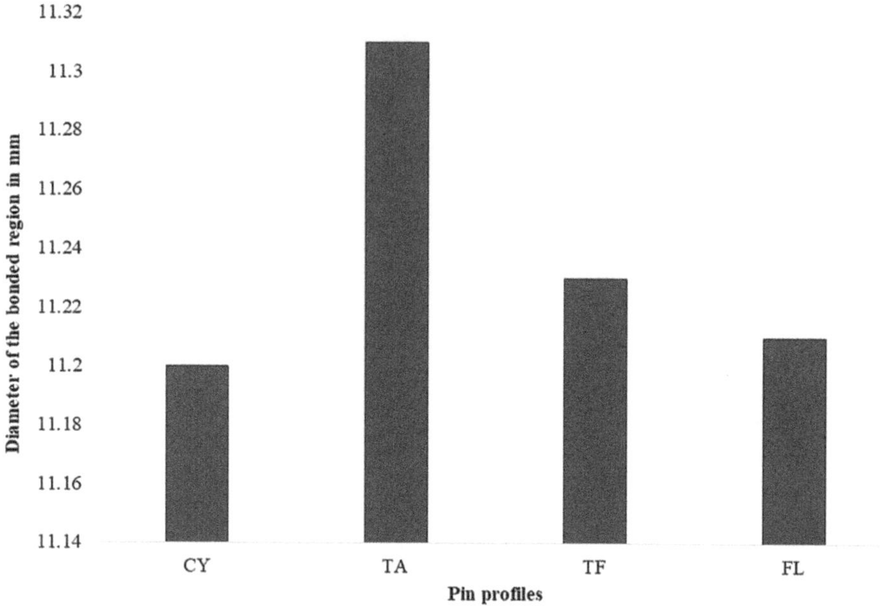

FIGURE 4.8 Diameter of the bonded region for the four tool pin profiles.

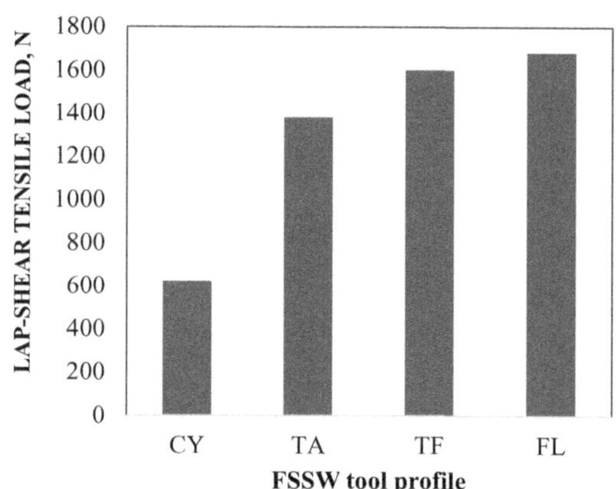

FIGURE 4.9 Lap-shear strength of ABS sheets welded with the four tool pin profiles.

the highest shear fracture force, but the sample with a cylindrical pin profile had poor mechanical characteristics. There is a vital observation that the lap-shear tensile load increases with bigger A_b, which means optimizing the FSSW tool pin geometry would effectively increase the A_b and enhance the quality of the weldment of ABS.

4.5 CONCLUDING REMARKS

The development of FSSW has been seamless and fruitfully progressing for the past two decades. FSSW is a solid-state welding process, and in this chapter, the various FSSW tools, FSSW tool geometry, FSSW tool materials, various tool geometrical factors, their application in various sectors, its advantages over fusion welding, along with application-based case studies like how tool geometry plays an important role in increasing the static strength of polyethylene sheets and how it improves the structural integrity of a thermoplastic material like ABS, were discussed. This chapter also publicizes the area of solid-state joining and will be able to distinguish between FSW and FSSW. Now coming to the futuristic scope of FSSW, a few areas of potential advancements and developments that could shape the future of FSSW are discussed below that might find their way to academic interest and industrial implementation.

- Tool development: Reducing wear and durability are the major areas of concentration in tool design. Future developments may focus on this area by investigating and modifying the shape, material, and surface coating. By concentrating on these areas, the desired results can be obtained.
- Integration of sensor: A sensor can be integrated into the welding process to monitor the process and provide real-time feedback, thereby helping in detecting defects and making adjustments to optimize the quality of the weld.
- Introduction of artificial intelligence: This technology can be applied to analyze vast amounts of data that are generated during FSSW operations. This can help in identifying the patterns, optimizing process parameters, and predicting the weld defects earlier.
- Dissimilar joining: Research and development efforts can focus on expanding the variety of materials that can be effectively spot-welded using friction stir techniques, including multiple combinations of metals, composites, and plastics.

These are the future areas on which the actual direction of future developments in FSSW will depend. This chapter has thus been able to strike a fine balance between the development of the FSSW field by discussing the earlier stages, and current development and providing the areas in which future research can be explored.

REFERENCES

1. Rizvi, S.A., 2010. *Advanced Welding Technology*. SK Kataria and Sons.
2. Procesov, E.P.O., 2011. Experimental comparison of resistance spot welding and friction-stir spot welding processes for the en aw 5005 aluminium alloy. *Materiali in Tehnologije*, 45(5), 395–399.
3. Shen, Z., Ding, Y. and Gerlich, A.P., 2020. Advances in friction stir spot welding. *Critical Reviews in Solid State and Materials Sciences*, 45(6), 457–534.

4. Mubiayi, M.P., Akinlabi, E.T. and Makhatha, M.E., 2018. Current state of friction stir spot welding between aluminium and copper. *Materials Today: Proceedings*, 5(9), 18633–18640.

5. Mubiayi, M.P., Akinlabi, E.T. and Makhatha, M.E., 2019. *Current Trends in Friction Stir Welding (FSW) and Friction Stir Spot Welding (FSSW)* (Vol. 6). Cham, Switzerland: Springer International Publishing.

6. Tozaki, Y., Uematsu, Y. and Tokaji, K., 2010. A newly developed tool without probe for friction stir spot welding and its performance. *Journal of Materials Processing Technology*, 210(6–7), 844–851.

7. Tweedy, B.M., Widener, C.A., Merry, J.D., Brown, J.M. and Burford, D.A., 2008. *Factors Affecting the Properties of Swept Friction Stir Spot Welds* (No. 2008-01-1135, SAE Technical Paper). SAE.

8. Lee, Y.W., Mowazzem Hossain, M.A.A., Hong, S.T., Yum, Y.J. and Park, K.Y., 2011, June. Characterization of friction stir spot welding of aluminium alloys using acoustic emissions. In: *ISOPE International Ocean and Polar Engineering Conference* (ISOPE-I). ISOPE.

9. Yuan, W., Shah, K., Ghaffari, B. and Badarinarayan, H., 2016. Friction stir welding of dissimilar lightweight metals with addition of an adhesive. *Friction Stir Welding and Processing*, VIII, 127–135.

10. Milon D Selvam, https://newsroom.mazda.com/en/publicity/release/2003/200302/0227e.html (assessed on 28/05/2023).

11. Aota, K. and Ikeuchi, K., 2009. Development of friction stir spot welding using rotating tool without probe and its application to low-carbon steel plates. *Welding International*, 23(8), 572–580.

12. Garg, A. and Bhattacharya, A., 2017. Strength and failure analysis of similar and dissimilar friction stir spot welds: Influence of different tools and pin geometries. *Materials & Design*, 127, 272–286.

13. Tier, M.D., Rosendo, T.S., Dos Santos, J.F., Huber, N., Mazzaferro, J.A., Mazzaferro, C.P. and Strohaecker, T.R., 2013. The influence of refill FSSW parameters on the microstructure and shear strength of 5042 aluminium welds. *Journal of Materials Processing Technology*, 213(6), 997–1005.

14. Aota, K., Takahashi, M. and Ikeuchi, K., 2010. Friction stir spot welding of aluminium to steel by rotating tool without probe. *Welding International*, 24(2), 96–104.

15. Bilici, M.K. and Yükler, A.I., 2012. Influence of tool geometry and process parameters on macrostructure and static strength in friction stir spot welded polyethylene sheets. *Materials & Design*, 33, 145–152.

16. Siddharth, S. and Senthilkumar, T., 2016. Optimization of friction stir spot welding process parameters of dissimilar Al 5083 and C 10100 joints using response surface methodology. *Russian Journal of Non-Ferrous Metals*, 57, 456–466.

17. Jannet, S., Mathews, P.K. and Raja, R., 2015. Optimization of process parameters of friction stir welded AA 5083-O aluminium alloy using response surface methodology. *Bulletin of the Polish Academy of Sciences. Technical Sciences*, 63(4), 851–855.

18. Habibizadeh, A., Honarpisheh, M. and Golabi, S.I., 2021. Determining optimum shear strength of friction stir spot welding parameters of AA1050/C10100 joints. *Manufacturing Technology*, 21(3), 315–329.

19. Karthikeyan, R. and Balasubramanian, V., 2010. Predictions of the optimized friction stir spot welding process parameters for joining AA2024 aluminium alloy using RSM. *The International Journal of Advanced Manufacturing Technology*, 51, 173–183.

20. Padmanaban, R., Vignesh, R.V., Arivarasu, M., Karthick, K.P. and Sundar, A.A., 2016. Process parameters effect on the strength of friction stir spot-welded AA6061. *ARPN Journal of Engineering and Applied Sciences*, 11(9), 6030–6035.

21. Manickam, S. and Balasubramanian, V., 2016. Optimizing the friction stir spot welding parameters to attain maximum strength in Al/Mg dissimilar joints. *Journal of Welding and Joining*, 34(3), 23–30.

22. Ibrahim, H.K., Khuder, A.W.H. and Muhammed, M.A.S., 2019. Effect of tool-pin geometry on microstructure and temperature distribution in friction stir spot welds of similar AA2024-T3 aluminium alloys. *International Journal of Mechanical and Mechatronics Engineering*, 19, 14–28.

23. Mubiayi, M.P. and Akinlabi, E.T., 2016. Evolving properties of friction stir spot welds between AA1060 and commercially pure copper C11000. *Transactions of Nonferrous Metals Society of China*, 26(7), 1852–1862.

24. Macías, E.J., Roca, A.S., Fals, H.C., Muro, J.C.S. and Fernández, J.B., 2015. Characterisation of friction stir spot welding process based on envelope analysis of vibro-acoustical signals. *Science and Technology of Welding and Joining*, 20(2), 172–180.

25. Özdemir, U., Sayer, S. and Yeni, Ç., 2012. Effect of pin penetration depth on the mechanical properties of friction stir spot welded aluminium and copper. *Materials Testing*, 54(4), 233–239.

26. Heideman, R., Johnson, C. and Kou, S., 2010. Metallurgical analysis of Al/Cu friction stir spot welding. *Science and Technology of Welding and Joining*, 15(7), 597–604.

27. Boucherit, A., Avettand-Fènoël, M.N. and Taillard, R., 2017. Effect of a Zn interlayer on dissimilar FSSW of Al and Cu. *Materials & Design*, 124, 87–99.

28. Sarila, V., Koneru, H.P., Cheepu, M., Chigilipalli, B.K., Kantumuchu, V.C. and Shanmugam, M., 2022. Microstructural and mechanical properties of AZ31B to AA6061 dissimilar joints fabricated by refill friction stir spot welding. *Journal of Manufacturing and Materials Processing*, 6(5), p.95.

29. Shen, Z., Li, W.Y., Ding, Y., Hou, W., Liu, X.C., Guo, W., Chen, H.Y., Liu, X., Yang, J. and Gerlich, A.P., 2020. Material flow during refill friction stir spot welded dissimilar Al alloys using a grooved tool. *Journal of Manufacturing Processes*, 49, 260–270.

30. Li, P., Chen, S., Dong, H., Ji, H., Li, Y., Guo, X., Yang, G., Zhang, X. and Han, X., 2020. Interfacial microstructure and mechanical properties of dissimilar aluminium/steel joint fabricated via refilled friction stir spot welding. *Journal of Manufacturing Processes*, 49, 385–396.

31. Ahmed, M.M., El-Sayed Seleman, M.M., Albaijan, I. and Abd El-Aty, A., 2023. Microstructure, texture, and mechanical properties of friction stir spot-welded AA5052-H32: Influence of tool rotation rate. *Materials*, 16(9), p.3423.

32. Cui, L., Zhang, C., Liu, Y.C., Liu, X.G., Wang, D.P. and Li, H.J., 2018. Recent progress in friction stir welding tools used for steels. *Journal of Iron and Steel Research International*, 25, 477–486.

33. Pan, T.Y., 2007. *Friction Stir Spot Welding (FSSW): A Literature Review*, SAE Technical Paper. SAE.

34. Tutar, M., Aydin, H., Yuce, C., Yavuz, N. and Bayram, A., 2014. The optimisation of process parameters for friction stir spot-welded AA3003-H12 aluminium alloy using a Taguchi orthogonal array. *Materials & Design*, 63, 789–797.

35. Pieta, G., Dos Santos, J., Strohaecker, T.R. and Clarke, T., 2014. Optimization of friction spot welding process parameters for AA2198-T8 sheets. *Materials and Manufacturing Processes*, 29(8), 934–940.

36. Plaine, A.H., Gonzalez, A.R., Suhuddin, U.F.H., Dos Santos, J.F. and Alcântara, N.G., 2015. The optimization of friction spot welding process parameters in AA6181-T4 and Ti6Al4V dissimilar joints. *Materials & Design*, 83, 36–41.

37. Gerlich, A., Yamamoto, M., Shibayanagi, T. and North, T.H., 2009. Selection of welding parameter during friction stir spot welding. *SAE International Journal of Materials and Manufacturing*, 1(1), 1–8.

38. Schmidt, H.B. and Hattel, J.H., 2008. Thermal modelling of friction stir welding. *Scripta Materialia*, 58(5), 332–337.

39. Cavaliere, P., Cabibbo, M., Panella, F. and Squillace, A., 2009. 2198 Al-Li plates joined by friction stir welding: Mechanical and microstructural behavior. *Materials & Design*, 30(9), 3622–3631.

40. Saunders, N., Miles, M., Hartman, T., Hovanski, Y., Hong, S.T. and Steel, R., 2014. Joint strength in high speed friction stir spot welded DP 980 steel. *International Journal of Precision Engineering and Manufacturing*, 15, 841–848.

41. Khosa, S.U., Weinberger, T. and Enzinger, N., 2010. Thermo-mechanical investigations during friction stir spot welding (FSSW) of AA6082-T6. *Welding in the World*, 54, R134–R146.

42. Su, P., Gerlich, A., North, T.H. and Bendzsak, G.J., 2006. Energy generation and stir zone dimensions in friction stir spot welds. *SAE Transactions*, 717–725.

43. Karthikeyan, R. and Balasubramanian, V., 2010. Predictions of the optimized friction stir spot welding process parameters for joining AA2024 aluminium alloy using RSM. *The International Journal of Advanced Manufacturing Technology*, 51, 173–183.

44. Gerlich, A., Su, P. and North, T.H., 2005. Tool penetration during friction stir spot welding of Al and Mg alloys. *Journal of Materials Science*, 40, 6473–6481.

45. Gerlich, A., Yamamoto, M. and North, T.H., 2008. Local melting and tool slippage during friction stir spot welding of Al-alloys. *Journal of Materials Science*, 43, 2–11.

46. Tozaki, Y., Uematsu, Y. and Tokaji, K., 2007. Effect of tool geometry on microstructure and static strength in friction stir spot welded aluminium alloys. *International Journal of Machine Tools and Manufacture*, 47(15), 2230–2236.

47. Manickam, S., Rajendran, C., Nathan, S.R., Sivamaran, V. and Balasubramanian, V., 2023. Assessment of the influence of FSSW parameters on shear strength of dissimilar materials joint (AA6061/AZ31B). *International Journal of Lightweight Materials and Manufacture*, 6(1), 33–45.

48. Cox, C.D., Gibson, B.T., Strauss, A.M. and Cook, G.E., 2012. Effect of pin length and rotation rate on the tensile strength of a friction stir spot-welded al alloy: A contribution to automated production. *Materials and Manufacturing Processes*, 27(4), 472–478.

49. Fujimoto, M., Inuzuka, M., Koga, S. and Seta, Y., 2005. Development of friction spot joining. *Welding in the World*, 49, 18–21.

50. Fujimoto, M., Koga, S., Abe, N., Sato, S.Y. and Kokawa, H., 2009. Analysis of plastic flow of the Al alloy joint produced by friction stir spot welding. *Welding International*, 23(8), 589–596.

51. Su, P., Gerlich, A., North, T.H. and Bendzsak, G.J., 2006. Energy utilisation and generation during friction stir spot welding. *Science and Technology of Welding and Joining*, 11(2), 163–169.

52. Yan, Y., Shen, Y., Lei, H. and Zhuang, J., 2019. Influence of welding parameters and tool geometry on the morphology and mechanical performance of ABS friction stir spot welds. *The International Journal of Advanced Manufacturing Technology*, 103, 2319–2330.

5 Interlayer Morphology of Friction Stir Spot Welding

V. Lakshmanan, P. Srinivasan,
M. Sanjay, and S. HariPrasadh

5.1 AN EXPLANATION OF THE PRIMARY PHASES

Friction stir welding can create both continuous and spot welds. Friction spot stir welding, or FSSW for short, is used in the latter situation. In order to attach the sheets when utilizing FSSW, a quick immersion of the sheets in a small amount of stirred material is all that is required. The friction spot stir welding process consists of these four steps: Figure 5.1 depicts diving, retracting, lingering in the dew, diving, and cooling. The tool is descending beside the upper layer, and probe tilt achieves the lower sheet during a descent phase, which spins at a predefined velocity [1]. Ejected material is the substance that the tool probe vertically extrudes. The descent process is stopped when the tool reaches the desired depth, and dwell is then started. The material close to the tool probe, which is where the bond among the sheets is established, is heated during this phase. At last, in-stay cooling is created. At this point, even tool rotation needs to stop. As a result, the materials might cool. The tool may be held vertically or under load control to complete this step. The gaps and porosities that frequently arise at the boundary between the agitated zone and the surrounding material can be eliminated using the second method, as has been demonstrated [2]. The welds may have better mechanical properties as a result. When the material reaches the necessary cooling temperature, the tool is removed. However, moving too quickly puts the neighbouring material at jeopardy [3].

5.2 INITIAL MATERIAL STREAM

In comparison to metals, polymers are significantly more temperature-sensitive and have a lower thermal diffusivity. As a result, substantial heat gradients can develop, and places with noticeably different mechanical behaviour can be discovered [2]. As shown in Figure 5.2a, an impacted zone appears beneath the tool probe as soon as it makes initial contact with a thermo-mechanical layer in the top layer.

Figure 5.2a shows the material temperature display thermo-mechanically affected zone (TMAZ), which is situated below the tool probe. It is significantly higher than the neighbourhood's surroundings and the ground level. As a result, it starts to reverse extrude vertically right away, as seen in Figure 5.2a. As the treatment goes

DOI: 10.1201/9781003432289-5

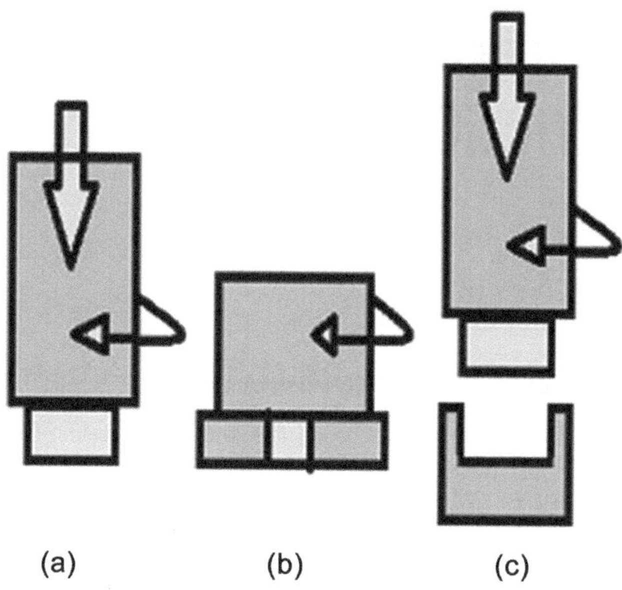

FIGURE 5.1 Foremost phases of FSSW improvement: (a) descent, (b) stay, and (c) tool renunciation.

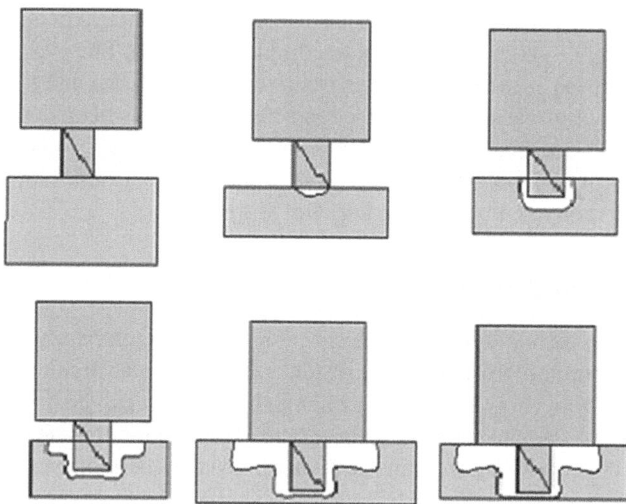

FIGURE 5.2 The flow of materials through the transparent polymer FSSW is shown in the diagram, including the initial position, chip generation, TMAZ formation beneath the tool pin, TAMZ expansion, deep dive, and material after dwell.

on and the interaction duration lengthens, the TMAZ zone keeps expanding. The region surrounding the tool shoulder experiences a substantial morphological shift, from "U-shape" to "V-shape," as shown in Figure 5.2a, as it approaches the upper face of the upper sheet. When a tool bear comes into contact with a polymer, it flows more easily and generates greater heat. The "V-shape" of the TMAZ adjustment significantly enhances the motorized deeds of the welds since it necessitates a longer expansion of the weld area [4]. If the TMAZ migrates evenly downward, the weld area can rise. The dwell phase's stretched contact time leads the welded zone to expand and produce additional frictional heat. Heat dispersion effects, which tend to elevate the local temperature, are what cause this. The process of cooling follows. The fundamental objective of this [5] is to allow the tool to retract as soon as the material regains strength.

As a result, the weld nugget, material reflow, and stirring zone can all be used to explain the geometry of friction spot stir welds. Typical machining methods like milling or drilling are frequently used to eliminate this final characteristic. On the other hand, the motorized deeds of friction spot stir welds depend on the size, shape, and stirring material of the weld nugget. And also, the instrument probe's encounters this situation when a blind hole (weld nugget) develops in the joint. In comparison to metals, polymers are significantly more temperature-sensitive and have a lower thermal diffusivity. Regions with clearly different mechanical behaviour can be detected as a result, and sizeable heat gradients can develop [6]. An affected zone is visible beneath the tool probe as it makes initial contact with a thermo-mechanical layer in the top layer. The diagram depicts the material flow through the transparent polymer FSSW, including the beginning position, chip creation, TMAZ formation underneath the tool pin, TAMZ expansion, deep dive, and material after dwell.

5.3 WELD MORPHOLOGY AND QUALITY EVALUATION

To better understand how progression parameters affect the motorized deeds of the welds, a thorough examination of morphology joints and their impact on failure mechanisms is vital. Figure 5.3a depicts the FSS weld joining the top sheet, bottom sheet, and agitated area. The top and bottom sheets retain their original compactness despite their thermo-mechanically altered characteristics [7]. The agitated area can occasionally have a lot of porosity. Moisture from the original substance and, in the case of particularly hygroscopic materials, air trapped inside the material during the process are the main causes of this [6]. In contrast, the top and bottom sheets of the agitated zone may exhibit porosity and concentrations of tightly packed bubbles. This is due to the material's thermal contraction as an outcome of cooling. Due to the limited thermal dispersion of polymers, there are large temperature variations between the agitated zone and its surroundings. Because it is significantly hotter than the top and bottom sheets, the agitated zone cools more slowly. This can result in a porous contact between the disturbed zone and its surroundings, as demonstrated in Figure 5.3a.

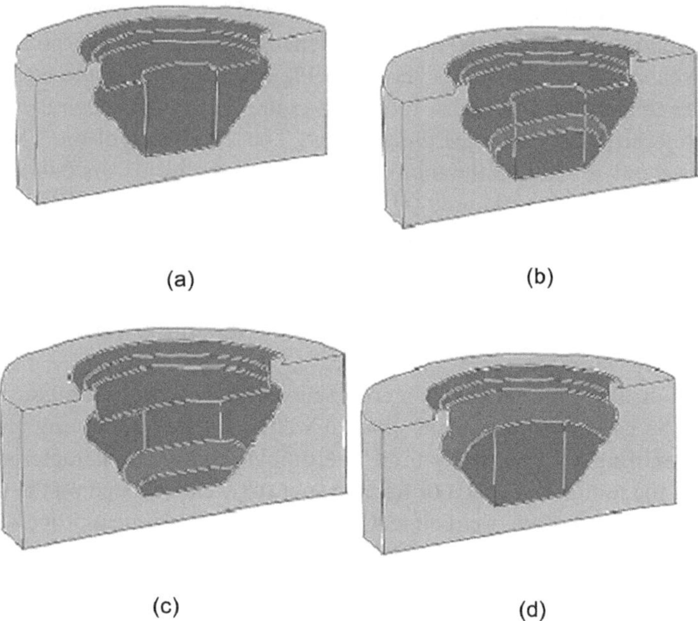

FIGURE 5.3   Friction stir spot weld interface characteristics [13]: (a) stirred zone, (b) Aweld, (c) Aup, and (d) Alow.

As shown, there are permeable zones where the top, lower, and agitated areas converge. Due to brittleness, the bottom, top, and bottom sheets all break. The SZ splits from the bottom sheet as a result of brittleness. Almost immediately after the peak stress, the fractures started. These joints are frequently robust and effective energy absorbers. Because the critical stress value is less than the material's strength in this situation, the fracture occurs fairly quickly. Large temperature gradients, residual tension caused by cooling, and porosities all contribute to the fracture. How brittle the collapse was is seen from the fracture surfaces. Failures, however, might result from the porous contact that was previously discussed that might happen between the agitated region and the surrounding material.

Numerous physical events, as well as morphological and geometrical characteristics, may have an impact on the occurrence of specific failure types. Mode failures usually result in a porous interlayer and cavities between the agitated zone and the foundation materials. A shear collapse could also happen if there are such defects in the agitated area. However, geometrical elements also play a big role. The common load-bearing locations were looked at in order to better understand how the right plunging depth affects friction spot stir welds. The agitated zone behaves almost identically like a rigid body in the failure modes and eventually splits lower and higher sheets. On the outer edge of the wounded area, the fracture first becomes visible. If there are significant contact regions between the agitated region and the top sheet, welds are consequently more likely to fail by detaching from it. Between the SR and the bottom layer, there are up and minuscule interface zones. In order to better understand how the proper plunging depth impacts friction spot stir welds, the typical load-bearing areas were examined. It acts virtually solidly after the bottom and top sheets gradually separate from the wounded area.

Starting at the margin of the agitated area, the fracture spreads along the base material's edge. Large interface zones are visible at the welds, where the top sheet and the agitated region combine. Therefore, it is more likely for small interface patches, which join the SR to the bottom layer, to fail by detaching from the top sheet [8].

5.4 PARAMETERS CONSEQUENCE PROCESS

The morphology of the welds is influenced by a number of parameter processes in FSW, including processing rates, phase lengths, and geometrical considerations. Later on, a thorough discussion of the friction spot stir welds' quality, shape, and mechanical behaviour will take place. It is crucial to keep in mind that occasionally there may be data gaps since process variables influence many research conclusions. Different processing windows, materials, analytical methods, and research environments could have an impact on this. The discrepancy in how a process parameter affects strength (particularly for the FSSW of polymers) is caused by the dearth of studies that have focused on a particular process parameter. As a result, extrapolating findings from a single study outside of the processing window that was the focus of that research is much more difficult.

The range is examined in several indications for the descent rate, tool spin velocity, and stay time. A straight link is typically unsuitable due to the differences in the materials and tools utilized for these projects (tool size, machine capabilities, etc.). By visualizing these ranges as graphs, it is possible to gain a deeper qualitative understanding of the investigated processing windows. The discovered range of the average diving speed is 20–40 mm/min. The two distinct panes have low tool spin rates (equipped with 500–2000 RPM) and high tool spin rates (2000–5500 RPM). Shows the short (20–60 s), medium (60–150 s), and long (>150 s) dwell time ranges that were finally selected. Frictional power and tool rotation speed are inversely correlated. Making joints requires a short dwell time because friction generates more heat when a tool rotates quickly. Using low or high tool spin rates, which correspond to differing power levels, can significantly change the temperature distribution.

5.5 REDUCTION RATE

The worth, shape, and motorized deeds of the friction spot stir welds will next be thoroughly discussed. It is critical to remember that because process variables affect many research outcomes, there may occasionally be data gaps. This could be affected by various processing windows, materials, analytical techniques, and research environments. The scarcity of studies that have concentrated on a certain

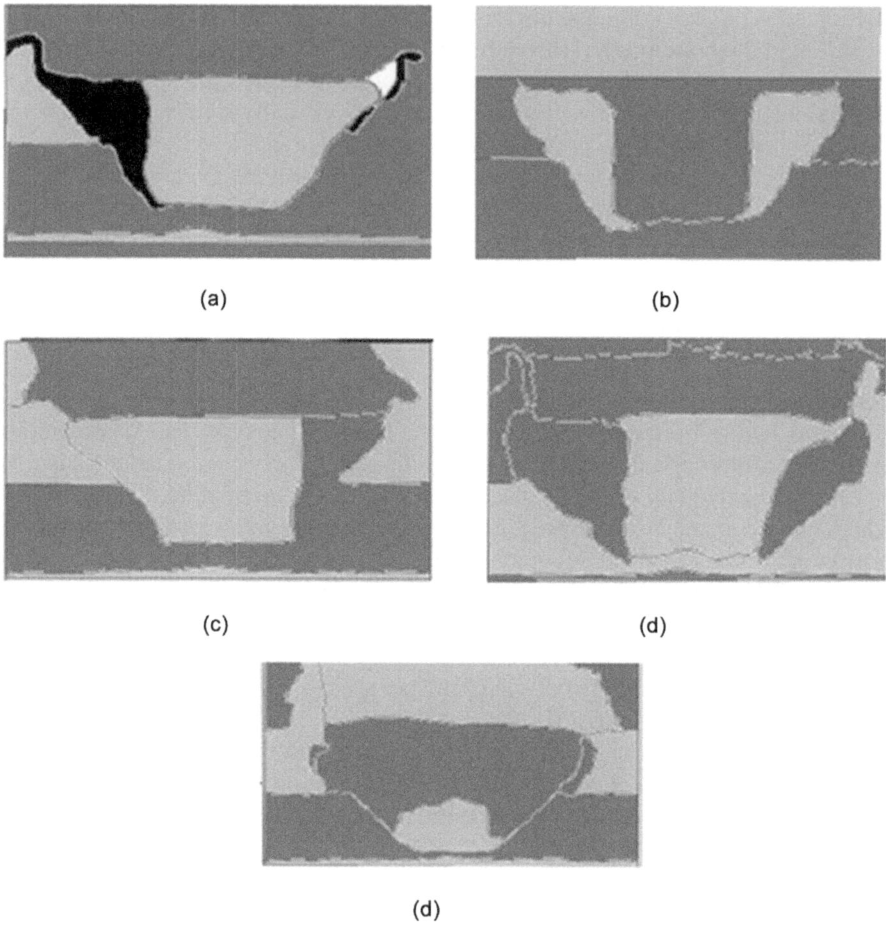

(a)

(b)

(c)

(d)

(d)

FIGURE 5.4 Process parameter effects on weld morphology: (a) indication form, (b) drop of tool spin velocity, (c) drop of descent velocity, (d) boost of stay time, and (e) drop of waiting time [13].

process parameter is what leads to the disagreement in how a process parameter impacts strength (especially for the FSSW of polymers). As a result, extrapolating results from a single study outside of the processing window that was the subject of that research is significantly more challenging, as demonstrated in Figure 5.4.

5.6 SPIN VELOCITY

Compares the ranges for the tool spin velocity, dwell time, and plunging rate that were looked at in various references. Due to the differences in the materials and tools used for these projects (tool size, machine capabilities, etc.), a direct comparison is often incorrect. It is possible to acquire a richer qualitative grasp of the examined processing windows by seeing these ranges as graphs. The tested diving speed ranges are shown, with the average dive speed being discovered to be between 20 and 40 mm/min. Two split window ranges for high tool spin rates (2000–5500 RPM) and low tool rotation rates (up to 500–2000 RPM) are shown in Figure 5.8b. The final decision was to choose three dwell time ranges: short (20–60 s), medium (60–150 s), and long (>150 s). Notably, rapid tool rotation rates were used in the articles that used shorter stay lengths [6,9]. It is not an accident. Yes, there is an inverse relationship between frictional power and tool rotation speed. Because friction produces greater heat when a tool rotates quickly, making joints necessitates a short dwell time. The temperature distribution can be considerably altered by using low or high tool spin rates, which correlate to a variety of power levels. The literature extensively discusses the effects of the three variables stay time, tool spin velocity, and descent rate, in that order, on the FSSW of polymers.

5.7 THE PRE-HEATING PHASE PERIOD

According to the findings, this characteristic has a negligible impact on the quality of the joints. The results demonstrate that when a pre-heating phase is added at the same time as connecting polymers, a localized warming of material takes place only, when the material is directly beneath the tool query warms up; the surroundings of the material remain untouched. This occurred as a result of the reduced heat diffusivity of polymers. As a result, neither the welds' strength nor their sizes are changed.

5.8 DWELL TIME

When the plunging depth is reached, this parameter helps determine how much frictional heat is given to the weld. Heat is currently causing the weld dimension to expand. Therefore, larger welds and ultimately higher load-bearing are connected to longer dwell durations, as demonstrated by the comparison presented. Extremely high dwell time counts, however, could be detrimental [6]. Accordingly, an excessively extended stay time may actually cause the welds to degrade [7]. This prevented the molecular chains from physically breaking down, which would have caused more material to be evacuated, the creation of a hollow, and a loss of molecular weight.

5.9 COOLING TIME

After the dwell time has passed, the substance in the afflicted area appears rubbery or pasty. To avoid ripping the SR, the tooltip requires plenty of time to cool. Due to the polymers' poor heat diffusivity, it is difficult for the agitated zone to rapidly cool down. This depends on a variety of elements, including the materials' thermal properties (such as the temperature at which glass transitions, for example), the operation's peak temperature, the area being stirred, and others. Currently, the device does two jobs. On the surfaces that come into contact with the agitated zone, it does apply some pressure. A temperature-based tool retraction trigger would be advantageous in this case since it also removes heat from the weld, allowing for faster cooling. Due to the thermal properties of polymers, a tool may be quickly pulled back as soon as the temperature falls below a specified threshold. The advantages in terms of productivity and process effectiveness would be clear to discern. This is nevertheless impractical since it is challenging to determine with accuracy the temperature of the agitated area when the therapy is being administered.

5.10　PLUNGE DEPTH

The size and placement of the chosen FSSW weld sections directly depend on the plunge depth. Figure 5.10 shows that the regions Aweld and Alow rise as the penetration depth (s) for welds increases, while the regions Aup (and Lup) fall. Because of the modest penetration depth, it was expected that the small weld zone would be weak. In reaction to the exceptionally high tool plunge depth values that separated the SR from the top layer, the Aup rapidly fell. The final shear force was also produced by these conditions at extremely low values. The shear force peaked at values of moderate penetration depth [10].

5.11　THE INSTRUMENTS' AND PROBES' SHOULDERS

The findings in Ref. [5] demonstrate that when tool probe diameter increased with increasing tool shoulder sizes, crucial tensile strain on UTS welds was reduced. The size of the agitated region and the differentiating zones had an impact on this. The diameter D of a tool shoulder is inversely related to the higher ideals of the feature regions. However, a diameter of the tool query had an effect at these spots and reduced the UTS of the welds. The joint strength of polycarbonate was capable of reaching 88% under the ideal production conditions [11].

5.12　DIMENSIONS OF THE TOOL PROBE

The tool probe dimension has a significant effect on strengthening the welds because the thermo-mechanical process that separates the two materials requires stirring. Lambiase, F et al.'s [12] original research was primarily concerned with how the tool fix affected the motorized deeds of FSS welds. Squares (SQ), triangular forms (TR), threaded cylinders (TH), tapered cylinders (TC), and hexagons (HG) were the six geometries that the researchers compared.

The results shown in Figure 5.5 demonstrate that, when contrasted to welds made using a straight container form under the indistinguishable progression conditions, the use of a threaded container probe increased the maximum shear force by 28%. To improve the flow of materials, this was taken care of. When the tool retracts from the welds, the probe profile also affects the size and shape of the weld nugget. As a result, this part of the material is incapable of supporting any weight. This concept provided the spark for more research into inventive probe

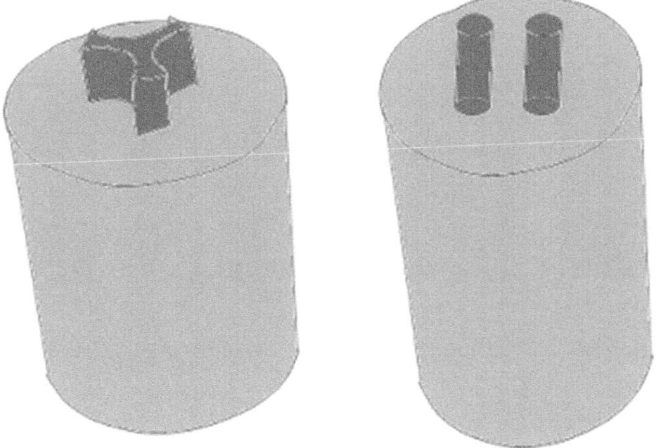

FIGURE 5.5   Triflute and dual fix utilized for FSSW of polymers representation.

types. In order to reduce the size of the weld nugget and the quantity of material expelled from the weld, [13] studied the use of a triflute tool query. This invention had a big impact.

5.13 PLUNGING FORCES' EFFECT

Most articles on friction spot stir welding of polymers use disarticulation management throughout the duration of the welding progression. As a result, the machines were altered to make setup easier. Additionally, the agitated zone must be surrounded by a permeable region [14]. Due to the polymer collecting moisture, air becoming trapped throughout the process, and thermal shrinkage after cooling, there were several bubbles in this location. By lowering the load-bearing area and functioning as stress raisers, these bubbles drastically changed how the welds behaved mechanically. Figure 5.6a–d demonstrates the material flow during transparent polycarbonate friction spot welding. In this instance, bubble formation amid the stirred and surrounding districts can be seen at the beginning of the stay period [1]. During the descending phase, the compressive tension produced by the axial force prevents the production of bubbles. But as the tool stops moving, the axial load drops, which makes it easier for bubbles to form [7].

Figure 5.6e–h depicts the stages of processing under load and displacement control in chronological order to prevent the formation of this porous zone during cooling. The porosities and cavities developed at the start of the stay period were able to close due to the compressive tension created by this. It is a very encouraging conclusion that 99% of the welds made under load control were joint-effective, according to the author's analysis.

Displacement control

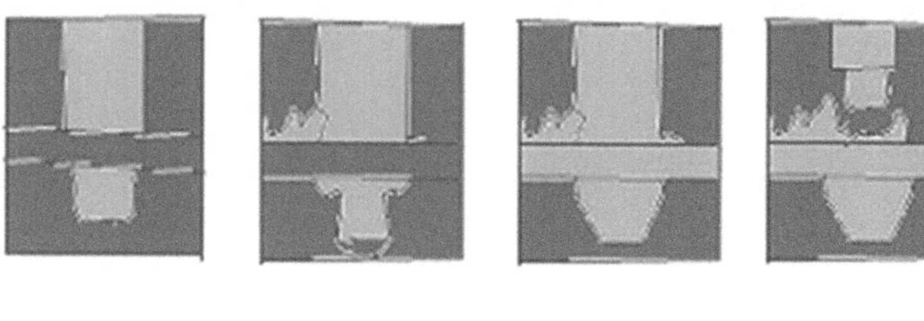

(a) Plunging   (b) Plunging (end)   (c) Stirring   (d) Tool retraction

Load control

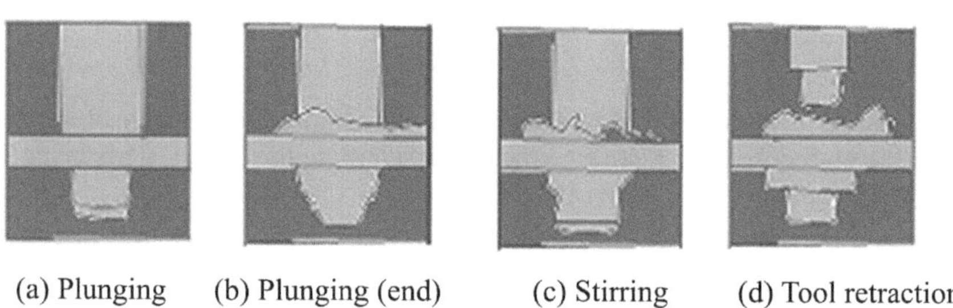

(a) Plunging   (b) Plunging (end)   (c) Stirring   (d) Tool retraction

FIGURE 5.6   Influence of freight management during the cooling phase [1]. Sequence of progression phases (a–d) disarticulation manage and (e–h) load manage.

5.14   HIGH-STRENGTH ALUMINIUM ALLOYS ARE WELDED TOGETHER VIA REFILL FRICTION STIR WELDING

It may be said that aluminium material is a feasible alternative to steel in many application domains for automotive and aircraft structures due to favourable material qualities and relatively low weight. Technically and financially viable lightweight solutions are produced when cutting-edge lightweight concepts are combined with well-suited manufacturing techniques, including joining. The joining of sheets in the shape of a spot is possible using the refill friction stir spot welding technique, which relies on two-sided friction [12]. It employs a spinning tool with a bear and a query that, through frictional warmth effort and axial proposition, mixes the material of the fusion cohort. The joint formation, in particular under lap shear strain, results in high weld potency and excellent load haulage ability. Refill FSSW creates high-quality welds in materials that are challenging to weld. Aluminium alloys with thicknesses of 2 mm for AA2219-O (upper) and 8 mm for AA2219-C10S (below) were employed as the study's foundation materials. According to Figure 5.7, the specimen

FIGURE 5.7 2-mm-thick AA2219-O (upper) and 8-mm-thick AA2219-C10S (lower) aluminium alloys.

had proportions of 120 mm by 30 mm and an overlap area of 30 mm by 30 mm. The tool for this experiment was made up of a pin with a 5.9 mm diameter, a sleeve with a 9 mm diameter, and a clamping ring.

A primary objective of this research is to understand how welding conditions impact the microstructure and mechanical assets of RFSSW connections of sheets with a high thickness ratio. Relationships among tensile characteristics and microstructure are referred to [15]. RFSSW joints typically include four microstructures: stir zone (SZ), thermo-mechanical affected zone (TMAZ), heat-affected zone (HAZ), and base metal (BM). These microstructures are depicted in Figure 5.7. They also have faults, including holes, frail bonding, adhesion ligaments, and hooks [16]. Increases in plunge depth result in larger amounts of material that are plastically deformed in both the thickness direction and joint width. Insufficient local material flow caused holes to form in the unit when the descent depth reached 2.7 mm, as shown in Figure 5.7. These holes will decrease the joint's actual load-bearing surface and encourage the spread of cracks, leading to early joint failure.

Figure 5.8 shows how altering the spin velocity from 1000 to 2500 rpm has an impact on the effective correlation length at a point where the TMAZ/SZ interface's quality is harmed. In this soldering state of affairs, the plasticized material streams into the gap amid the tools [17]. This has to do with how important the joint's effective connection length is to the mechanical properties of the joint in the thickness direction. In this case of a weak joint, the effective connection length is shortened. The hook faults seize on an inverted V shape when the stirring tool is inserted into the lower plate, suggesting that interface bending may be the origin of these flaws.

The elevated-slant granule margins (HAGBs) bigger than 15° are shown in black lines in Figure 5.8, and the short-slant granule margins (LAGBs) relating to 2°–15° are shown in white lines. Blue squares in Figure 5.9 indicate the discovered areas. Figure 5.10a demonstrates how refined and equiaxed the grains are in SZ. SZ is exemplified by HAGBs as a result of the vibrant re-crystallization caused by the stirring effect of pin and sleeve. Sub-grains form inside dislocations that were originally polygonalized during the welding process. The neighboring sub-crystals are stirred by the welding tool, and as they spin and continuously absorb pattern warp, they renovate into HAGBs and produce exquisite equiaxed crystals. The grains in the TMAZ have undergone severe torsional distortion as a result of pin and sleeve churning, and

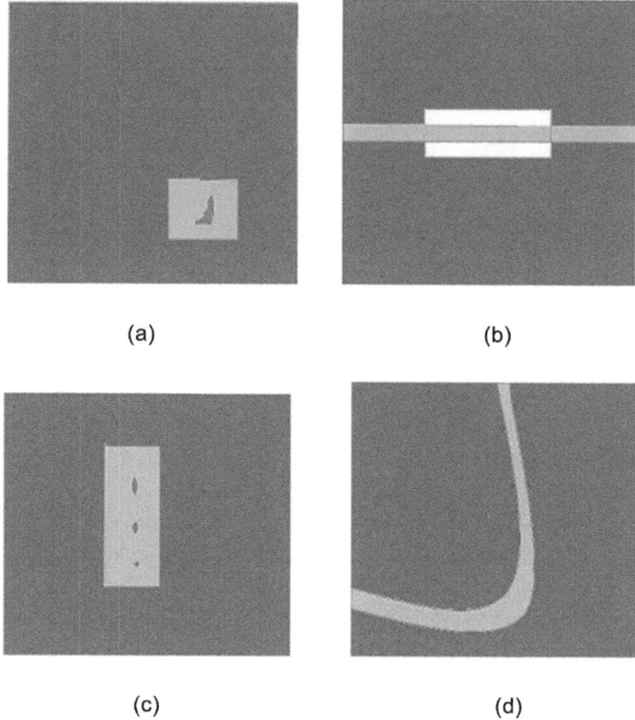

FIGURE 5.8 RFSSW joints commonly include the following flaws: (a) hole, (b) adhesion ligament, (c) weak bonding, and (d) hook.

numerous white lines can be visible inside the grains. In grain margins are expands' sizes and slant in the HAZ. Heat in this area only occurs during welding. In HAZ, the HAGB to HAZ ratio is 40.95%. The re-crystallization distribution plan of the SZ can be used to quantitatively describe the level of re-crystallization in various joint locations. Deformed grains are most noticeable in this region, followed by grains that have undergone re-crystallization and the substructure. The SZ is exposed to mechanical stirring and temperature cycling during the welding process, which results in constant dynamic re-crystallization [18]. The dislocation density then decreased. However, because recovery re-crystallization and plastic deformation take place at the same time, the re-crystallized grains are also subject to plastic deformation, which causes the dislocations to grow and re-agglomerate. As a result, even in the SZ, there are more malformed grains in addition to re-crystallized grains [19].

The re-crystallization ratio in TMAZ is lower than in SZ; however, there is a minor rise in the fraction of deformed grains and sub-grains. This is due to the fact that during the welding process, TMAZ's thermal cycling and mechanical effects are frailer than those of SZ. Therefore, the percentage of re-crystallization is low. In addition, HAZ's side re-crystallization occured. The temperature cycle only has an impact on this area during the welding process. As a result, the distorted grains are primarily sub-grains and much smaller than those in TMAZ. The SZ, TMAZ, and

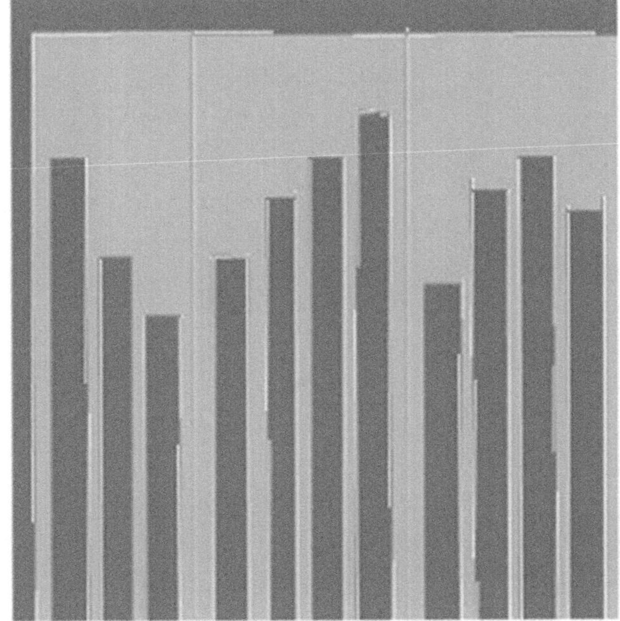

(a) Plunge depth   (b) Welding time (c) Rotational speed

FIGURE 5.9 Tensile-clip potency for variation welding parameters: (a) plunge depth, (b) welding time, and (c) rotational speed.

HAZ in the conventional RFSSW joint are locally misaligned. The gradient of the compass reading change inside the grain can be described by the confined compass reading variation by contrast. As can be observed, TMAZ has a substantially larger local compass reading variation than HAZ and SZ. It is because TMAZ is subjected to enthusiastic stirring exploited by the stirring tool during the welding process, which results in significant plastic deformation. However, during the welding progression, SZ was exposed to high temperatures and significant plastic deformation, which resulted in the production of fine and uniform grains as a result of dynamic re-crystallization. The stored deformation energy was thus completely released. Even though the compass reading variation between HAZ grains and TMAZ grains is smaller, it is still slightly higher than that between SZ grains. This is because during welding, some warp vigour cargo space is unconfined and HAZ is recovered by heat, causing the local compass reading variation to be amid TMAZ and SZ.

The RFSSW joints' hardness distribution under various processing conditions. An M-shaped distribution is seen in all hardness profiles. The AA2219-O BM has a hardness of roughly 45 HV, and from the BM to the TMAZ, the hardness increases. However, due to vibrant re-crystallization and a decrease in dislocation density, the hardness in the SZ marginally drops. LAGBs are created by the intensive mechanical stirring of the TMAZ. The sub-grain boundaries, which make the material

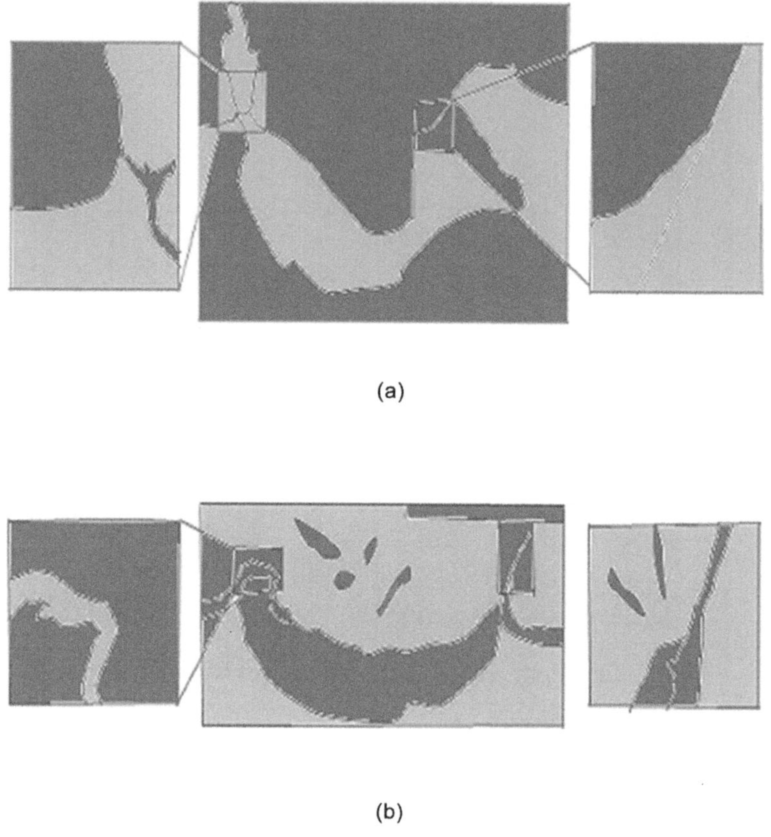

(a)

(b)

FIGURE 5.10   (a) Conventional FSW joint and (b) Ni interlayered joint [22].

harder, prevent dislocations. There is a proportionate link between spin velocity and SZ hardness, with utmost stiffness for a spin velocity of 2000 rpm, as can be seen when comparing hardness for various joints. After this, increased rotation speed causes SZ to become less hard. Similar to how rotational speed affects hardness, welding duration does as well. In addition, the joint's hardness reduces as the dive depth rises.

Figure 5.9 displays tensile-clip potency of RFSSW joints. The maximal potency is 7.2 kN at 2000 rpm, 3.6 seconds for welding, and 2.2 mm for the depth of the plunge. The strength of the plunge is strongly inversely proportionate to its depth. This finding relates greater mechanical strength to shallower plunge depths. Longer welding periods, on the other hand, are related to better mechanical characteristics of the joints for a given plunge depth and rotating speed. In addition, as spin velocity increases, the tensile-clip force grows as well. The descent depth has a significant impact on the mechanical potency of RFSSW joints, according to Yang Moochani et al. [20]. It was discovered that the pattern of the hook defect had a significant impact on the final weld strength. In this investigation, it was discovered that the potency of the joint is closely linked to the spin velocity and welding time, in addition to the dive depth.

Based on the results of the investigation into the impacts of the above-mentioned parameters, can be concluded. The relationship between RFSS mechanical qualities and macro/microstructure aspects fusion joints.

1. Defects, including frail bonding, hooks, and holes, are detected in joints with high depth proportions when certain process parameters are fulfilled. In a typical joint, the SZ showed superior and equiaxed crumbs as a result of vibrant re-crystallization, whereas the TMAZ's principal microstructure features were elongated crumbs with a high fraction of LAGBs as an outcome of inadequate heat input and deformation.
2. The upper sheet of the joint's hardness distribution was M-shaped. Maximum hardness was attained at 2000 rpm of rotation. Additionally, welding time has a similar impact on stiffness allotment to rotating speed. Though, as the depth of the drop increases, the joint's hardness decreases.
3. With a spin velocity of 2000 rpm, a welding period of 3.6 s, and a descent depth of 2.2 mm, the utmost tensile-clip potency was 7.2 kN. Additionally, the twisting altitude of the hook effect and frail bonding length at the TMAZ/SZ crossing point were found to have the burliest effects on the tensile-clip potency [20].

5.15 CONCEPT INTERLAYER

A foreign metal (different from the base metal) with a melting point close to the base metal is sandwiched between the base metal to recover the microstructural and mechanical qualities of welded joints; this foreign material is referred to as an interlayer. Interlayers are becoming more useful in welding applications recently. While interlayers are advantageous in joints, improperly chosen interlayers can considerably worsen the mechanical and microstructural characteristics of weld joints [11].

5.15.1 Applications

Due to its low weight and great strength, the combination of aluminium and magnesium is a commonly used material. Compared to standard welding, this welding offers a number of advantages and a quick procedure that can handle high volume manufacturing. This, when joined without defects, saves a lot of time [9]. The FSW has superior mechanical qualities, is resilient, and can be automated. Industries began utilizing FSW in a variety of fields after realizing its economic potential.

5.16 FORMATION AND GROWTH OF THE INTERMETALLIC LAYER

The creation of IMCs is critical and requires special attention in welding dissimilar alloys so as to generate weld joints of the highest quality, which highlights the impossibility of completely eliminating IMCs. Large IMC accumulations in the weld local areas promote crack propagation and brittle fracture failure [21]. Because intermetallic compounds (IMCs) occur during friction stir welding, it can be difficult to create high potency joints of Al/Mg [15]. It is possible to get joints with greater strength by removing Al-Mg IMCs. It is also possible to stop or reduce the dissemination of Al and Mg by sandwiching a metallic interlayer among the metals. This interlayer might act as a blockade to prevent Al and Mg atoms from combining and moving around. As a result, Al-Mg IMCs can be removed, leading to joints with increased potency. We suggest sandwiching a Ni interlayer amid Mg and Al plates in an effort to reduce or completely remove IMCs. Using a Ni interlayer during welding can effectively reduce solidification cracks. The distinctive IMCs Al_3Mg_2 and $Al_{12}Mg_{17}$ are abolished by the Ni interlayer, which can significantly stop the molecular dissemination of Mg and Al diagonally at their interfaces. The weld zone exhibits significant Ni dispersion with Al and Mg, resulting in the formation of Al_3Ni and Mg_2Ni compounds .

Reactive diffusion governed the development of the intermetallic layer (IMC), which showed a negative correlation between its thickness and the joint's tensile strength. The mechanical characteristics are improved by IMC formation [15].

A joint made with a Ni interlayer is shown in Figure 5.10a and b without any visible fracture or groove issues. It suggests the Ni interlayer may include suppressed flaws by decreasing the heat interaction and IMCs. Figure 5.10 (a): Ni interlayered joint (b) Ni without interlayer joint [22].

Ni has a melting point of 1453°C, which is far higher than the FSW process's claimed maximum temperature range of 450°C–550°C. The clipped interlayer becomes broken up into petite crumbles in the weld zone during the FSLW process. Under the influence of the tool and pin, the Ni interlayer thins and disintegrates into tiny flakes, which subsequently mix with Mg and travel upward with the perpendicular stream of the material [23]. Ni flakes are not expected to melt, as was previously described; nonetheless, their division into small hews and consequent mixing of Al and Mg substrates can be inferred.

5.17   CONCLUSION

In this chapter, interlayer morphologies in FSSW are studied in detail.

- Keyholes are avoided by using an interlayer material without the need for any additional post-processing steps.
- The interlayer material additionally improves ductility and offers a joint with a maximum lap shear load.
- An increase in interlayer morphology will also result in a rise in the weld nugget zone and the effective bearing area.
- The mechanical characteristics are improved by interlayer morphologies.

REFERENCES

1. Bilici, M.K YAkler, A. Ao.; Kurtulmu Å, M. The optimization of welding parameters for friction stir spot welding of high-density polyethylene sheets. Mater. Des. 32 (2011), 4074–4079.

2. Bilici, M.K. Application of Taguchi approach to optimize friction stir spot welding parameters of polypropylene. Mater. Des. 35 (2012), 113–119.

3. Paoletti, A.; Lambiase, F.; Di Ilio, A. Optimization of Friction Stir Welding of Thermoplastics. Procedia CIRP 33(2015), 563–568.

4. Memduh, K. Friction stir spot welding parameters for polypropylene sheets. Sci. Res. Essays 7 (2012), 947–956.

5. Bilici, M.K.; Yükler, A. Ao. Influence of tool geometry and process parameters on macrostructure and static strength in friction stir spot welded polyethylene sheets. Mater. Des. 33 (2012), 145–152

6. Y. C. Liu, J. J. Lin, B. Y. Lin, C. M. Lin, and H. L. Tsai, "Effects of process parameters on strength of Mg alloy AZ61 friction stir spot welds," Materials and Design, vol. 35 (2012), 350–357.

7. M. K. Bilici and A. I. Yükler, "Influence of tool geometry and process parameters on macrostructure and static strength in friction stir spot welded polyethylene sheets," Materials and Design, 33, no. 1(2012), 145–152.

8. Lambiase, F.; Paoletti, A.; Di Ilio, A. Mechanical behaviour of friction stir spot welds of polycarbonate sheets. Int. J. Adv. Manuf. Technol. 80,(2015), 301–314.

9. Lambiase, F.; Paoletti, A.; Di Ilio, A. Effect of tool geometry on mechanical behavior of friction stir spot welds of polycarbonate sheets. Int. J. Adv. Manuf. Technol. 88 (2017), 3005–3016.

10. Yan, Y.; Shen, Y.; Zhang, W.; Hou, W. Friction stir spot welding ABS using triflute-pin tool: Effect of process parameters on joint morphology, dimension and mechanical property. J. Manuf. Process. N 32(2018), 269–279.

11. Yan, Y.; Shen, Y.; Hou, W.; Li, J. Friction stir spot welding thin acrylonitrile butadiene styrene sheets using pinless tool. Int. J. Adv. Manuf. Technol. 97(2018), 2749–2755.

12. Lambiase, F.; Paoletti, A.; Di Ilio, A. Friction spot stir welding of polymers: control of plunging force. Int. J. Adv. Manuf. Technol. 90 (2017), 2827–2837.

13. Hui Shi, Ke Chen, Zhiyuan Liang , Fengbo Dong , Taiwu Yu , Xianping Dong , Lanting Zhang , Aidang Shan Intermetallic Compounds in the Banded Structure and Their Effect on Mechanical Properties of Al/Mg Dissimilar Friction Stir Welding Joints, Journal of Materials Science & Technology (2016)

14. Wu YN, Liao HC, Yang J, Zhou KX (2014) Effect of Si content on dynamic recrystallization of Al-Si-Mg alloys during hot extrusion. J Mater Sci Technol 30, 1271–1277.

15. Alireza S, Hamidreza E , Moslem P (2022) ,Protrusion friction stir spot welding of dissimilar joints of 6061 aluminum alloy/Copper sheets with Zn interlayer, Materials Letters, 328, 133–137.

16. Yan, Y.; Shen, Y.; Zhang, W.; Guan, W. Effects of friction stir spot welding parameters on morphology and mechanical property of modified cast nylon 6 joints produced by double-pin tool. Int. J. Adv. Manuf. Technol. 92 (2017), 2511–2523

17. Paoletti, A.; Lambiase, F.; Di Ilio, A. Analysis of forces and temperatures in friction spot stir welding of thermoplastic polymers. Int. J. Adv. Manuf. Technol. 83, (2016), 1395–1407

18. Derazkola, H.A.; Simchi, A. Experimental and thermomechanical analysis of the effect of tool pin profile on the friction stir welding of poly (methyl methacrylate) sheets. J. Manuf. Process. 34 (2018), 412–423.

19. Yang C, Wu CS, Shi L Phase-field modelling of dynamic recrystallization process during friction stir welding of aluminium alloys. Sci Technol Weld Join 25 (2019), 345–358

20. Moochani, A.; Omidvar, H.; Ghaffarian, S.R.; Goushegir, S.M. Friction stir welding of thermoplastics with a new heat-assisted tool design: mechanical properties and microstructure. Weld. World 63 (2019), 181–190.

21. Dashatan, S.H.; Azdast, T.; Ahmadi, S.R.; Bagheri, A. Friction stir spot welding of dissimilar polymethyl methacrylate and acrylonitrile butadiene styrene sheets. Mater. Des. 45 (2013), 135–141.

22. Silva BH, Zepon G, Bolfarini C, dos Santos JF, Refill friction stir spot welding of AA6082-T6 alloy: hook defect formation and its influence on the mechanical properties and fracture behavior. Mater Sci Eng A 773 (2020), 138–724

23. Lambiase, F.; Paoletti, A.; Di Ilio, A. Effect of tool geometry on loads developing in friction stir spot welds of polycarbonate sheets. Int. J. Adv. Manuf. Technol. 87, (2016), 2293–2303.

6 Mechanical Properties Evolution in FSSW

*V. Lakshmanan, T. Arunnellaiappan, and
G. PaulRaj*

6.1 INTRODUCTION

The friction stir spot welding (FSSW) method modifies the friction stir welding (FSW) method; the tool does not move linearly. The process of FSSW results in a weld that is localized. The technique of FSSW welding, originally created by TWI in the UK, is now used to join aluminium alloys. This method is also utilized in manufacturing automotive parts like vehicle masks, decklids, and doors made of aluminium. The application of FSSW has several advantages over alternatives. Resistance spot welding (RSW) and riveting are examples of aluminium spot joining techniques. FSSW has great scope in the near future because of its cost savings in terms of operation and investment, reduction in weight, enhanced repeatability and consistency, minimal maintenance requirements, improved work environment, and the potential to recycle the materials. The FSSW process involves using an elongated tool with different shapes and sticks to pierce through sheets. Underneath the lower layer, a backing plate that supports the downhill might be applied. The tools maintained for explicit duration spawn heat in the course of chafing, known as the dwell phase. This causes the material beside the tool to soften and deform plastically, resulting in a bond between the upper and lower sheet surfaces, forming a solid-state bond. To finish the process, the utensil is removed, leaving a distinct hole where the pin sticks out in the joint's middle. FSSW has a number of advantages over conventional joining techniques, including reduced distortion, increased fatigue life, no consumables required, and environmental friendliness. A specific kind of solid-state joining procedure called FSSW has the ability to replace single-point joining methods such as riveting. For years, the aeronautical and automobile industries have been intrigued by friction-based welding because of its ability to save weight. Unlike fusion-based welding, it can avoid specific problems and even join materials that are typically difficult to weld using other technologies.

Figure 6.1 displays a diagram of FSSW. The plates are securely clamped together with a suitable fixture and placed above the backing plate. The FSSW tool is rotated using a spindle while simultaneously applying downward force.

Typically, an aperture is left after FSSW is completed. There are process variants such as replenish friction stir spot welding (RFSSW), which enables the material to occupy the cavity created by the process [1], RFSSW, has garnered considerable interest due to its surface without holes, reliable mechanical characteristics, and lack of infill material requirements. It has been utilized primarily for welding similar and dissimilar

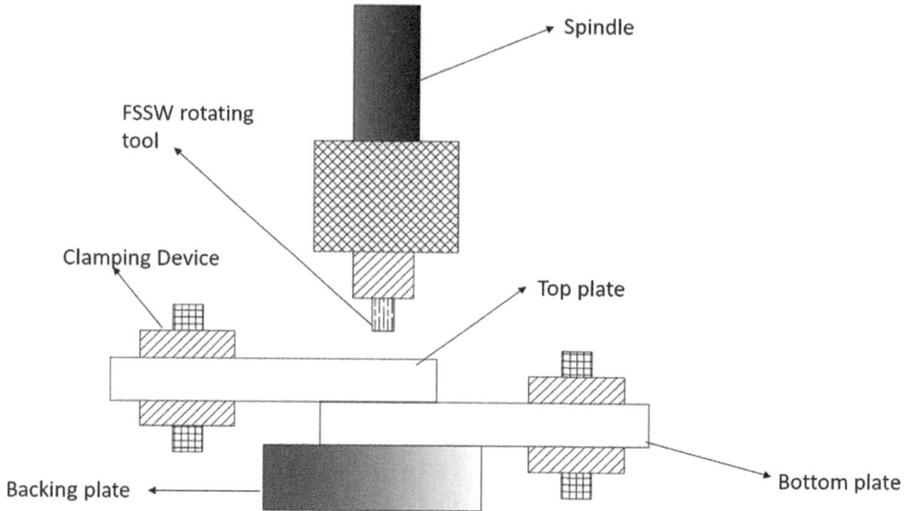

FIGURE 6.1   Schematic representation of friction stir spot welding (FSSW).

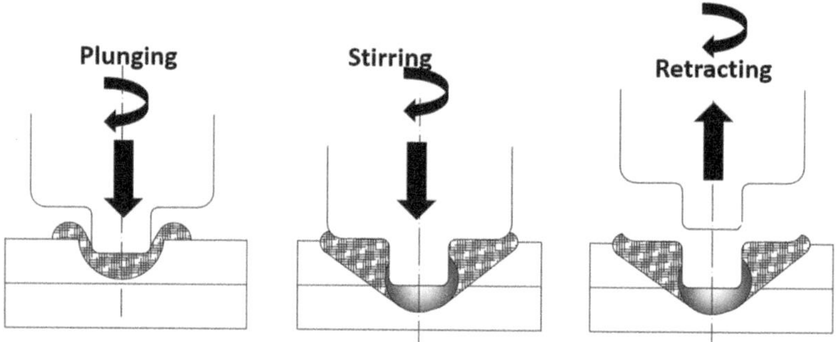

FIGURE 6.2   Dissimilar stages of friction stir spot welding process.

aluminium (Al)-based combinations [2]. The solid-state joining process, FSSW, is effectively utilized in automobile, railway, high-speed railway, and aircraft industries. This method depends on the heat produced by mechanical friction, the plastic flow of metals that are softening, and the axial forging pressure exerted by the machine spindle. Thus, in order to carry out local metallic bonding, it combines mechanical and thermal qualities, avoiding a number of problems associated with metal melting [3].

6.2   STAGES OF FSSW

Three steps make up the FSSW process: sloping, shaking, and retreating. Schematic views of FSSW stages are shown in Figure 6.2.

6.2.1 Plunging

Initially, gizmo rotates at a sky-scraping gaunt speed. Next, the utensil is pushed onto the upper sheet, awaiting the contrivances to touch the crest exterior, creating a weld spot. This plunging motion results in material being expelled.

6.2.2 Stirring

Once the contrivance reaches a set vigour, the stirring juncture begins following the plunge. During this stage, the tool continuously rotates within the workpiece, generating frictional heat. This heat softens and mixes the materials surrounding the utensil, ultimately creating a solid-state joint.

6.2.3 Retracting

After achieving the acceptable level of bonding, the tool has been removed from the workpieces, and there is a centre in the joint's corkscrew. Keyholes may affect the automatic properties of the welded joint.

6.3 FUNCTION OF FSSW IN WELDING WITH DIFFERENT SUBSTANCES

Due to metallurgical incompatibility, a contradictory weld in an industrial constitution is the primary concern of proposal and welding engineers. However, it has many benefits, including deterioration, confrontation, and a soaring potency-to-power quotient. The main issues in dissimilar welding that pose considerable challenges in traditional welding include fragile alloying elements, intermetallic chemicals, hardening, splitting porousness, and evaporation are among the factors that cause grain coarsening. Consequently, an alternative technology, solid-state welding, FSSW, is utilized to address these issues. Due to the inadequate solubility created by minimal ‚Compared to fusion-welded joints, overall joint efficiency is improved a little more.

6.4 FSSW PARAMETERS

The weld quality in FSSW is significantly impacted by the subsequent process parameter.

6.4.1 Territorial Dimensions

The superiority of a mutual is affected by various geometrical variables such as slip breadth, integer of row and rivet, spot distance of rivets from sheet boundaries, nail ground and diameter, and rivet pattern. All these factors contribute to the joint's mechanical qualities, as shown in Figure 6.3.

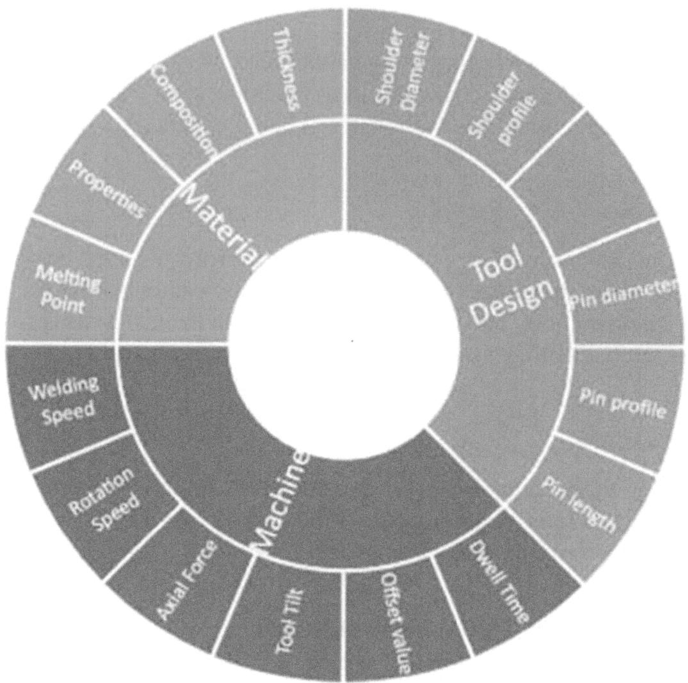

FIGURE 6.3　Factors affecting mechanical properties of FSSW.

6.4.2 ROTATIONAL SPEED

It has been observed that the turning speed (RS) affects the stiffness of the weld. The ceiling inflexibility is located near the keyhole, regardless of the RS used. The frontier between the TMAZ and HAZ has the tiniest stiffness. The grain size in each of the three weld zones is also impacted by the RS. Thermal flux in the crystalline zone converts the elongated and coarse grains in the BMs into fine, equivalent granules. All zones have a decrease in grain size as RS rises.

The revised breakage exterior has revealed that the RS plays a significant role in determining the failure approach of FSSW joints. The amount of frictional heat generated is directly linked to RS, which subsequently leads to an increase in the magnitude of the chunk neighbourhood. However, auxiliary increases in the RS do not have a significant impact on the nugget zone's volume. The privileged RS has little impact on heat generation as the torque decreases. Less heat input and partial recrystallization in the nugget zone due to less stirring activity arise from the lower peak temperature and faster cooling rate at lower RS values. On the other hand, higher peak warmth at privileged RS levels and slower cool times produce grain in the hunk zone [4].

6.4.3 PLUNGE DEPTH

Plunge depth must be above a crucial value that ensures bonding and mechanical interlocking between foils and the lower sheet ratio (opt) in order to achieve sky-high shear strength. In this study, the given equation has been used to determine the option.

$$\text{opt} = Q + (QA3 - Q) + (QB1 - Q) + (QC3 - Q)$$

The mean *SN* ratio of many studies measuring peak tensile strength is represented by Q in the expression above. The *SN* ratios for rotating speed and plunge depth are $QA3$ and $QB1$, respectively. The *SN* ratio for inspiring time is QC3. The charge of tensile potency at the pinnacle can be considered by,

$$y_{opt}y_{opt_2} = (10)^{\eta_{opt}} / 10 = (10)^{\eta_{opt}} / 10$$

Due to frictional heat transfer during the pin tool's plunge depth, the top and bottom plates' microhardness levels changed from what they were originally. This caused the material to disperse its grains between the layers [5].

6.4.4 PLUNGE SPEED

Based on micrographs, it has been well known that the granule mass that formed in the nugget is influenced by the plunge speed; the chunk zone of joint fictitious at other plunge speeds has a slightly coarser grain size [6]. In contrast to the other factors, the velocity of heat intake is negatively related to plunge speed. Higher plunge speeds result in reduced cycle time, whereas lower plunge speeds increase cycle time. The nugget zone coarse grain formation is an end result of excessive mixing action brought on by prolonged cycle periods. FSSW joints have a keyhole blemish at the floor sheet, and edgy materials may be debarred as flare. However, at low plunge speeds, low stirring motion causes a decrease, which leads to poor bonding and a lower SFL of FSSW joints. Less heat input and insufficient churning occur as a result of lower peak temperatures, which can result in the nugget zone having a coarse grain structure. Grain quality is improved in the hunk region as a result of a higher peak temperature and longer axial force extent. Due to the much reduced material, the nugget zone has a higher rate of dislocation and a different distribution of precipitation than the BM, which is significantly softer. Plunge speed changes result in different materials mixing and swirling around the spinning pin, varying the creation of the peak temperature, and applying pressure to the welding material sheets. Plunge speed is indirectly affected by the material's hardness, which has a significant impact on joint performance. Higher plunge speeds result in a reduced incidence of forging force, leading to a condensed expanse of full metallurgical tie and insufficient textile consolidation between the crest and floor sheets in the weld piece. Joint potency is influenced by the maximum heat produced by the plunge rate and the preservation of a constant torque magnitude determined by the nugget zone's hardness. Varying joint strengths are produced by changing the cycle time in combination with varying heat input and axial force on the spot.

6.4.5 RESIDE TIME

Dwell time refers to the amount of time it takes a device to swivel and create a common. The fracture morphology shows that joints made with a shorter dwell time and a medium dwell time have a "Partially Curved" approach of stoppage, while joints made with a longer dwell time have a "Nugget Pullout" category of collapse. This indicates that dwell time significantly affects the type of malfunction of FSSW joints. As dwell time increases, the cycle time also increases. Lower dwell times cause not enough plasticization, resulting in a partially recrystallized constitution in the nugget sector. Higher dwell times cause excessive plasticization, leading to coarser grain due to slow cooling rates. Reasonably lofty temperature and axial force extent result in well grain in the piece region. Longer dwell times reduce forging force due to higher heat input. Moreover, longer dwell times generate more frictional heat, resulting in a larger weld diameter and increased joint strength. During the FSSW process, weld joints are not created at the keyhole region. Welds with dispensation times greater than the lowest amount doling out time array contained hole, whereas weld with dispensation times shorter than maximum giving out time array did not. The holes were caused by decidedly plasticized substance being dragged out by the tool throughout tool retraction.

6.5 MECHANICAL PROPERTIES OF FSSW JOINTS

Various factors can affect the unconscious properties of FSSW-welded joints, including tool geometry, pin profile, instrument rotation speed, downward force, settle time, and utensil transverse speed. Other factors that can influence these properties include base material composition, post-weld heat treatments, and stress relief processes. Selecting the appropriate welding methods, materials, and post-weld treatments can help to optimize the mechanical characteristics such as tensile strength, yield strength, elongation, fatigue strength, fracture toughness, crash toughness, and residual stress of welded joints for specific applications. As soon as the tool pin pierces the bottom layer, it starts to strengthen the connection and contribute to its formation. The mechanical properties are impacted by the keyhole that is made during penetration.

6.5.1 SHEAR TENSILE STRENGTH

Tensile strength is the highest load that a weld can withstand without fracturing. It is a key indicator of the joint's load-bearing capacity and resistance to external forces. The main drawback is the remaining eyehole in the solder's middle nuggets after welding. The presence of a keyhole in FSSW functions as a notch during loading and concentrates too much tension. This would result in the beginning and growth of the crack, which would result in poor mechanical properties. Another major issue in dissimilar welding, FSSW, is the arrangement of components in the weld region. As the IMC layer's thickness increases, the weld's lap-shear tensile strength drops. According to a paper, shear tensile analysis may reveal cracks if intermetallic compounds are present at the joint's interface [7].

In the FSSW of AA6K21 and AZ31, the lap-shear strength of the welds dropped as the IMC layer thickness increased. This decrease in vigour was observed as a result of crack initiation and propagation in IMCs due to the frictional warmth created by the high contrivance rotation rate dwell time [8]. Similar results were obtained by FSSW of AZ31B and AA5083 and showed an inversely proportional connection between the thickness and shear strength of intermetallic compounds (IMCs) [9]. Typically, lap-shear stress samples are prepared as per the dimensions shown in Figure 6.4. The breakdown load of FSSW dissimilar welds was influenced by the welds' geometrical characteristics that affect the width of the bond and interlock, as well as the configuration of IMCs in the blend zone. When an unremitting stratum of IMCs is present, it weakens the weld, but when IMCs are discontinuous, the weld strength improves [10].

Regardless of the gizmo rotation speed, the tensile cut-off vigour increases as the product span increases. At the same probe length, the tensile shear strength starts to increase as stay moment and device revolving haste are increased. Figure 6.5 shows the liaison linking tool hustles and clip tensile power [11]. According to reports, the

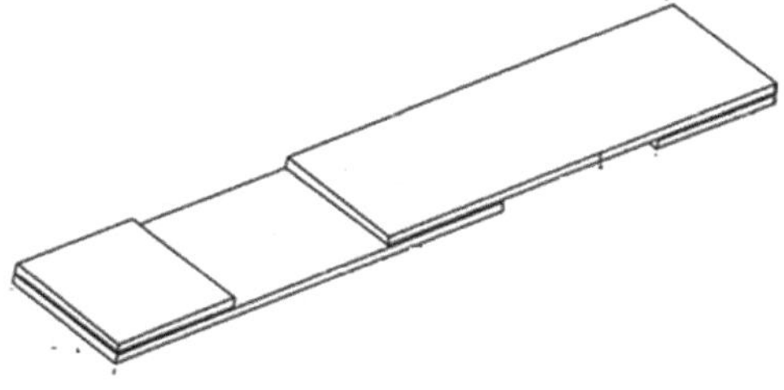

FIGURE 6.4 Schematic representation of lap-shear stress sample.

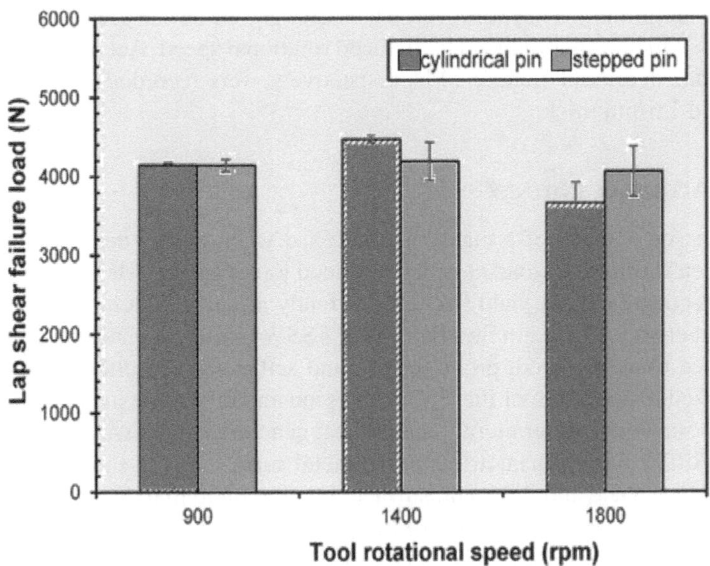

FIGURE 6.5 Impact of contrivance speed on lap trim tensile strong point.

probe length had no effect on cross-tension potency. Stagnant strength dependence on survey length is accurately reflected by considering the upper lamina's thickness beneath the take-on indentation and the nugget size [12]. With growing tool rotation haste, joint elongation and the critical tensile strength have both been seen to rise. Additionally, gizmo negotiated and gyratory velocities improved. The heat-impacted zone expanded as the nugget's diameter shrank [13]. Successful weld width and weld strength were directly correlated, and escalating utensil rotating pace and dwell occasion increased the tensile shear vigour.

Siddharth et al. [14] elaborately impact process parameters such as tool rotation, plunge depth, and dwell time on lap-shear tensile strength. Their results revealed that no welding occurred in the FSSW joints. The low heat generation in the weld zone was attributed to tool spinning speeds under 1000 rpm. At the same time, welding did not proceed properly at a tool rotation speed greater than 1600 r/min because of the undue heat produced in the weld zone and the melt of the base material. They also found that insufficient penetration and the stirring effect at 0.75 plunge depth produce poor quality welds. On the erstwhile tender, a thrust depth greater than 2.25 results in excess penetration, increases keyhole size, and produces an improper weld. Lack of heat generation was observed at 10 s dwell time, and the dual was not twisted. In the folder of 18 s dwell time, excess heat generation, more molten metal at the weld zone, and also rigorous stirring produce the improper weld.

A design-of-experiment (DOE) methodology was used in a work by Henrichs et al. [15] to explore the effects of rotating speed, plunge depth, plunge rate, and two tool geometries on the tensile lap-shear strength of 1.1 mm 5754-O material. The research discovered that a lower rotating speed (1500–1750 rpm) was required to achieve the maximal failure load of 2.6 kN. Similar research was conducted by Freeney, Sharma, and Mishra [16] using a tapered-pin tool to examine the influence of rotating speed, plunge rate, and depth on the tensile shear failure stress of 5052 H32 aluminium. They discovered that stronger welds had greater weld interface diameters, which were produced by reduced rotational speed. According to the study, failure loads of around 4.2 and 5 kN, respectively, were recorded using samples that were 1 and 1.6 mm thick.

6.5.2　Acquiesce Potency

The acquiesce potency of a material is stressed when plastic deformation starts. It helps in determining the load at which a welded joint begins to deform irreversibly. It is a well-known fact that yield strength is greatly affected by temperature. The frictional heat energy created in the stir zone of FSSW mainly depends on utensil geometry, device rotation speed, probe length, and settle time. Frictional heat increases the localized temperature of the stir zone region and drives the metals to flow plastically. The schematic diagram of the total heat generated at the weldment is shown in Figure 6.6. The friction heat arises at the metal surface due to the interaction of the rotating contrivance and the metal surface. In addition to that, heat is also generated by the plastic flow of metal in the stir zone.

The total heat generated at the surface is the sum of the warmth generated at the tool/work piece interface by plasticity. If the stress developed on the whisk sector

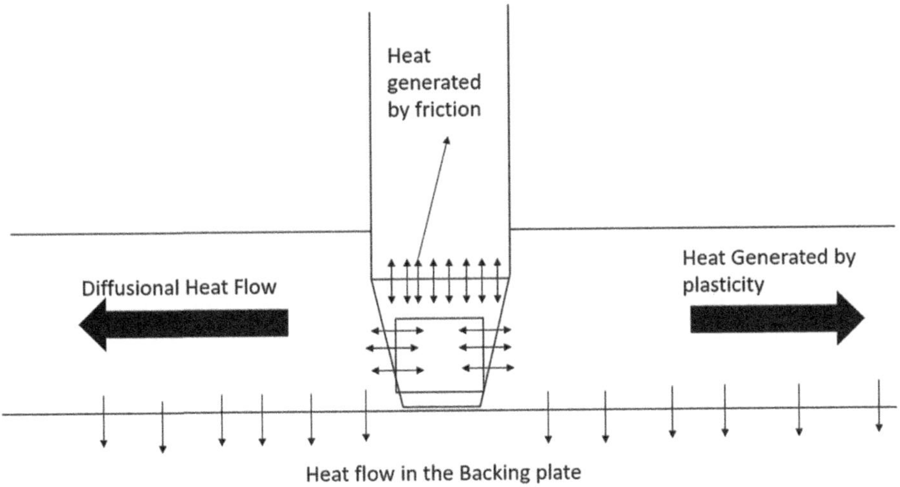

FIGURE 6.6 Schematic diagram of heat generation at the tool and weld interface.

is greater than the tensile stress of the fabric, failure may occur. In some cases, tool material reacts with the base metal at elevated temperatures to produce undesirable phases and degrade the tool. During the welding process, grain growth in the HAZ region produces coarse grains. By having a coarse grain structure, there is less pile-up at the grain boundary, and the HAZ requires less stress to deform. This causes the weldment's yield strength to decrease at the HAZ, ultimately allowing for ductile failure [17]. The grain growth at the HAZ is significantly suppressed by the low tool rotation speed. Consequently, produce a fine-grain structure in the HAZ region that enhances the yield potency of the weld region [18].

Manickam et al. [19] predicted the liaison between lap-shear strength and plunge rate. Their results revealed that the lap-shear vigour increases with increasing plunge rate, with the greatest value of 1.6 cm/min. Tool speed has a major impact on the shear might of the FSSW weld sample. Shear strength increases with increasing the contrivance hustle up to an optimum value; after that, it starts to decrease with the tool speed. The quality of the weld was controlled by the heating and cooling cycles. Tool rotation speed and downward plunge rate increase the heat at the stir zone that experienced server plastic deformation and produced fine and recrystallized grains, thereby increasing the shear strength. However, the decrease in strength against shear could be attributed to the excessive heat input that promotes the large grains and decreases their strength.

6.5.3 Fatigue Strength

Fatigue strength describes the resistance of a welded joint to repeated cyclic loading, which can contribute to failure over time. It is essential in structures subject to varying masses. Few studies have been reported related to the fatigue behaviour of FSSW weld. The exhaustion of similar FSSW repairs was superior to that of dissimilar welds. The lower fatigue life of dissimilar welds can be attributed to the presence of intermetallics at the interface between two different metals. In similar welds, it was observed that the cyclic load level led to two distinct failure scenarios. In contrast, under lower cyclic loads, the keyhole opened due to fracture start in the TMAZ and HAZ, which caused weakness stoppage, making a corner to the load direction. Due to fatigue fracture propagation around the nugget, nugget pullout failure happened at greater cycle loads. The growth of fatigue cracks was mostly governed by the creation of fatigue striations. The dissimilar weld showed a nugget deboning failure mechanism because of the attendance of an intermetallic compound layer [20].

Although scatter increases as fatigue increases. The resources in the beat zone were compressed and microstructural changes occurred as a result. The specimen's and the tool's fatigue lives appeared to be prolonged by increasing the thickness of the sheet at the junction [21]. Using a horizontal tool in lap-shear specimens, Lin et al.'s [22] investigation looked into failure modes of spot rasping welds on aluminium 6111-T4 sheets. The FSSW analytical approach for fatigue cracking onset and propagation resembles RSW. Welds made with a plane gizmo have further intricate failure modes with a bowl-shaped utensil. The primary kind, as it does under quasi-static loading circumstances, expands close to the greater surface of the sheet inside the weld, towards the edge of the beat zone. The second type begins and develops outside the whip zone in the foundation expanse. The third form starts at the upper sheet's bent surface outside of the weld. Wang and Chen [23] investigated the weariness lives of FSSW in Al 6061-T6 lap-shear specimens under cyclic loading conditions. They discovered that welds created using a concave tool have different failure mechanisms than those created with a flat tool. According to experimental findings, the fatigue crack in spot friction welds appears to start close to the original notch tip in the stir zone when they are subjected to cyclic loading. After then, it extends around the nugget's perimeter through the sheet's thickness and eventually develops in the width direction, resulting in the final fracture. Lap-shear and cross-tension specimens are frequently used for shear and opening dominating loading conditions, respectively, to evaluate the mechanical properties of spot friction welds. The majority of earlier research, however, concentrated on the fatigue behaviour of spot friction welds in lap-shear specimens. The failure modes of FSSW in cross-tension specimens of Al 6061-T6 sheets were initially examined in the work of Lin et al. [22,24]. They discovered that under low-cycle loading conditions, the upper nugget pullout failure mechanism is caused by kinked fractures that propagate into the greater sheet from the fracture points. The lower chunk pullout failure mechanism is caused by kinked fatigue cracks that propagate into the lower sheet from the crack points under high-cycle loading conditions. The behavior of FSSW joint fracture is significantly influenced by the concavity of the tool shoulder surface. According to findings, failures below stagnant tensile lap-shear stress situations started close to the whip zone in the cetre of the nugget and spread around their circumferences until they finally fractured. According to Pan et al.'s research [8,9], the microstructure and geometry of welds affected fracture processes under static tensile shear tests. But another element that had an impact on the fatigue fracture mechanisms during the cyclic fatigue test was the cyclic load amplitude. The fatigue of FSSW joints appears to be amenable based on the inclusive and local hassle greatness components of the conflict spot weld, the fatigue crack propagation model.

6.5.4 HARDNESS

The term "hardness" refers to how much material is able to withstand deformation, wear, or indentation. In welding, the stability of both the heat up-posh zone (HAZ) and weld metal can be influenced, which can influence the overall performance of the joint. The FSSW's microhardness profile had a W-shaped look with increasing resistance in the knothole bearing. Furthermore, it was observed that TMAZ and HAZ had softer hardness values. Due to the presence of an intermetallic compound layer, the stir zone in the divergent weld exhibited much superior stiffness principles. The inflexibility morals decrease with the ever-increasing distance from the weld midpoint to the base metal. The chief hardness principles were obtained at the stir zone (SZ), which can be ascribed to the grain modification at the SZ. The lowest hardness values were recorded at HAZ, which can be attributed to grain growth along with thermal softening. It has been reported that the lower hardness of HAZ can be associated with the dissolution of intermetallic clusters during the temperature rise [13]. The dispersion of fine precipitates in the area has been linked to the higher hardness values of the swirl sector and the surface slam to the spyhole. The greater hardness values of SZ [25] were therefore a result of both the grain refining and the dispersion strengthening processes working together.

The chart above shows how the hardness values vary from the swirl region to the base metal. The stir zone had the highest recorded hardness value, while the lowest value was found at HAZ.

Due to the HAZ's heat cycle without deformation, the particle size of the HAZ is coarser than that of the BM in Figure 6.7. It was found that as the current increases, the particle size in the HAZ grows. The HAZ's peak temperature rises, and its high temperature dwell time lengthens as a result of the increase in current. The average particle size of TMAZ falls when the current is increased from 100 to 200 A because the dynamic recrystallization behaviour and the quantity of tiny grains both increase as the current increases [1].

6.5.5 RESIDUAL STRESSES

When welding, residual stresses are induced in the joint as a result of heating and cooling cycles. These stresses can impact the joint's mechanical behaviour, potentially leading to distortion or cracking. Due to the absence of melting, the liveliness of participation for FSSW is extremely stumpy. Heat-affected zones and residual strains related to welds can thus be relatively insignificant. Consequently, friction stir spot welding provides substantial performance benefits. However, the measurement of residual stresses is very important in FSSW. The purpose of measuring outstanding stress after weld is to investigate how multiple spots interact and how the lingering strain pasture affects propagation of weariness cracks. A Both a zone of enhanced compressive residual strain between the spots and a zone of enhanced tensile residual pressure above and around the spot have been observed. The tensile residual tension surrounding the site is now speeding up the crack. As a result, the crack grows in a circular manner all the way around the affected area until the external load takes over. The crack then starts to grow horizontally. It is palpable from examining the

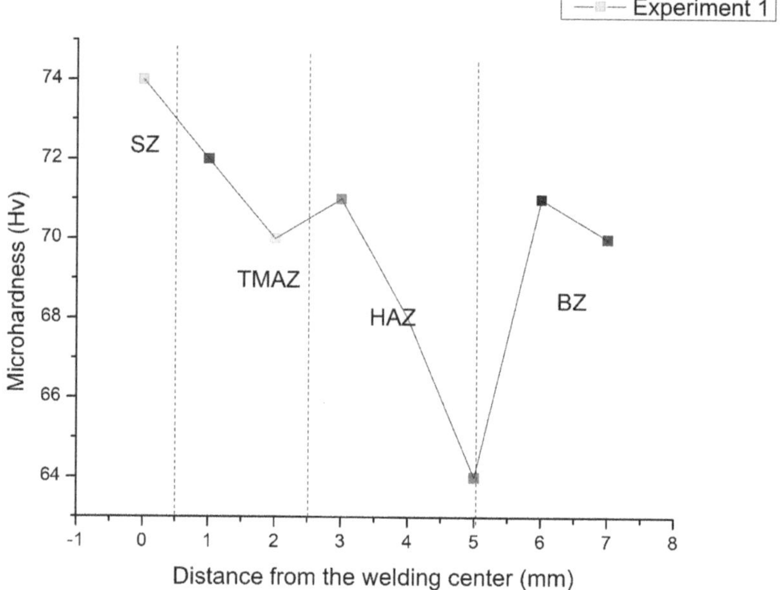

FIGURE 6.7   Microhardness profile of the FSSW.

tensile lingering stress on bad skin, as seen on the residual stress maps. Tensile residual stresses around the spot accelerate the development of cracks. But the compressive space between each point inhibits the rate of crack expansion [26,27].

Bernardi et al. [26] prove that the fatigue properties of the FSSW joints have improved. A fatigue test campaign was run on joints with and without sealant to assess any potential synergies between the refill FSSW and sealant. Statistics are applied, and analyses of tiredness data are performed. A thorough investigation that takes into account macro- and microstructures, the deformation strain field, the distribution of residual stress, and the behaviour of fatigue crack propagation reveals the underlying history of multi-spot joint fatigue damage. The findings show that adding sealant to the FSSW greatly enhances fatigue performance while creating lap joints.

6.6 SUMMARY

Achieving optimal mechanical properties for the FSSW requires precise control over welding parameters. This encompasses factors such as rotation speed, stab intensity, careful selection of utensil materials, and geometrizes. Optimal weld parameters depend on the metallurgical and reflex properties of the base substance and the tool geometry of the FSSW process. A concentrated area of greater hardness at the weld joint results from the stir zone of FSSW's intense plastic deformation and grain refining. Formation of a keyhole is inevitable in FSSW, which can introduce residual stresses and alter the microstructure, ultimately reducing the material's fatigue performance. Moreover, the mechanical properties of a weld can be either positively or negatively impacted by the residual stresses produced during welding. Residual stress can exacerbate the corrosion properties of a weld, particularly in cases of stress corrosion cracking. Post-weld heat treatment or other treatments can enhance microstructure and mechanical properties, but their impact depends on welding conditions and materials.

REFERENCES

1. H Yang. Ji. Xiaoqing, C. Shujun, Y. Tao, Z. Hongwei, B. Yafeng, Xi. Yiming, Li. Xiaoxu. Microstructure and mechanical properties of electrically assisted friction stir welded AZ31B alloy joints, Journal of Manufacturing Processes, 43, (2019), 26–3.
2. F. Banglong, J. Shena, Uceu F.H.R. Suhuddin, T. Chena, F. Jorge, D. dos Santos, B. Klusemann, M. Rethmeier, Improved mechanical properties of cast Mg alloy welds via texture weakening by differential rotation refill friction stir spot welding, Scripta Materialia, 203, (2021), 114–113.
3. B. Zhang, Xi. Chen, K. Pan, Li. Ming , W. Jianing, Thermo-mechanical simulation using microstructure-based modeling of friction stir spot welded AA 6061-T6, Journal of Manufacturing Processes, 37, (2019), 71–81.
4. S. Manickam, C. Rajendran, V. Balasubramanian, Investigation of FSSW parameters on shear fracture load of AA6061 and copper alloy joints, Heliyon 6 (2020) 407.
5. R. Kumar, R. Singh, Ahuja, A. Fortunato, Thermo-mechanical investigations for the joining of thermoplastic composite structures via friction stir spot welding, Composite Structures 253 (2020) 11277.
6. S Ragunath, N Radhika, SA Krishna, N Jeyaprakash Enhancing microstructural, mechanical and tribological behaviour of AlSiBeTiV high entropy alloy reinforced SS410 through friction stir processing, Tribology International, 188 (2023), 108840.
7. U. Özdemir, S. Sayer, C. Yeni, I. Bornova-Izmir. Effect of pin penetration depth on the mechanical properties of friction stir spot welded aluminum and copper, Materials Testing in Joining Technology, 54 (2012) 233–239.
8. D-H Choi, B-W Ahn, C-Y Lee, Y-M, Yeon K Song, S-B Jung. Formation of intermetallic compounds in Al and Mg alloy interface during friction stir spot welding. Intermetallics 19 (2011) 125–30.
9. N Yamamoto, J Liao, S Watanabe, K Nakata. Effect of intermetallic compound layer on tensile strength of dissimilar friction-stir weld of a high strength Mg alloy and Al alloy. Mater Trans 50 (2009) 2833–2839.
10. H.M. Rao, W. Yuan, H. Badarinarayan, Effect of process parameters on mechanical properties of friction stir spot welded magnesium to aluminum alloys. Materials & Design. 66 (2015), 235–245.
11. Y. Tozaki, Y. Uematsu, K. Tokaji, Effect of processing parameters on static strength of dissimilar friction stir spot welds between different aluminium alloys, Fatigue and Fracture of Engineering Materials and Structures 30 (2007) 143–148.
12. T. Yasunari, Y. Uematsu, and K. Tokaji. Effect of Tool Geometry on Microstructure and Static Strength in Friction Stir Spot Welded Aluminium Alloys." International Journal of Machine Tools and Manufacture 47 (2007): 2230–2236.
13. G. D'Urso, C. Giardini, Thermo-mechanical characterization of friction stir spot welded AA7050 sheets by means of experimental and FEM analyses, Materials 9 (2016) 1–14.
14. S. Siddharth, T. Senthilkumar, and M. Chandrasekar. "Development of Processing Windows for Friction Stir Spot Welding of Aluminium Al5052/Copper C27200 Dissimilar Materials." Transactions of Nonferrous Metals Society of China 27, (2017) 1273–1284.

15. J.F Henrichs, C.B Smith, B.F Orsini, R.J DeGeorge, B.J Smale, P.C Ruehl, Friction Stir Welding for the 21st Century Automotive Industry", Proceedings of the 5th International Symposium of Friction Stir Welding, 10 (2004) 14–16.

16. T. A Freeney, S. R. Sharma, R. S Mishra, Effect of Welding Parameters on Properties of 5052 Al Friction Stir Spot Welds, SAE Technical Paper 2006-01-0969, SAE 2006 World Congress, (2006) 3–6.

17. M.M.Z Ahmed, E. Ahmed, A.S. Hamada, S.A. Khodir, M.M. El-Sayed Seleman, and B.P. Wynne. "Microstructure and Mechanical Properties Evolution of Friction Stir Spot Welded High-Mn Twinning-Induced Plasticity Steel." *Materials & Design* 91 (2016): 378–87.

18. W. Xiaopei, Y. Morisada, and F. Hidetoshi , Interface Development and Microstructure Evolution during Double-Sided Friction Stir Spot Welding of Magnesium Alloy by Adjustable Probes and Their Effects on Mechanical Properties of the Joint." Journal of Materials Processing Technology 294 (2021) 117104.

19. S. Manickam, C. Rajendran, S. Ragu Nathan, V. Sivamaran, and V. Balasubramanian. "Assessment of the Influence of FSSW Parameters on Shear Strength of Dissimilar Materials Joint (AA6061/AZ31B)." International Journal of Lightweight Materials and Manufacture, 6, (2023): 33–45.

20. H Chowdhury, D.L. Chen, S.D. Bhole, X. Cao, and P. Wanjara. Lap Shear Strength and Fatigue Life of Friction Stir Spot Welded AZ31 Magnesium and 5754 Aluminum Alloys. Materials Science and Engineering: A 556 (2012) 500–509.

21. K. Anton Savio, J. Lewisea, D Edwin Raja, R. Pandiyarajan, S. Sabarish, Metallurgical and mechanical investigation on FSSWed dissimilar aluminum alloy, Journal of Alloys and Metallurgical Systems 2 (2023) 100010.

22. P. Lin, J. Pan, and T. Pan, Failure modes and fatigue life estimations of spot friction welds in lap-shear specimens of aluminum 6111-T4 sheets. Part 2: welds made by a flat tool, Int. J. Fatigue. 30, (2008) 90.

23. D., A Wang, and C., H. Chen, Fatigue lives of friction stir spot welds in aluminium 6061-T6 sheets, Journal of Material Processing Technology 209 (2009), 367.

24. S. Arul, T. Pan, P.C. Lin, J. Pan, Z. Feng, and M. Santella, Microstructures and failure mechanisms of spot friction welds in lap-shear specimens of aluminum 5754 sheets, SAE Technical Paper. 1 (2005).

25. B. Soumyabrata, M Mondal, S. Y. Anaman, K. Gao, Sung-Tae Hong, and Hoon-Hwe Cho. Gas Pocket-Assisted Underwater Friction Stir Spot Welding. Journal of Materials Processing Technology 320 (2023): 118100.

26. B. Matteo, F.H Uceu , FU. Suhuddin, Banglong, P. Juliano. Gerber, Mateus Bianchi, Ilya Ostrovsky, Bjoern Sievers, et al. "Fatigue Behaviour of Multi-Spot Joints of 2024-T3 Aluminium Sheets Obtained by Refill Friction Stir Spot Welding with Polysulfide Sealant." International Journal of Fatigue 172 (2023): 107539.

27. Y. Sharma, A. Mehta, H. Vasudev, N. Jeyaprakash, G. Prashar, C. prakash et al. Analysis of friction stir welds using numerical modelling approach: a comprehensive review. International journal interact design and manufacturing (2023). https://doi.org/10.1007/s12008-023-01324-6.

7 Metallurgical Transition

*S. Sathish, V. Anandakrishnan, M. Baskaran,
M. Prabu, and S. Senthilraja*

7.1 INTRODUCTION

The need for improved performance, reduction in weight, increased fuel economy, and enhanced strength focused the research on lightweight materials to facilitate the requirements of aerospace, automotive, and defence applications [1]. In particular, aluminium alloys are the most attractive lightweight material owing to their excellent properties such as a greater strength-to-weight ratio, high resistance to corrosion, and better thermal and electrical conductivity [2]. In most cases, the materials are used in the form sheets, which need to be joined together in order to meet the required geometries. Conventional mechanical fastening such as riveting, tightening with screws, bolt, and nuts can be used. Such mechanical fastening needs a hole that is made with the help of drilling, which leads to a stress concentration on the drilled zone that ends up being reduced in strength. Besides, such mechanical fastenings result in increased cost and an increase in the weight of the system [3,4]. The joining of materials with conventional welding is found to be difficult owing to their higher thermal conductivities and high affinity for oxidation. Besides, the welds are experienced with several defects such as voids, pores, distortions, cracks, and residual stress formation. Thus, an advanced welding technique is required to meet the above requirements. Friction stir spot welding (FSSW) is a solid-state welding method in which the tool is where the spinning tool is used to stir, and it is plunged in the workpiece to create the weld between the two plates. The stirring and plunging develop a frictional heat that causes the material to soften, and the stirring causes the plastic to deform. The FSSW has a distinct keyhole at the weld zone, and several enhancements to the welding have been made to enhance the properties and remove the keyhole. The research accomplishments give rise to different FSSW processes, namely pinless, refill, two-step spot welding, protrusion welding, and weld with reinforcement inclusion. Such modifications in the process and its associated process parameters result in the variation of microstructure with and without the formation of intermetallic compounds (IMCs). The mechanical, tribological, and corrosion properties vary according to the microstructure of the weld.

7.2 MACROSTRUCTURE AND MICROSTRUCTURE CHARACTERISTICS

The macrostructure of the weld joint displays the key hole at the centre with the ring at the interface approximately equal to the shoulder and pin. The profile of the keyhole may be cylindrical or conical, depending on the type of pin used to create

DOI: 10.1201/9781003432289-7

the weld. The sectional view of the weld joint identifies the different zones, namely the base metal zone (BM), heat-affected zone (HAZ), thermomechanically affected zone (TMAZ), and stir zone (SZ). Besides, the sectional view displays the formation of hooks and rings. The SZ is the zone in which the tool has direct contact and where severe plastic deformation occurs. TMAZ is the zone followed by the SZ, where it is subjected to heat and distortion. Next to the TMAZ is the HAZ, where there is no deformation or distortion and only thermal effects. The size of the above-mentioned zones is subject to variation with respect to the process parameters, which determine the material strength and metallurgical changes. The microstructure of the weld joint is highly dependent on the process parameters of the FSSW. In specific, the critical process parameters are rotational speed, tool profile, pin length, dwell time, plunge rate, and plunge depth, which have a significant influence on the microstructure. As mentioned, the different zones, namely the SZ, TMAZ, and the heat-affected exhibit different microstructures, which are responsible for the variations in the microstructure. The microstructure of the SZ exhibits dynamic recrystallization of grains due to the higher plastic deformation, which gives rise to finer grains. The SZ displays the fully bonded regions, which leads to effective bonding strength. Owing to the deformation and rapid heating, there will be a modification in the grain structure, crystallographic textures, and formation of precipitates. The TMAZ exhibits partially modified microstructure grains owing to the thermal effect and partially deformation. On the contrary, in the HAZ, there won't be any significant changes in the grain size, which mostly resembles the base metal microstructure.

7.3 METALLURGICAL INSIGHT ON THE JOINING OF SIMILAR METALS

The joining of two aluminium metal sheets of the AA5182 alloy is performed with the FSSW technique by considering the 6 kN clamping force, 1300 rpm rotational speed, and 2.5 s plunging time [5]. The Z38CDV5 steel tool with a threaded pin is used to join the 2 and 1.2 mm upper and lower aluminium sheets, respectively. The welded sheets were sectioned longitudinally (schematic sketch is shown in Figure 7.1), and their microstructure and textures were analysed with scanning electron microscopy and EBSD techniques, respectively. The micrograph of the welded sheet shows the development of the hook at the fused crossing point and flash formation, as visualized in Figure 7.1. The microstructure of the welded sheets displayed the recrystallized and

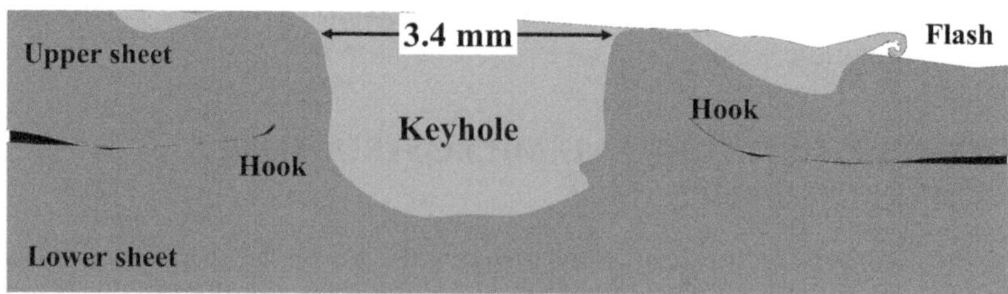

FIGURE 7.1 Typical cross-sectional view of the welded joints.

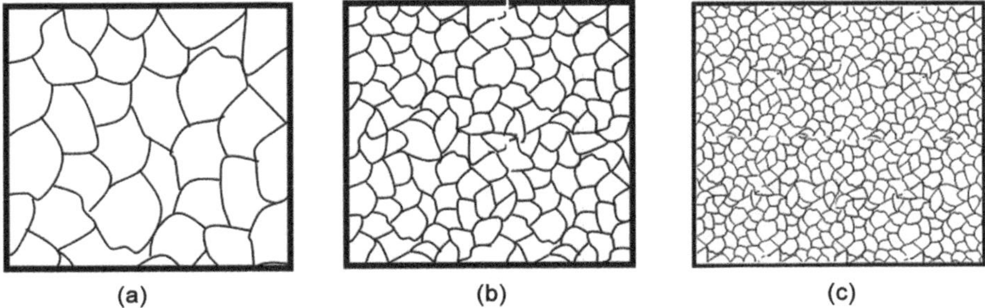

FIGURE 7.2 Illustration of grain size as a function of distance from the key hole on the left side: (a) 5 mm, (b) 3 mm, and (c) 2 mm.

equiaxed grain in both sheets with a mean grain size of 16.6 and 19.5 mm for the top and bottom sheets, respectively. Figure 7.2 shows the schematic view of the grain size variation from the weld centre. The pole figure and ODF of the base material (BM) display recrystallization textures that resemble those of with the face-centred cubic alloy, with the dominance of static recrystallization and cubic components for the bottom sheet and top sheet, respectively. In general, the grains are found to be fine in the stirred zone, and the grain size tends to increase with the increased distance from the hole. The deformed fine grains in TMAZ, heavier grains in HAZ, and fully recrystallized grains in the base metal zone are observed in the grain orientation spread maps. After the welding, the sheets displayed HAZ, TMAZ, and SZ with 8.4%, 41.6%, and 50% for the top sheet and HAZ, TMAZ, SZ, and base metal of 41.6%, 25%, 16.7%, and 16.7% for the bottom sheet. The major observations are that the microstructures of the left-hand side and right-hand side are different, and the deformation in the top sheet is more distinct compared to the bottom sheet.

The AA 5052 sheet (wt% of Mg-2.35, Si-0.16%, Fe-0.20%, Mn-0.02%, Cr-0.11%, Cu-0.02%, Zn-0.01%, Ti-0.01%) is welded with AA 6061-T6 sheet (wt% of Mg-0.7, Si-0.4%, Fe-0.5%, Mn-0.09%, Cr-0.15%, Cu-0.2%, Zn-0.04%, Ti-0.0495%) using the friction stir welding (FSW) process [6]. During this welding method, the specimen was prepared at 100 mm × 30 mm. The two sheets are first cut into 100 and 30 mm lengths and widths, respectively, before being welded together with a 30 mm overlap length by inserting the AA 6061-T6 sheet at the bottom and the AA 5052 sheet on top. A tool with a 5 and 18 mm pin diameter and shoulder was used to accomplish spot welding, respectively. The FSW was completed with a plunge speed of 30 mm/min, a plunge depth of 0.3 mm, a toll speed of 900 rpm, and a dwell time of 2 s. After the welding process, the microstructural and mechanical properties were studied using a scanning electron microscope, a universal tensile machine, and a hardness testing machine. The results of the microstructure analysis revealed that the layer of AA 5052 material has been deformed near the keyhole area by tool pressure. During the welding process, more heat is generated because of pin rotation and friction. This causes the temperature to rise and the material to soften. Finally, the material is joined due to tool pressure and plastic deformation. According to the hardness test findings, the top plate HAZ has the lowest hardness, while the bottom plate has the maximum hardness.

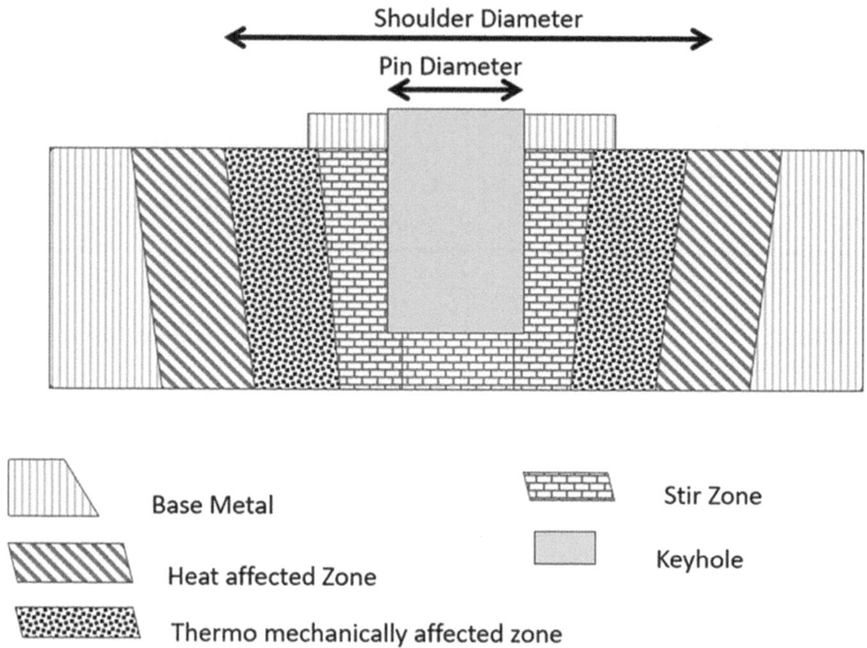

FIGURE 7.3 Metallographic structure of FSW joint.

Two comparable 3-mm thick AA6082-T6 were welded utilizing a vertical milling machine with a rotating velocity of 2000 rpm, a dwell duration of 15 sec, and a tool shoulder diameter of 16 mm [7]. A 100 kN load and a 2 mm/min crosshead speed were considered in the universal testing machine, and the test was conducted at room temperature. The schematic representation of the metallographic structure of FSW is shown in Figure 7.3. Due to the higher tool rotational speed, more compressive loads are applied near the SZ. This leads to more crystal structure changes in the SZ. At the same time, partial crystallization and no crystallization happen in the TMAZ and base metal zones. To study the impact of welding on crystal structure, XRD was carried out at a voltage of 45 kV and 40 mA. The XRD analysis results confirmed the (hkl) values (111), (200), (220), (311), and (222) with the standard peak values of face-centred cubic structure aluminium.

The role of geometry of the tool and rotation speed on the microstructure and mechanical properties of dissimilar welding of AA2024-O/AA6061-T6 sheets is examined [8]. Initially, two 3 mm thick A2024-O and AA6061-T6 aluminium composition sheets were overlapped together and welded using a vertical universal milling machine. During this study, the tool rotation was varied from 900 to 1800 rpm and maintained at three speeds, such as 900, 1400, and 1800 rpm. Similarly, the dwell time and shoulder plunge depth were maintained at 5 s and 0.1 mm, respectively. A plain cylinder-shaped pin (5 mm diameter and 5 mm length) and a stepped pin (8, 6, and 4 mm diameters and 5 mm length) were used to study the impact of mechanical properties. Vickers hardness machine and universal testing machine were used to

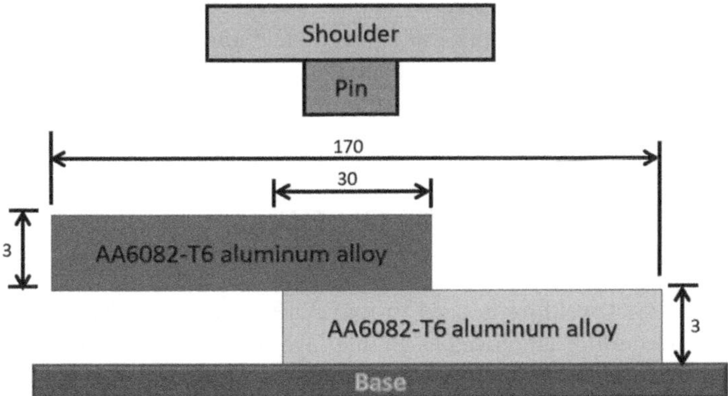

FIGURE 7.4   Schematic diagram of the setup.

study the mechanical properties of the welded specimen. Owing to plunging and stirring, the sheets bent upwards towards the upper sheet. The results of the EDX analysis show that the welded region consists of a reasonable amount of oxygen, around 4 wt% and 1.5 wt% of Cu and Mg, respectively. But in the SZ, the wt% of copper and magnesium decreased to 1.8 and 0.8, respectively. More heat is produced while increasing the tool's rotation speed. This leads to increased grain growth during the welding process. Based on this study, the optimum tool rotation speed is obtained at 1350 rpm to obtain a proper hook width, hardening, and grain size.

A welding of two similar AA6082-T6 Al alloy sheets using the FSW process was carried out by Ahmed et al. [9]. Figure 7.4 shows the experimental arrangements of two sheets of 1 and 2 mm thickness, a 3 s dwell time is maintained, and the rpm of tool rotation varied from 400 to 1000 rotations per minute during this process. In the beginning of the research, aluminium alloy sheets were cut into two pieces of 100×30 mm ($L \times B$) and overlapped together with a 30 mm overlap. This welding process uses AISI H13 steel with 20 and 5 mm shoulder and pin diameter, respectively. FSW is a combination of mechanical and thermal processes. The rpm of tool motion produces mechanical energy, which is converted into heat through friction. Hence, the rpm of tool rotation is primary factor for heat generation in this welding process. During this process, maximum of about 3 kJ of heat is obtained for a 1000 rpm tool rotation. The results of the microstructure analysis revealed that the thickness reduction is triggered by strong plastic deformation. This leads to an average grain size of 6.6 ± 3 and 14.0 ± 2 μm for 1 and 2 mm thickness, respectively.

Two different thicknesses of 2219 aluminium alloy sheets were welded using Refill Friction Stir Spot Welding (RFSSW) and characterization study was conducted by Zou et al. [10]. In this research, a 2 mm thickness of 2219-O alloy is put on top, and three various thicknesses (4, 10, and 14 mm) of 2219-C10S are placed at the bottom and welded using an H13 steel tool. During this work, the rpm of the tool, time of welding, and depth were maintained at 2000 rpm, 6 s, and 2.2 mm, respectively. In order to measure the temperature distribution, K type thermocouple was placed 8 mm from the weld axis. After welding, the microstructure and mechanical behaviour were analysed by using an optical microscope, SEM, and shear testing machine. The thickness of the plate and heat distribution effect were analysed using Abacus software, and the temperature increased by reducing the space from the centre of the joint. Also, it was noticed that the highest temperature increases with decreasing the width of the bottom plate.

7.4 METALLURGICAL INSIGHT ON THE JOINING OF DISSIMILAR METALS

Pinless friction spot welding is used to join pure copper and the aluminium alloy AA2024, and zinc foil is employed as an interlayer in lapped welding zones [11]. The welding speed is set at 2000 and 2350 rpm with dwell times of 1, 3, and 5 s to achieve optimal characteristics. ASTM-E3–112 and ASTM-E112 are used to identify the grain size in welding zones using the optical microstructure. The HAZ is affected by thermal stress without any plastic deformation; the SZ is subjected to extremely high temperatures and deformation; and the TMAZ is subjected to very high temperatures and deformation. Figure 7.5 shows the typical view of the welded joint microstructure of the aluminium side with a HAZ layer and base metal. As the aluminium is the bottom plate, it does not undergo severe plastic deformations. High temperature, stir action from the top copper plate, and the HAZ have an impact on each other. The HAZ is affected by high temperatures and the stirring motion from the top copper plate. Pure copper has a lower hardness than aluminium, which causes copper to flow more quickly. The rpm and dwell duration directly affect the dimension of the grain. Due to the increased heat generated by mechanical extrusion and the reduced cooling rate, recrystallized grains develop with sufficient energy and time. Gaps develop due to insufficient dilution between the aluminium and copper plates. Therefore, choosing the right filler metal is crucial for FSSW. The high heat input, increased rotating speed, and tool pressure result in very high mechanical extrusion. Direct contact between the aluminium and zinc occurs because of the distance between the welding centre and the zinc interlayer metal. The strength of the joint is affected as it leads to brittle aluminium-copper IMCs with low zinc content.

Friction stir solid-liquid spot welding can improve the extremely poor efficiency of an aluminium-copper weld junction [12]. FSSW is used to attach two thin plates,

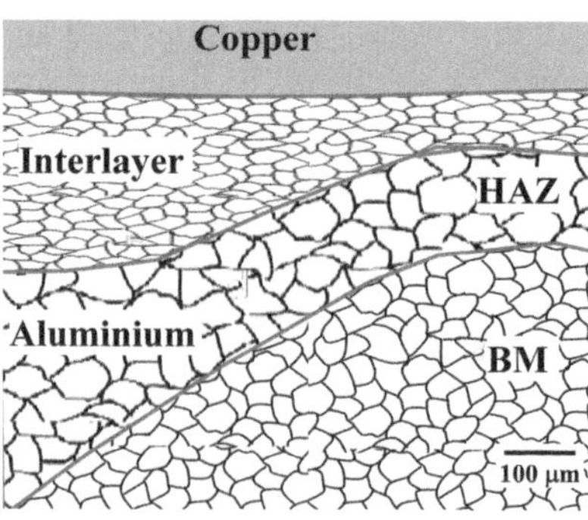

FIGURE 7.5 Typical microstructure of the welded joint of the aluminium side.

one made of 2 mm thick 1060 Al and the other of T2 copper. The foundation material, with dimensions of $105 \times 30 \times 2$ mm, consists of pure copper and 1060 aluminium. The aluminium sheet, held below the copper sheet, overlaps by 35 mm. The interlayer sheet, made of 99.99% pure zinc, measures 30 mm $\times$ 35 mm and has thickness options of 0.11, 0.06, and 0.021 mm. During FSSW welding, the pin does not penetrate the copper sheet. The joint quality mainly depends on the aluminium-copper interface, and the zinc interlayer sheet has no adverse effect on temperature variation. The microstructure of the lap interface can be divided into two areas, A and B. The interface microstructure differs in areas A and B. The thickness of the zinc-added joint in area A is 79 µm, while in area B it is 124 µm. In comparison, the conventional welding thickness at joints A and B is 100 and 144 µm, respectively. EDS analysis is used to determine the composition of IMCs. The microstructure mainly consists of lamellar eutectic and dendrite eutectic. The lap interface contains elements of copper, zinc, and aluminium. The aluminium substrates have a lower zinc content, while the copper substrates have a higher zinc content. However, the copper substrates still contain more zinc than the aluminium substrates. The zinc also melts into the copper and aluminium substrates due to its penetration. The eutectic layer thickness is dominated by liquid zinc, with lower thicknesses resulting from less zinc and higher thicknesses resulting from a larger volume of zinc.

A keyhole-defect-free Al/Ti-welded junction was created through refill RFSSW [13]. During welding, a solid solution layer is produced due to the dispersion of titanium and aluminium atoms at the contact. The Ti6Al4V and AA6061 plates used for RFSSW were 76 mm by 30 mm by 1.5 mm and 76 mm by 30 mm by 2 mm in size, respectively. RFSSW effectively bonded the different Al/Ti alloys, with titanium and aluminium separating from one another. The edge layer is not caught due to the thinness of the IMC layer. The atomic ratio of titanium to aluminium is approximately 3:1. Mapping data confirms that atomic diffusion creates the interfacial layer at the edge. The development of the interface layer primarily depends on atom diffusion during the heating process since the tools do not come into direct contact with the titanium alloy. It is evident that as the scanning distance increases, the titanium content increases while the aluminium content decreases. An intermediate layer with a thickness of 50 nm was made possible by the diffusion of elements along the concentration gradient. The Al-to-Ti ratio was found to be roughly 3:1, and an IMC layer was generated at the interface, as indicated by the line scanning of IMCs. The obvious interface, produced by the diffusion layer, is located at the centre of the picture. Through calibration of the electron diffraction pattern and comparison with the reference card, the IMC was identified as TiAl3. It exhibited a clear diffusion layer between aluminium and titanium with a constant thickness of less than 50 nm and a distribution resembling a ribbon shape. In RFSSW, the titanium alloy does not come into direct contact with the welding tool; instead, it is agitated against the aluminium alloy. The main factors influencing the formation of the interface are the plastic flow and heat input of the aluminium alloy, as well as the distribution of contact between aluminium and titanium. The little heat input also prevents the formation of the IMC.

Numerous experimental investigations have been conducted to explore different process parameters for joining the incompatible AA 7075-T651 and Ti-6Al-4V alloys using FSSW [14]. The welding factors, such as dwell time and rotating speed, affected the mechanical and microstructural properties of the weld joints. The experiments utilized AA7075-T651 and Ti-6Al-4V alloys, each with a width of 4 mm. The plates are fabricated according to the dimensions (100 mm × 35 mm × 4 mm) specified in the JIS Z3136 Standard. The microstructure was examined using optical and SEM-EDS techniques under different welding conditions. The faying surfaces of the FSSW joints exhibited a distinctive geometric feature known as a "hook." This hook formation is attributed to trapped oxide layers, insufficient metallic connection between the two sheets, and upper displacement during the lower penetration of the welding tool. The fractured oxide layers and plasticized material effectively combine within the SZ, resulting in a continuous joint. However, outside of the SZ, the oxide layer fragments are not thoroughly mixed due to limited swirling, resulting in a curved profile referred to as a "hook." The size and dispersion of the hook and oxide film depend on the welding parameters. Specifically, weld specimens with shorter dwell times exhibited larger bond widths at 1000 and 1400 rpm with a 5-s rest duration. Longer dwell times, on the other hand, are important for softening the materials at 1000 rpm and generating an upward flow towards the pinhole, resulting in smaller hook heights, narrower bonds, and stronger joints. However, flaws were observed in these small juncture hook areas, which may be attributed to inadequate metallurgical attachment during the welding process. These juncture hooks act as pre-existing fissures, leading to the failure of the weld joint. Microcracks were identified in the hook area, possibly due to fragmented and unmixed titanium particles trapped in the oxide layer during material mixing under specific welding conditions. Additionally, elongated titanium alloy grain particles were observed in the hook region, indicating partial recrystallization under specific welding conditions. The SZ and TMAZ of the welded structure clearly demonstrate the influence of welding parameters. The optical microstructure study revealed unmixed areas in the SZ joint interface, which can be attributed to low rotating speeds that produce insufficient heat and short dwell times that hinder proper stirring for strong weld connections. The presence of visible fissures in the SZ region leads to decreased joint strength. Similar results indicate the impact of dwell duration on the microstructure, with smaller hook areas observed at a rotating speed of 1000 rpm and a dwell time of 10 s equated to a dwell time of 5 s. Longer dwell times are essential for achieving high strength, as they facilitate material stirring and enhance material flow when the tool is inserted. Increasing the rotating speed to 2000 rpm in the compound area results in changes in the microstructure due to increased heat production. Grains in the SZ region are refined due to mechanical stresses and friction in the applied force direction. However, the TMAZ zone exhibits elongated grains parallel to the boundary, indicating material movement. The wideness of intermetallic composite layers formed at the weld juncture is the primary factor influencing the mechanical characteristics of FW joints between different titanium and aluminium alloys. The IMC layer is determined to have a critical thickness of 5 μm. During FSSW of aluminium/titanium joints, it has been identified that even small modifications in the dwell time significantly impact the diffusion process.

The objective of the reading was to judge the feasibility of utilizing FSSW to create 4 mm thick connections from comparable spot lap junctions using 2.2 mm wide AA5052-H32 aluminium composite plates [15]. The experiments employed lower dwell periods of 1, 2, and 3 s, along with a fixed tool turning speed of 500 rotations per minute. The input heat utilized during the FSSW process was estimated based on the parameters used. All spot welds were examined for their appearance, macrostructures, and microstructure characteristics. The material used in the study was an AA5052-H32 composite sheet with measurements of 1000 mm × 1000 mm × 2 mm. Strips of AA5052-H32 were obtained from the sheets and initially cut into dimensions of 10 cm by 3 cm by 0.2 cm. These strips were then subjected to FSSW, with dwell periods of 1, 2, and 3 s and a constant velocity of 500 rpm. The processing of AA5052-H32 in 4 mm dense lap joints was found to be satisfactory under the specified spot welding conditions, resulting in distortion-free joints. The overall appearance of the FSSWed joints was similar, characterized by circular indentations from the shoulder projection. The grain size generally averaged around 40 μm, with some smaller grains and equiaxed coarse grains comprising the majority of the grain structure. Due to their distinctive qualities, ultra-lightweight alloys such as AA6061 and AZ31B are widely used in numerous significant applications [16]. Solid-state technology is used to join various materials to address the problems with fusion welding. Examining how different process factors in FSSW affect shear fracture stress is the goal. These factors include dwell time, rotation speed, plunge rate, and the proportion of the tool's shoulder-to-pin diameter. The findings show that the FSSW joint exhibits a larger shear fracture strain at a rotational speed of 1000 rpm, a plunge rate of 16 mm/min, a dwell period of 5 minutes, and a shoulder-to-pin diameter ratio of 2.5.

The BMs for the investigation were sheets of AZ31B and AA6061 that were 2 mm thick. The grain size in the SZ, TMAZ, and HAZ was strongly influenced by the FSW tool's rotation speed, according to optical micrographs of symmetrical sections. While the TMAZ featured longer, coarser granules, the SZ consistently displayed finer grains. Both equiaxed and elongated grains were seen in the HAZ, possibly as a result of insufficient heat input close to the weld nugget, which caused the grains to partially anneal. As a result, at a rotational speed of 1000 rpm, the grain size fluctuation at the interface was decreased as compared to other joints. Joints made with a diameter ratio of 2.5, a rotation speed of 1000 rpm, a plunge rate of 16 mm/min, and a dwell time of 5 minutes demonstrated greater lap shear strength. Combining these two elements produced welds with fewer flaws and ideal heat and material flow. The coarse, elongated grains changed into fine, equiaxed grains in the Nugget Zone (NZ) of the joints. The tool's axial and rotational speeds had an impact on the grain morphology in the TMAZ, which led to the development of elongated and slanted grains. There were a considerable number of equiaxed grains in the HAZ (HAZ). Joints made in the NZ at a dive rate of 16 mm/min had smaller grains, but joints made at higher plunge rates had larger grain sizes. In contrast to lower dive rates of 8 and 12 mm/min, fractured joint surfaces at 20 and 24 mm/min revealed a "partially curved" failure pattern. The amount of heat input was decreased as a result of shorter cycle times and higher plunge rates. Lower plunge rates caused longer cycle durations, which raised peak temperatures and required more heat input. Long cycle times resulted in excessive stirring, which made the NZ's grain structure coarser.

In contrast, the NZ produced coarser grains as a result of insufficient stirring and lower heat input brought on by shorter cycle times, lower peak temperatures, and faster plunge speeds. The maximum peak temperature, which was determined by the axial force and its duration, led to the formation of finer grains in the NZ. Dislocation density and precipitate dispersion significantly decreased in the NZ material, which was softer. The metallurgical bond area in the NZ decreased due to lower material mergers between the top and bottom sheets, which also decreased forging power. Due to differences in peak temperature, torque retention, and magnitude, variations in plunge rates have an impact on the SZ's hardness. Axial force, heat input, and cycle duration all had an impact on joint strength.

Protrusion friction stir spot welding (PFSSW) and zinc foil were used to join various AA6061/pure copper sheets together [17]. Pure copper (1 mm thick), pure zinc (100 m), and AA6061-T6 alloy (1.5 mm thick) were the three different types of raw materials that were used. A probe-less apparatus with a 14 mm shoulder diameter was used for the surgery. On the backing plate's surface, a protrusion with a diameter of 10 mm and a height of 0.4 mm was produced. The procedure's variable parameter was the tools' rate of rotation. The PFSSW sample's macrostructure displayed a continuous interface at 2500 rpm, with a few discontinuous zones caused by voids. Non-wettability of the zinc interlayer matrix led to the formation of black holes during non-equilibrium solidification. The zinc melted at the interface between the aluminium and copper due to frictional heat input during the joining operation. After the molten zinc solidified, gaps were left behind that allowed for the interface's bonding to fail and aided in the spread of fractures. A tighter interface made up of IMCs with a smaller grain size is formed at higher rotational speeds. This was attributable to the greater heat input brought on by faster rotation speeds, higher tool pressure, and better material mixing. At 1600 rpm, there were three separate parts in the brazed zone. Close to the Al sheet, a dendritic structure formed by the Al-rich phase was observed.

Due to the proximity of the melting temperatures of Zn and Al to the process temperature, the second area showed a dendritic structure. A spontaneously formed Al-Zn-Cu lamellar structure (76.23Al-2.2Cu-21.57Zn in wt%) was found between these regions and dendritic structures. Based on the linear speed of the tools, the tool pressure in the brazed zone at 1600 rpm was not significant, but the process temperature had a significant effect. As a result, melting predominated in this zone and mixing was minimal. At 2500 rpm, the brazed zone displayed a dense covering of IMCs, similar to the stirring zone, but at 1600 rpm, the zone lacked the dendritic structure due to the increased temperature and pressure. The brazed zone was affected by the tools' rotational speed, showing a dendritic structure at 1600 rpm and a noticeable interlayer of IMCs at 2500 rpm. The study used threaded pin tools manufactured from pre-hardened H13 tool steel to perform FSSW between 1.5 mm 6061 Al-T6 and 1.5 mm oxygen-free pure copper sheets [18]. A study on weak welds was done to verify the relevance of the copper ring and the absence of a continuous intermetallic layer in the fracture zone. A Cu ring was frequently absent in subpar weld cross-sections, which may have prevented the two sheets from engaging and joining. Instead, the joint was only supported by a thin layer of aluminium. SEM images show that both strong and weak welds had layered intermetallic structures that were discernible in the SZ particles. The schematic sectional view of the micrograph (Figure 7.6) shows that a strong weld had a prominent Cu ring that extended from the lower copper sheet into the upper aluminium sheet, aiding interlocking and bonding.

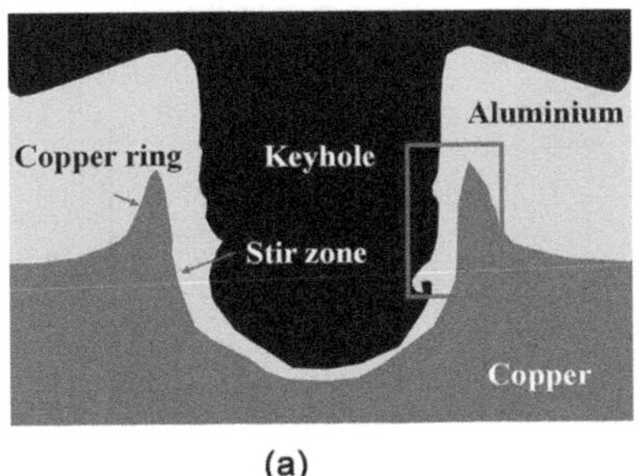

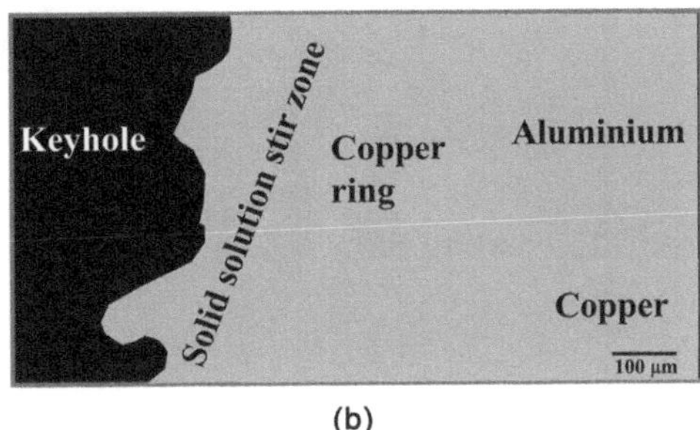

FIGURE 7.6 Typical cross section of friction stir spot welded aluminium copper: (a) sectional view and (b) magnified view.

Additionally, no continuous brittle IMCs were discovered at the junction of the copper ring and the surrounding aluminium, which would have reduced the weld strength. The production of IMCs along the interface, positioned away from the revolving tool, was less likely due to the substantial temperature gradients at the pin/Al contact during FSSW. It was found that there was a thicker layer of aluminium between the top of the Cu ring and the keyhole in welds where the Cu ring was positioned higher.

7.5 METALLURGICAL INSIGHT ON THE JOINING OF METALS WITH ADDED REINFORCEMENTS

The aluminium alloy AA5083 and pure copper are joined together through the FSSW using H13 steel with and without the addition of nano-silicon carbide particles [19]. The silicon carbide particles are packed in the bottom plate by creating a cube groove of 2×2 mm before starting the welding process. The welding of two plates is performed at 1200 rpm with a 5 s pause time and a tilt angle of 1 degree. The microscopic examinations of the welded region show three distinct regions: severe plastic deformed regions (stirred zone), followed by TMAZ and HAZ. The silicon carbide-added weld sample displayed fine grains compared to the weld sample without the silicon carbide addition. The addition of silicon carbide in the weld regions significantly reduces the grain size in all three zones in both the aluminium and copper plates. The addition of silicon carbide particles and effective stirring lead to grain filtration in the SZ. Furthermore, the addition of silicon carbide reduces the interface layer thickness to 14.6 mm, compared to 76.1 mm for the base metal weld. In the interface layer, Al_2Cu and $AlCu$ intermetallic complexes are seen along with copper granules in the aluminium matrix. The production of IMCs is also confirmed by x-ray diffraction studies. The TEM analysis of the SZ of the silicon carbide-added weld sample displayed the uniform dispersion of particles.

The aluminium alloy AA2024 and pure copper are joined together through the FSSW using H13 steel with and without the addition of nano-silicon carbide particles of 50 nm [20]. The silicon carbide particles are packed in the bottom aluminium plate by creating a cube groove of 2×2 mm before starting the weld procedure.

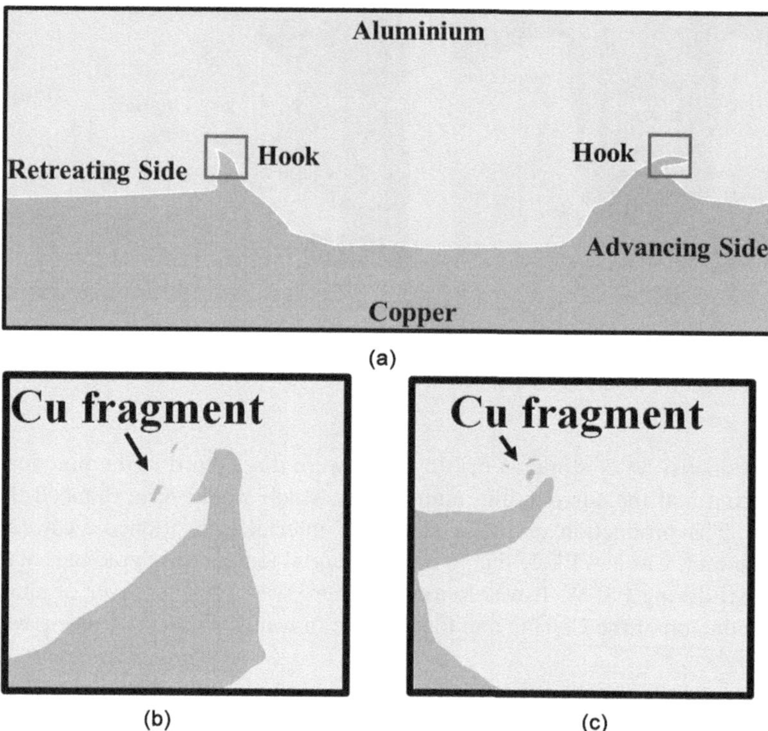

FIGURE 7.7 Typical sectional view of the aluminium-copper FSSW joint: (a) sectional view and (b, c) magnified view.

At a 2000 rpm rotating speed and a 4 s pause period, two separate plates are welded, both with and without the inclusion of silicon carbide. The macroscopic examination of the weld joint (Figure 7.7) shows the weld zones with the formation of hooks on either side, free from tunnels and cavities. The observation of copper ring formation on the advancing side shows that the copper material diffuses in the aluminium, and the length of the ring is higher on the advancing side compared to the retreating side. Besides the fine copper fragments, they are also noticed in the aluminium matrix. The microscopic observation of the weld sample with and without silicon carbide addition shows the voids, porosity, and lack of diffusion in the weld without silicon carbide addition. The addition of silicon carbide develops the frictional heat, which supports material flow with enhanced wettability. Due to the significant plastic deformation, it is also seen that the interface region is wavy in character. In addition, the SZ contrasts the HAZ and TMAZs by showing the equiaxed fine grains. It is common to note that the grain size of the advancing side is finer than the grain size of the retreating side. Additionally, in all three places, it was found that the grain size of the aluminium side was finer than the grain size of the copper side. The aluminium side exhibits the 57% HAGB, 32% LAGB, and 11% twin boundary, and copper side exhibits 56% HAGB, 29.5% LAGB, and 24.5% twin boundary. The shear texture is observed on both the aluminium and copper sides from the pole figure analysis.

The addition of silicon carbide particles to the stirring leads to a reduction in grain size. The width of the layers of IMCs is observed higher in the silicon carbide-added weld sample with a thickness of 3.1 μm compared to the conventional weld sample with a thickness of 1.9 μm. In the SZ, Al, Cu, and Al_2Cu IMC are observed in the conventional weld sample from the x-ray diffraction analysis. In addition to the above intermetallic Al_4Cu_9, I observed its presence in the SZ for the silicon carbide-added weld sample.

The aluminium alloy AA 6061-T6 is joined together through the FSSW using steel with a high level of chromium and carbon varied percentage of silicon carbide particles of 45 μm [21]. The welding was performed at different parameter combinations, namely, guiding hole diameter, rotational speed, pre-dwell time, dwell time, plunge rate, and plunge depth. The parameters are varied at three levels: 0, 2.5, 3, and 3.5 mm guiding hole diameter; 900, 1300, 1700, and 2100 rpm rotational speed; and 2, 6, 10, and 14 s of pre-dwell time. By contrast, the parameters dwell time, plunge rate, and plunge depth are kept constant as 20, 15, and 4.8, respectively. The microscopic examinations on the weld sample performed at 6 s of pre-dwell time, 900 rpm rotational speed, and 3 mm guiding hole diameter show the unsymmetrical structures from the weld centre with three distinct regions, namely, bonded, unbounded, and partly bonded regions. Besides, it exhibits voids on either side of the hole in the partially bonded regions and the formation of hooks initiated from the right side void. Where the sample was welded in other combinations, it displayed a perfectly bonded region without any visual defects. The microscopic examinations on the weld sample performed at 6 s of pre-dwell time, 1700 rpm rotational speed, and 3 mm guiding hole diameter show the perfect interface between TMAZ and SZ. Besides, silicon carbide particles are observed to be distributed homogenously in the SZ with an exceptional bonding with matrix. Furthermore, the energy dispersive analysis validates the silicon carbide presence. Whereas the microscopic examinations on the weld sample performed at 6 s of pre-dwell time, 1300 rpm rotational speed, and 3 mm guiding hole diameter shows the silicon carbide clusters owed to the improper flow of materials and mixing of reinforcements.

In another attempt, the aluminium alloy AA 6061-T6 and pure copper are joined together through the FSSW using high-carbon high-chromium steel with a varied percentage of silicon carbide particles of 45 μm [22]. The welding was performed at different parameter combinations, namely, guiding hole diameter, rotational speed, pre-dwell time, dwell time, plunge rate, and plunge depth. The parameters are varied at three levels: 2.5, 3, and 3.5 mm guiding hole diameter; 1300, 1700, and 2100 rpm rotational speed; and 6, 10, and 14 s of pre-dwell time. Whereas the parameters dwell time, plunge rate, and plunge depth are kept constant at 20, 15, and 4.8, respectively. The experiments were performed based on the Hybrid WASPAS–Taguchi Technique method, and the optimized combination of parameters is obtained at 3.5 mm guiding hole diameter, 1700 rpm rotational speed, and 14 s of pre-dwell time. The microscopic examination of the weld sample performed at the optimized parameters displayed hook formation on either side of the key hole, and there was no existence of voids or partially bounded regions. The SZ in the weld is clear and displays the homogenous dispersion of silicon carbide particles with a complete bonding with the matrix. Similar to the work mentioned above, AA 6061-T6 aluminium alloy is used for the 45 m silicon carbide dispersion FSSW [23]. Tool pin shape, shoulder plunge depth, shoulder diameter, pre-dwell time, dwell time, plunge rate, and guiding hole diameter were some of the parameter combinations used for welding. Tool pin shapes of round, triangular, and square; shoulder plunge depths of 0.36 and 0.6 mm; and shoulder diameters of 10, 13, and 16 mm are among the three degrees of parameter variation. Pre-dwell time, dwell time, plunge rate, and guiding hole diameter, on the other hand, are all held constant at 5 s, 20 s, 20 mm/min, and 2 mm, respectively.

The production of hooks may be seen in the typical depiction of the weld sample macrograph (Figure 7.8) at a shoulder plunge depth of 0.3 mm and a shoulder plunge diameter of 13 with square, triangular, and round tools. It's interesting to see that the tool profile modification from the microstructure analysis causes the material flow (shown by the dashed line) to alter. The square pin exhibits higher material flow, while the circular pin exhibits lower material flow, which significantly affects the distribution of silicon carbide particles. The ratio of the dynamic (sheared work-piece material volume) and static (difference in tool pin volume that is in contact with workpiece material) volumes is what determines how much material flows. Additionally, as illustrated in Figure 7.9, the tool pin profile's change in the direction of force varies. The forces can be broken down into two parts: normal force and shear force. Normal force pushes the material away from pin position 1, while

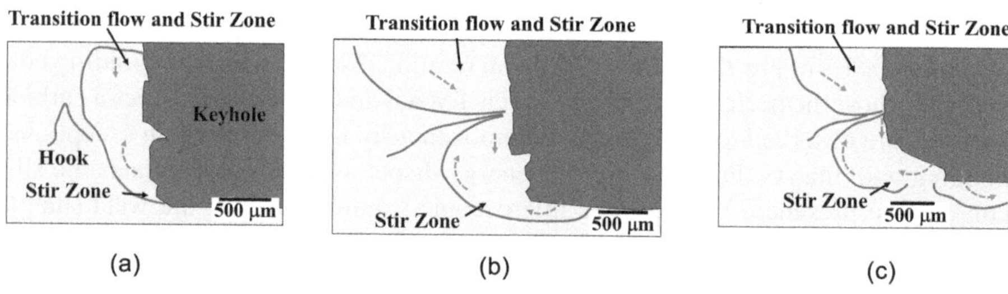

FIGURE 7.8   Typical illustration of the weld sample macrograph with different tool profile: (a) circular, (b) triangular, and (c) square.

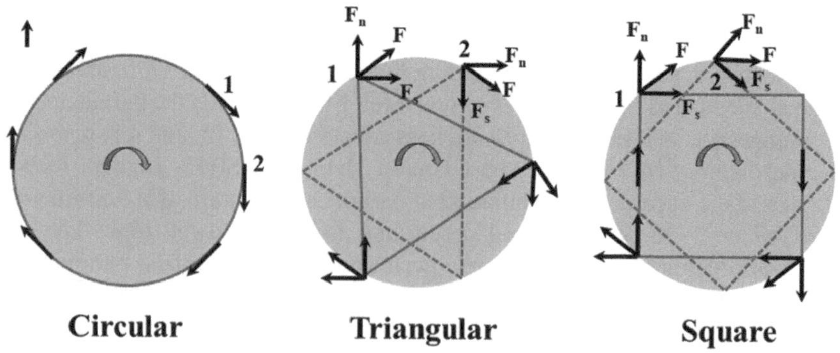

FIGURE 7.9   Comparison of the direction of forces acting on the periphery of tool profile.

shear force causes shear deformation. There is a decrease in normal force when the pushed material reaches point 2, which causes it to flow inward by the tool pin walls before being pushed outward once more by the tool pin edge. As a result, the material flow increased, increasing the SZ's width. In contrast, the square pin's greater number of edges improves material flow as compared to the triangular profile. The material flow will be induced closer to the profile alone in the circular profile due to the tangential pressures operating on the perimeter, resulting in a smaller SZ width. Because of this, the square pin profile produces a stronger weld zone than a triangular or round pin. The possible reason is that agitation, shearing, and friction are the mechanisms behind the deformation of material in the triangular and square pin profiles, whereas friction is the only mechanism in the circular profile. These variations in materials give rise to variations in hook length of 614.77, 300.51, and 216.21 μm for the circular, triangular, and square tool profiles.

The aluminium alloy AA 5083-H116 is joined together through the FSSW using H13 tool steel with a varied percentage of silicon carbide particles of 2 μm and 40 nm [24]. From the pilot experiments, the welding parameter are established at 1300 rpm rotational speed, a plunge rate of 25 mm/min, and a 10 s pause time, and spot welding is performed. The welding was performed with two different nano-silicon carbide particles and a threaded profile tool. The macroscopic examination of the welded joints shows a smooth crown appearance without any defects. The microscopic examination of the weld joint of micro- and nano-added samples exhibited the fine equiaxed grains in the SZ and the grain size of the are finer in the nano-silicon carbide-added sample compared to the micron-sized silicon carbide-added weld. The reason behind the fine grain formation is the pinning effect, which inhibits the growth of grains by reducing grain boundary sliding. The TMAZ displays fine grains, and the microstructure of the HAZ is similar to that of the base substance. The microstructure examination of the micro- and nano-silicon carbide-added weld joints in SZ displays the homogenous distribution of particles over the matrix, but the nano-silicon carbide-added weld joint shows a better dispersion than the micron particle added weld joint. Besides the few agglomerates, the micron particle added weld contains few fragmentation and clusters.

7.6 CONCLUSION

FSSW is a unique method used to join two plates of similar or dissimilar material with varied parameters. The FSSW has different modified variant methods to reduce or remove the keyhole after the welding process. The variations in parameters such as rotational speed, tool profile, pin length, dwell time, plunge rate, and plunge depth have a significant influence on the metallurgy of the welded material. The metallurgy of the weld zone differs even with the joining of similar materials. The metallurgy of the weld zone displays the transformation in grain size, crystallographic texture, and dynamic recrystallization. Besides, the weld microstructure is observed to be non-symmetric on either side, i.e., the right side and the left side. FSSW of dissimilar materials results in the development of IMCs with interface layers and grain refinement. It is also obvious that adding reinforcements to the friction stir spot weld zone improves deformation, reduces flaws, and refines grain.

REFERENCES

1. Y. Tozaki, Y. Uematsu, K. Tokaji, A newly developed tool without probe for friction stir spot welding and its performance, *J. Mater. Process. Technol.* 210 (2010) 844–851. https://doi.org/10.1016/j.jmatprotec.2010.01.015.
2. R.P. Mahto, R. Kumar, S.K. Pal, Characterizations of weld defects, intermetallic compounds and mechanical properties of friction stir lap welded dissimilar alloys, *Mater. Charact.* 160 (2020) 110115. https://doi.org/10.1016/j.matchar.2019.110115.
3. N.T. Nguyen, D.Y. Kim, H.Y. Kim, Assessment of the failure load for an AA6061-T6 friction stir spot welding joint, *Proc. Inst. Mech. Eng. Part B J. Eng. Manuf.* 225 (2011) 1746–1756. https://doi.org/10.1177/0954405411405911.
4. X.W. Yang, T. Fu, W.Y. Li, Friction stir spot welding: A review on joint macro- and microstructure, property, and process modelling, *Adv. Mater. Sci. Eng.* 2014 (2014). https://doi.org/10.1155/2014/697170.
5. T. Baudin, S. Bozzi, F. Brisset, H. Azzeddine, Local microstructure and texture development during friction stir spot of 5182 aluminum alloy, *Crystals* 13 (2023). https://doi.org/10.3390/cryst13030540.
6. S. Venukumar, K.H. Phanindra, B. Venkatesh, M. Sameer, N. Likhith, M.S. Johith, E.V. Kondaiah, M. Cheepu, Microstructure and mechanical properties of similar and dissimilar friction stir spot welded AA 5052 and AA 6061-T6 sheets, *AIP Conf. Proc.* 2648 (2022). https://doi.org/10.1063/5.0114180.
7. S. Kumar, S. Jambhale, M. Maurya, S. Kumar, S. Pandey, Evaluation of shear force and fractography of friction stir spot welded joints of AA 6082-T6 alloy, *J. Eng. Res.* 10 (2022) 124–144. https://doi.org/10.36909/jer.10265.
8. Tiwan, M.N. Ilman, K. Kusmono, S. Sehono, Microstructure and mechanical performance of dissimilar friction stir spot welded AA2024-O/AA6061-T6 sheets: Effects of tool rotation speed and pin geometry, *Int. J. Light. Mater. Manuf.* 6 (2023) 1–14. https://doi.org/10.1016/j.ijlmm.2022.07.004.
9. M.M.Z. Ahmed, M.M.E.S. Seleman, E. Ahmed, H.A. Reyad, K. Touileb, I. Albaijan, Friction stir spot welding of different thickness sheets of aluminum alloy AA6082-T6, *Materials (Basel)* 15 (2022) 1–20. https://doi.org/10.3390/ma15092971.
10. Y. Zou, W. Li, X. Yang, V. Patel, Z. Shen, Q. Chu, F. Wang, H. Tang, F. Cui, M. Chi, Characterizations of dissimilar refill friction stir spot welding 2219 aluminum alloy joints of unequal thickness, *J. Manuf. Process.* 79 (2022) 91–101. https://doi.org/10.1016/j.jmapro.2022.04.062.
11. A.H. Vaneghi, B. Bagheri, A. Shamsipur, S.E. Mirsalehi, A. Abdollahzadeh, Investigations into the formation of intermetallic compounds during pinless friction stir spot welding of AA2024-Zn-pure copper dissimilar joints, *Weld. World* 66 (2022) 2351–2369. https://doi.org/10.1007/s40194-022-01366-6.
12. H. Liu, Y. Zuo, S. Ji, J. Dong, H. Zhao, Friction stir solid-liquid spot welding of Cu to Al assisted by Zn interlayer, *J. Mater. Res. Technol.* 18 (2022) 85–95. https://doi.org/10.1016/j.jmrt.2022.02.067.
13. X. Nan, H. Zhao, C. Ma, S. Sun, G. Sun, Z. Xu, L. Zhou, R. Wang, X. Song, Interface characterization and formation mechanism of Al/Ti dissimilar joints of refill friction stir spot welding, *Int. J. Adv. Manuf. Technol.* (2023) 1539–1551. https://doi.org/10.1007/s00170-023-11226-2.
14. M. Asmael, T. Nasir, Q. Zeeshan, B. Safaei, O. Kalaf, A. Motallebzadeh, G. Hussain, Prediction of properties of friction stir spot welded joints of AA7075-T651/Ti-6Al-4V alloy using machine learning algorithms, *Arch. Civ. Mech. Eng.* 22 (2022) 1–19. https://doi.org/10.1007/s43452-022-00411-x.

15. M.M.Z. Ahmed, M.M. El-Sayed Seleman, A.M.E.S. Sobih, A. Bakkar, I. Albaijan, K. Touileb, A. Abd El-Aty, Friction stir-spot welding of AA5052-H32 alloy sheets: Effects of dwell time on mechanical properties and microstructural evolution, *Materials (Basel)* 16 (2023). https://doi.org/10.3390/ma16072818.

16. S. Manickam, C. Rajendran, S. Ragu Nathan, V. Sivamaran, V. Balasubramanian, Assessment of the influence of FSSW parameters on shear strength of dissimilar materials joint (AA6061/AZ31B), *Int. J. Light. Mater. Manuf.* 6 (2023) 33–45. https://doi.org/10.1016/j.ijlmm.2022.07.005.

17. A. Shahrabadi, H. Ezatpour, M. Paidar, Protrusion friction stir spot welding of dissimilar joints of 6061 aluminum alloy/Copper sheets with Zn interlayer, *Mater. Lett.* 328 (2022) 133107. https://doi.org/10.1016/j.matlet.2022.133107.

18. R. Heideman, C. Johnson, S. Kou, Metallurgical analysis of Al/Cu friction stir spot welding, *Sci. Technol. Weld. Join.* 15 (2010) 597–604. https://doi.org/10.1179/136217110X12785889549985.

19. B. Bagheri, A. Abdollahzadeh, A. Shamsipur, A different attempt to analysis friction stir spot welding of AA5083-copper alloys, *Mater. Sci. Technol.* 39 (2022) 1083–1089. https://doi.org/10.1080/02670836.2022.2159633.

20. B. Bagheri, M. Alizadeh, S.E. Mirsalehi, A. Shamsipur, A. Abdollahzadeh, Nanoparticles addition in AA2024 aluminum/pure copper plate: FSSW approach, *Microstruc. Evol. Text. Study Mech. Prop.* 74 (2022) 4420–4433. https://doi.org/10.1007/s11837-022-05481-z.

21. N. Chaudhary, S. Singh, Development of in situ MMC joint using friction stir spot welding of Al6061-T6, *Int. J. Adv. Manuf. Technol.* 123 (2022) 3633–3646. https://doi.org/10.1007/s00170-022-10490-y.

22. N. Chaudhary, S. Singh, M.P. Garg, H.K. Garg, S. Sharma, C. Li, E.M. Tag Eldin, S. El-Khatib, Parametric optimisation of friction-stir-spot-welded Al 6061-T6 incorporated with silicon carbide using a hybrid WASPAS-Taguchi technique, *Materials (Basel)* 15 (2022) 6427. https://doi.org/10.3390/ma15186427.

23. N. Chaudhary, S. Singh, Experimental investigation on microstructural and mechanical properties of in situ SiC-reinforced friction stir spot weld of Al 6061-T6, *J. Mater. Sci.* 58 (2023) 1849–1868. https://doi.org/10.1007/s10853-022-08110-x.

24. S. Suresh, E. Natarajan, G. Franz, S. Rajesh, Differentiation in the SiC filler size effect in the mechanical and tribological properties of friction-spot-welded AA5083-H116 alloy, *Fibers* 10 (2022). https://doi.org/10.3390/fib10120109.

8 Thermo-mechanical Analysis

K. Ananthakumar, S. Sathish, P. Parthiban,
N. Tiruvenkadam, and M. Baskaran

8.1 INTRODUCTION

Friction stir welding (FSW) is a solid-state joining process used as a successful method for joining alloy parts in industrial sectors such as aerospace, high-speed rail, and automobiles. Besides the axial forging force generated by the machine spindle, this process also depends on plastic flow and heat generation in the metal due to mechanical friction that is softening. In order to establish localized metallurgical bonding, FSW combines the dual functions of mechanical and thermal energy, overcoming a number of flaws brought on by metal melting. [1]. It has been established that welded joints possess improved mechanical properties [2]. Friction stir spot welding (FSSW) is a different technology first developed by Mazda Corporation for welding automobile parts made up of sheet metal, and it is like FSW apart from tool motion [3]. This innovative technique is especially well suited for forming lap joints in the laminated edges of automobile panels or structural components.

Additionally, in the present situation of widespread utilization of lightweight metals (primarily aluminum and magnesium alloys) in the automobile sector, which are frequently referred to as hard-to-weld metals because of their simple oxidisability and high heat conductivity [4], in order to reduce emissions and adapt to global energy conservation in terms of post-weld joint integrity and weldability, FSSW has benefits over other joining methods [5]. As a result, this upgraded technology gradually replaces traditional riveting, resistance spot welding (RSW), and threaded connections, which are fragile points in the strength and fatigue resistance of a vehicle's construction. A white auto body contains hundreds of joints, so mechanically improving each joint through appropriate process control will significantly contribute to the optimization of overall performance, including stiffness, collision resistance, and other factors [6].

8.2 CONVENTIONAL FSSW

A typical FSSW procedure involves inserting the rotating tool into the lapped metal plates to a preset depth, followed by a brief duration of coupling action between the shoulder and tool pin (corresponding to the shoulder pressed depth) before the shoulder touches the upper surface of the workpiece. As indicated in Figure 8.1, the usual FSSW process can be classified into three parts [7,8]. As shown in Figure 8.1, the spinning tool with a connected pin descends through the plates at a predetermined rate

DOI: 10.1201/9781003432289-8

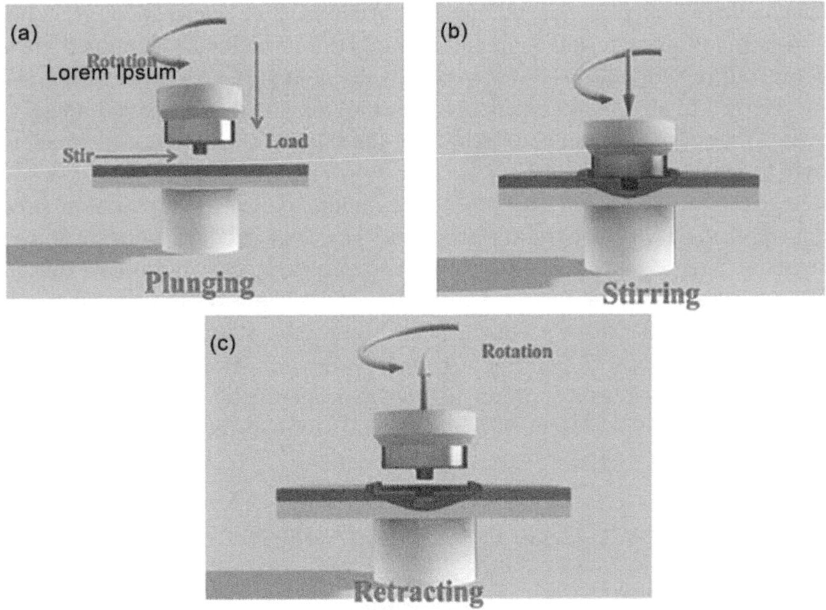

FIGURE 8.1 FSSW process steps (a) plunging, (b) dwell, and (c) retraction.

(Figure 8.1a) until the shoulder reaches the predetermined depth. The plunge rate (p_r) refers to how quickly the tool descends into the workpiece, and the plunge depth (d_z) describes how far the tool's shoulder enters the workpiece. These are the terms used in relation to FSSW. The mechanical contact between the plates and the tool, like in FSW, encourages frictional heating, which softens the base material [8]. A backing plate, or anvil, is placed at the lower surface of the workpiece to support the downward force that the tool exerts on it during the process [9–11]. After the plunge depth is reached, the dwell stage (Figure 8.1b) begins. It lasts for a predetermined amount of time, during which the tool encourages the joining of the two distinct sheets to form a solid-state junction [12,13]. The dwell time, also known as the maintenance time, is the length of time the tool is submerged at this point (t_{man}). After this time has passed, the tool denies (Figure 8.1c), leaving an unfavorable keyhole that greatly reduces the joint strength by functioning as a stress concentration flaw. As the tool contacts with the raw material in the FSSW are more intricate than they initially seem to be due to the highly temporary welding circumstances, as opposed to the actual FSW, which achieves a stable state after the first plunging and dwell times, the FSSW is a deceptively simple process.

8.3 FSSW'S THERMO-MECHANICAL PROCESS

A thermomechanically affected zone (TMAZ), a stirred zone (SZ), a heat-impacted zone, and a surplus base zone (BZ) can all be considered sub-regions of the solid bond region. Investigations on different material and alloy joints have revealed that the mechanical properties depend on microstructural characterization, grain boundary, including grain size, and degree of precipitation, which may be altered via welding temperature and fluidity [14,15]. Additionally, a weakly bonded transition section (often referred to as a hook) exists between the unbonded and bonded areas, and it is created due to the plastic deformation of the bottom sheet metal. Its incidence is thought to represent the FSSW joint's crack-start position.

It is vital to investigate the physical relation to heat generation in order to comprehend the FSSW operating principle. The heat produced during the initial mechanical interaction between the tool and the surface to be joined started the FSSW process. The work material is rotated and advanced by the rotating tool in a continuous motion, which produces heat through the plastic distortion of sheared and strained layers of soft material near the revolving tool surfaces. Throughout the welding process, various heat production devices operate intermittently or in cycles. This chapter deals with the thermo-mechanical study of the FSSW process in different materials.

8.4　CASE STUDY 1: FSSW OF AA6060 SHEETS

8.4.1　Process Conditions

The FSSW process was completed in the CNC machine tool with the processed sample lab configuration of 2 mm thick AA6060-T6 aluminum alloy sheets. AISI 1040 steel is used as tool material with cylindrical sections with dimensions of 12, 4, and 3.4 mm of shoulder diameter, pin length, and pin head diameter, respectively. The experimental study shows how the welding process parameters (shown in Table 8.1) influence the mechanical behavior of the processed samples.

8.4.2　Thermal Characterization

The test specimens were created as a single piece by machining the plate through milling in order to prevent adverse effects from the resistance to thermal contact among the plates and to provide precise and consistent temperature measurements during the tests. On each sample, 1 mm diameter holes were drilled in the width direction, and each hole was placed 4, 5, 7, 8, and 10 mm from the specimen's center. The depth of the holes was made at half the specimen width (15 mm) and half the height of the overlapping portion (2 mm) [16].

TABLE 8.1

Parameters Considered for Experimentation

Rotational Speed, S (rpm)	Plunging Depth, Z (mm)	Feed Rate, F (mm/min)	Dwell Time, t (s)
1000	3.6	10	0.5
3000	3.7	20	1
5000	3.8	30	1.5

8.4.3 WELD PRESENCE

It is clear how different values of rotating speed lead to an inadequate mixing of materials and an impressive and erratic flash creation. On the other hand, the intermediate circumstances result in both good material mixing and a small amount of regular flash. The sectional view of the joints that were formed by the upper and lower boundary conditions for revolving speed and feed rate is shown in Figure 8.2 after alkaline etching. Two distinct joint morphologies can be seen. The first is the perfectly straight joint line with a hint of curvature just next to the hole where the pin is left behind during the lower rotational speed. For high revolving speed, the connection line is curved, and the hook is evident. Finally, the varying feed rate may cause a change in bonded zone breadth, especially for lower rotational speed values.

8.4.4 WELDING TEMPERATURE

In all instances, excellent overall repeatability with little data scatter was seen from the analysis of variance. The p-values are extremely small (below the α-value set to 0.05), indicating that both rotating speed (S) and feed rate (F) have an impact on welding force and temperature. Plunging depth (Z) can be partially attributed to this influence, although dwell duration (s) did not significantly affect the output parameters under consideration. Only the parameters that were found to be significant based on these factors were included in the subsequent study. Figure 8.3 shows the temperature distribution at varied rotational speeds and feed rates for the measured temperatures T1 and T3. The temperature distribution exhibits systematic behavior, as detailed. For the same rotational speed, the change in feed rate gives rise to a higher

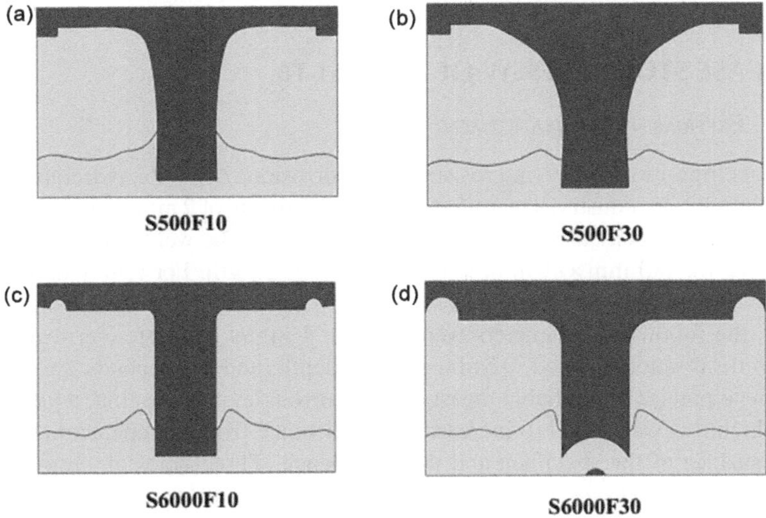

FIGURE 8.2 Macrograph of weld sections at different combination of rotational speed and feed rate.

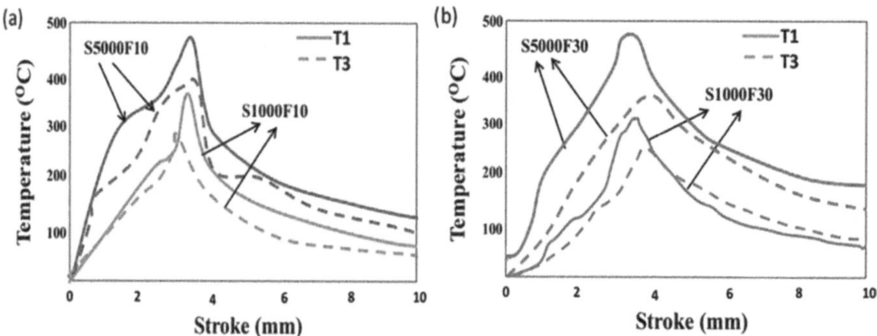

FIGURE 8.3 Temperature distribution [°C] as a function of S and F, measured by the thermocouples T1 and T3.

and sharper peak. Independent of the position of thermocouples, the heat distribution is found to be alike at the initial phases for the lower feed rate. By contrast, at a higher feed rate, the curves seem to overlap at the end phase. Besides, the rise in the rotational speed gives rise to the surge in temperature in a faster manner.

The outcome of revolving speed and feed rate on the heat distribution and welding force is plotted and observed. The rise in revolving speed increased the temperature and decreased the welding force. The rise in the feed rate reduced the temperature and increased the welding force. This increase in temperature has a substantial contribution to the material flow with the reduced resistance to penetration and welding force. Also, the time it takes for the thermal dispersion to occur at the tool-workpiece interface and within the workpiece itself is shortened, which in turn limits the temperature increase that would otherwise lead to a larger welding force.

8.5 CASE STUDY 2: FSSW OF AA 6061-T6

8.5.1 Experimental Arrangement

Due to its high formability and resistance to corrosion, AA 6061 is frequently used in the automobile industry. The rolled AA6061-T6 sheets of 2 mm width were lapped together with the conical HS6-5-2C HSS tool pin [17]. The welded work piece measurements are 180 mm × 40 mm, with a 40 mm lap length. For removing the oxide layers from lapping surfaces, wire brushing was utilized. To achieve a higher weld quality, the following parameters were chosen: 1 mm/s feed rate (vertical plunge), 1200 r/min rotational speed, 0.3 mm pressed depth, and a 3 seconds dwell period. To achieve precise and reliable measurements throughout the testing, a plate with a 2 mm thickness was added to increase heat resistance from the contact between the lowest surface of the specimen and the workbench. Three holes of 1 mm diameter were drilled at intervals of 9, 12, and 15 mm from the weld center over the depth equivalent to half the thickness of the plate to eliminate the temperature inaccuracy. The temperature is measured using K-type thermocouples, and the axial weld force is measured with the force sensor, which is mounted on the spindle.

8.5.2 MICROSTRUCTURE

The schematic sectional view of the weld joint is represented in Figure 8.4, which shows the flat bottom surface and keyhole formation in the weld joint. In addition, the formation of burrs made the surface irregular, which ends up in the complex stress distribution. Besides, there is a gap of 0.16 mm between the two plates with a sharp notch at the bonding interface, which is susceptible to producing a certain amount of stress concentration. According to subsequent analysis of the SEM image, the lap boundary in a hook-shaped upward curvature with a significant amount of internal micro-voids along the path is observed. Under stress, these micro-voids will quickly deform, spread, and eventually polymerize, forming a crack band.

8.5.3 THERMAL BACKGROUND

An FEM is developed to investigate the temperature distribution with the Gaussian surface and double ellipsoidal heat source. The analysis was compared with the experimental values obtained with the measurements by thermocouples. The temperature curves obtained through the simulation and experimentation have a good correlation. From the analysis, the inferred mechanism of the welding is as follows. First, heat is produced due to friction when the tool and workpiece come into contact, heat is produced by plastic deformation in flowing metal, and some of the heat produced is also lost through a variety of channels. The temperature at the weld center increases quickly during the beginning stages of welding because it is believed that the sum of heat produced is higher than heat dissipation. Furthermore, as the temperature rises, the yield stress decreases, thereby lowering the frictional heat. The actual and simulated temperature curves show that the rate at which the temperature rises is found to decrease during the steady welding stage. Finally, the temperature at the weld center reaches a maximum before the equilibrium between heat production and dissipation is achieved. Thus, it is perceived that the negative change in yield stress and the ensuing adjustment to heat produced due to friction are the major factors controlling the welding temperature.

8.5.4 THE FEATURES OF RESIDUAL STRESS

The residual stress that remains in the workpiece after the welding significantly affects the mechanical characteristics of the weld. The excessive tensile stress at the

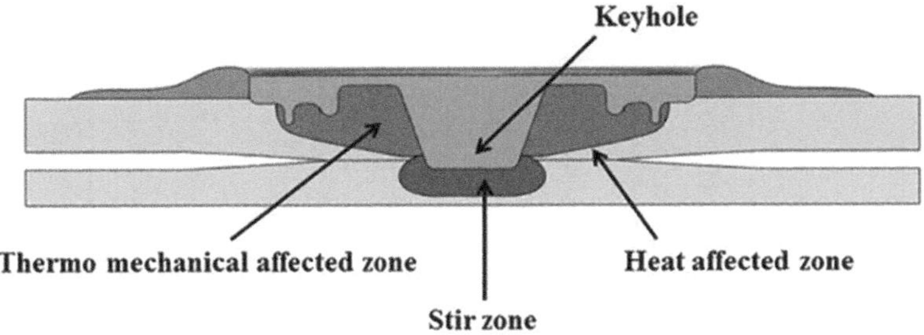

FIGURE 8.4 Sectional view of weld joint.

hook's tip accelerates the evolution of the crack. The residual stress analysis along the longitudinal (*x*-axis) and lateral directions (*y*-axis) showed the unsymmetrical profiles with the observation of compressive stress in the hook tip which ends up in a positive effect on the joints.[17]

8.6　CASE STUDY 3: FSSW OF ALUMINUM ALLOY 2024-O WITH ALUMINUM ALLOY 6061-T6

8.6.1　Process Conditions

The 3 mm thick aluminum alloy 2024-O and 6061-T6 sheets are joined together by the FSSW with the AISI H13 tool [18]. The tool is made with two different pin profiles, namely cylindrical and stepped, as depicted in Figure 8.5. The welding is performed at different rpms (900, 1400, and 1800 rpm) with a constant plunge depth and dwell time of 0.1 mm and 5 seconds, respectively. The samples are overlapped by keeping the aluminum alloy 6061 as the bottom plate and 2024 as the upper plate. To observe the thermal history, K-type thermocouples are positioned at four places from the weld center over a distance of 0, 4, 6, and 8 mm.

8.6.2　Thermal History

It is widely known that temperature dispersion has a significant impact on the metallurgical and mechanical characteristics of the weld material. Measurements are made of the temperature fluctuations in the weld samples at various rotational speeds using the cylindrical and stepped pins. Figure 8.6 shows the typical variations in the temperature. The rate of heating is observed to be higher owing to the rubbing heat developed by the tool rotation and plunge. During the dwell time, the temperature is observed to be constant, and after the retraction, it undergoes a cooling. Two major observations are reported as increased peak temperature with increased rotational

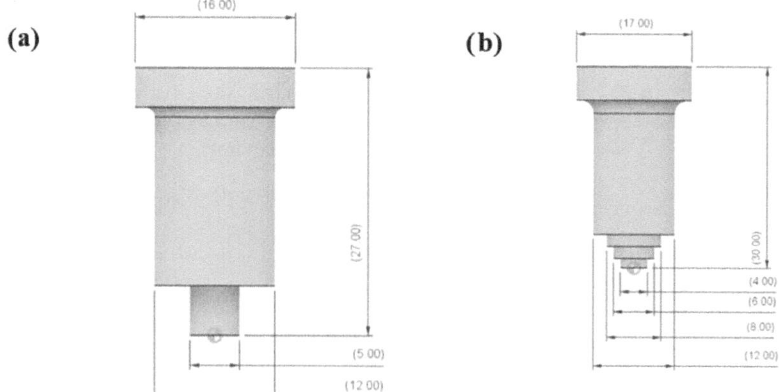

FIGURE 8.5　Typical sketch of pin profile: (a) cylindrical and (b) stepped.

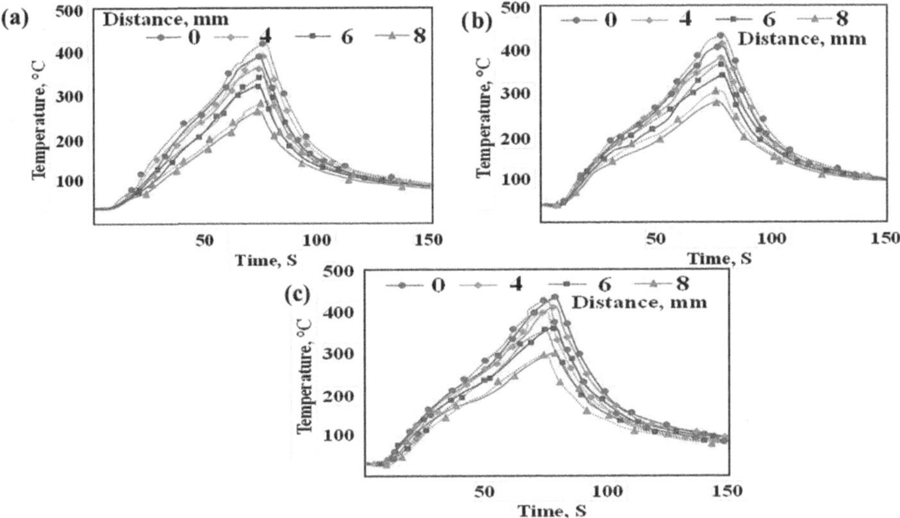

FIGURE 8.6 Temperature distribution: (a) 900 rpm, (b) 1400 rpm, and (c) 1800 rpm (continuous line-cylindrical pin; dotted line–step pin).

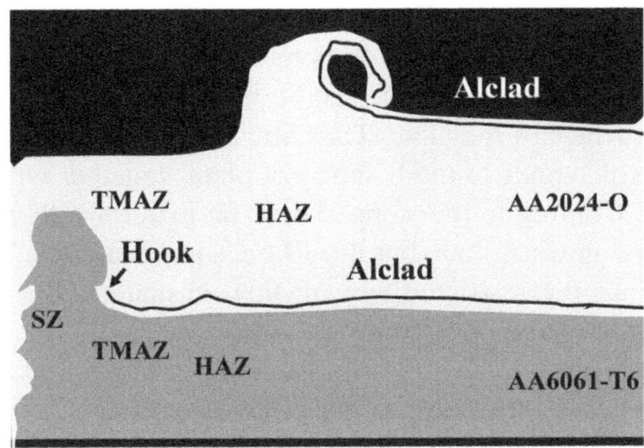

FIGURE 8.7 Illustration of weld appearance with different zones.

speed and variation in pin profile reflected in the temperature, i.e., a higher peak temperature is achieved with the stepped pin.

8.6.3 METALLURGICAL CHARACTERISTICS

The weld samples were sectioned and etched with Keller's reagent to explore their metallurgical features. The weld appearance, as shown in Figure 8.7, displays the five zones, namely the unbonded zone, partially bonded zone, HAZ, TMAZ, and SZ. Nearer to the SZ, a hook formation is observed owing to the plunge and stir action. Also, the formation of a thin Alclad layer of 200 mm thickness helps in the enhancement of corrosion resistance. The microstructure of the base metal 6061 aluminum alloy displays the elongated grains oriented toward the roll direction, with the existence of coarse intermetallic precipitates in the borders of the grains and inside the grains. The microstructure of the base metal 2024 aluminum alloy also displays the elongated grains with the existence of the fine and coarse intermetallic precipitates of Al_2Cu, Al_2CuMg, and $Al_6(Fe, Mn)$. The HAZ displays the coarse grains owing to the grain expansion during welding, and the TMAZ displays the uncrystallized grains in both aluminum alloys. The only difference observed in the alloys 2024 and 6061 is the change in distribution of particles and their size: coarse grains of 0.5–10 mm in the 6061 alloy and finer grain grains in the 2024 alloy.

8.6.4 CHARACTERISTICS OF MATERIAL FLOW

The flow of material during the stirring will result in variations in the metallurgy of the weld zone. The pace of spinning has increased the degree of extrusion of material from the bottom plate to the top plate at both pin profiles, whereas the width of bonding is observed to decrease when the rotating speed rises. The comparison of the degrees of extrusion shows the lower extrusion for the stepped pin profile owing to the restriction of upward material flow by the shoulder surface of the step being horizontal. The direction of flow material during the process of stirring is illustrated schematically in Figure 8.8a. In the case of a circular, the action of stirring induces the material to flow in a descending direction from the bottom of the shoulder toward the pin bottom. Next, the material from the bottom of the pin is pushed upward toward the upper surface of the plate. Thus, the material movement caused the mixing of materials, which ends in the bonding of plates together with the formation of hooks and flashes. Coming to the stepped pin, the material's flow follows a similar trend of inward and upward flow, but it will be separated at each step as proposed in Figure 8.8b. Thus, the restricted upward flow of material by the step shoulders reduces the extrusion of the bottommost layer of material to the top surface. Based on the speed and pin profile, the bond width and hook geometry vary. The rise in the revolving speed increased the hook height and reduced the bond width owing to the

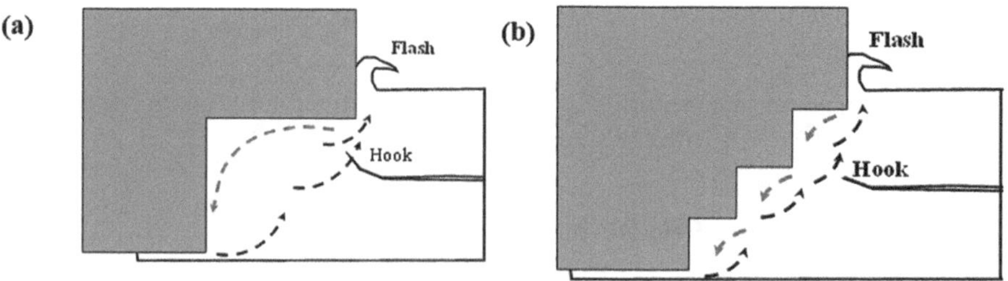

FIGURE 8.8   Material flow: (a) circular in profile and (b) stepped pin profile.

difference in material flow. Due to the restriction of material flow, the steeped pin weld specimens showed a short hook height and a wide bandwidth when weighed against cylindrical pins. On the contrary, at the higher revolving speed of 1800 rpm, the bond width is observed to increase, which shows that the increased stir force pushes the material upward.

8.6.5 Mechanical Properties

The mechanical behavior of the weld models is examined with the hardness and lap shear tests. The hardness test performed on the weld sample displayed a symmetrical pattern of hardness distribution in all the samples, with an average value of 57.3 and 76.8 Hv for the AA2024 and AA6061 alloys, respectively. The AA2024 alloy's hardness value is discovered to increase exponentially toward the key hole due to precipitation hardening, decrease in the stir zone due to lower dislocation density, and then abruptly increase closer to the key hole due to precipitation and work hardening. By contrast, in the aluminum 6061 alloy, the value decreases steadily toward the key hole, and in the underneath region, the maximum value is observed at the weld center, and it reduces outward. The decrease in the rotational speed results in increased hardness in the top sheet and decreased hardness in the lower sheet. Besides, the geometry of the pin significantly influences the hardness, and the larger hardness is found in the weld done using the cylindrical pin compared to the step pin in the stir zone. Examining the relationship between pin profile and rotating speed and the weld's shear strength, it was found that, for cylindrical pins, a higher rotational speed increases lap shear load up to 1400 rpm but then decreases it at 1800 rpm. A similar observation of variation in shear strength is reported with the step pin profile. Overall, the sample welded with the cylindrical profile is found to be feasible compared with the step pin profile weld.

8.7 CASE STUDY 4: FSSW OF ALUMINUM ALLOY 6061 WITH CARBON FIBER-REINFORCED POLYMER

8.7.1 Process Conditions

The aluminum alloy 6061 in T6 condition is joined with the carbon fiber-reinforced polymer composite (CFRP) through FSSW using the SKD tool [19]. The aluminum alloy plates of 100.0 mm length, 50.0 mm width, 2.0 mm depth, and 3 mm thick polymer composite are taken and overlapped over 50 mm. The factors such as plunge depth, plunge speed, and rotation speeds are varied, and the same is named as cases 1, 2, and 3. Case 1 has the parameter combination: 0.3 mm plunge depth, 0.1 mm/s plunge speed, and 1500 rpm rotation speed; case 2 has the parameter combination: 0.6 mm plunge depth, 0.1 mm/s plunge speed, and 1500 rpm rotation speed; case 3 has the parameter combination: 0.3 mm plunge depth, 0.1 mm/s plunge speed, and 1000 rpm rotation speed. The variations in temperature during the FSSW process are captured with the infrared thermal image system, and the recorded data is used to validate the simulated model.

8.7.2 Thermal History

At four distinct points away from the tool edge, at distances of 0.0 mm, 2.50 mm, 5.0 mm, and 7.50 mm, the temperature was measured. In case 1, the temperature of the upper surface of the aluminum alloy next to the tool is found to have increased drastically in a short period of time. There was an instant decrease in temperature with the tool extraction after 3 seconds. The recorded temperature as a function of time shows the fluctuation at the initial time, and a higher temperature is observed at the point closer to the tool edge. In the simulated model, the temperature dispersal of the weld is explored. The temperature dispersal at different times, namely 1, 2, 3, 4, 5, 6, 7, and 8 seconds, from the initial stage to the cooling stage is modeled for the aluminum alloy top surface and interface of the aluminum alloy and CFRP processed under the case 2 condition. The modeled temperature distribution is divided into two stages: the heating stage from 1 to 6 seconds and then the cooling stage. The maximum temperature at the tool and aluminum alloy contact surface was observed to be 480°C–582°C, which was lower compared to the melting temperature of aluminum alloy. The highest temperature at the aluminum alloy and CFRP contact surface was observed to be 575°C, which was obtained after 6 seconds. After a 5 second time period, it is observed that there is a slighter temperature variation at the contact surface owing to the dynamic balancing of heat generation and heat transfer among the surroundings. Similar observations of temperature variation in the contours are observed at the aluminum alloy and CFRP contact surfaces with an increased time span [19].

The variation of temperature when using the FSSW plate's sectional view over various time intervals. When contrasted with the tool center and the aluminum alloy contact surface, the temperature of the surface near the edge of the tool rises quickly in the beginning, or 0.8 seconds, and the trend entirely reverses after 1 second. The temperature variation between the aluminum alloy and the interface of aluminum alloy CFRP is higher compared to the central part. Moreover, the temperature distribution among the aluminum alloy CFRP is found to be symmetric with respect to the axis line. Though the increase in plunge depth caused more heat transfer to the CRFP side, the maximum temperature of 410°C is only achieved in case conditions. The outcomes attained from the established model are equated with the experimental results, and the profiles match each other and show it can be used efficiently to understand the thermal history and deformation.

8.7.3 Mechanical Properties

To comprehend the significance of the process factors, the tensile shear strength of the welded junction between the aluminum alloy and CFRP composite is evaluated. The fractured surface of the weld sample displays the fracture at the aluminum alloy and CFRP interface. Figure 8.9 displays the observations of tensile shear strength, tensile force, and bonded area in all three cases. The maximum shear strength is observed at the sample welded in case 3, whereas the bonded area and ultimate tensile force are found to be minimum at this condition. The main cause is that the

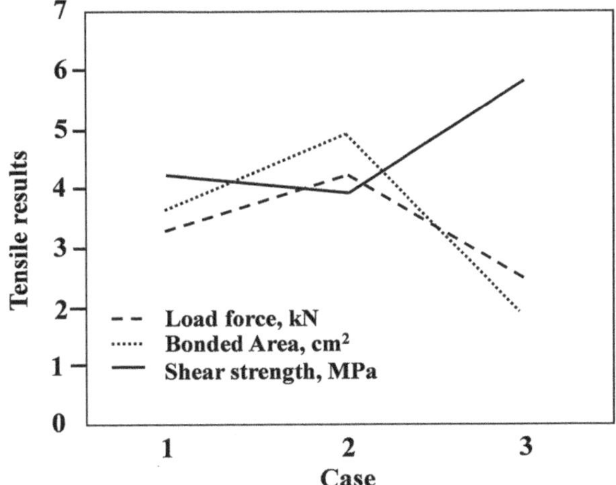

FIGURE 8.9 Comparison of tensile results.

adhesion bonding region is more prominent due to the feasible temperature range of 220°C–340°C. The maximum bonded area is observed at the sample welded in case 2, as more CFRP material is deformed into the aluminum alloy.

8.7.4 Characteristics of Fracture Surface

The edges of the welded aluminum alloy and CFRP composite plates show burrs and the appearance of indentation, and the geometry of the indentation resembles the shape of a tool. There will be a distinct region on the fracture surface of the metal and polymer-welded samples, namely the adhesion, transition, and plastically deformed zones. In the deformation zone, the aluminum alloy gets deformed and forms the bowl-shaped surface on the CFRP side, where the resin matrix underwent deformation by melting. The temperature has a substantial effect on the material flow during the deformation. But if the temperature goes above the decomposition temperature, it results in defect formation owing to the release of gases such as NH_3, CO_2, H_2O, and hydrocarbons. Deformed CFRP materials with aluminum alloy are perceived in the deformed and transition zones. In order to have adhesion bonding, a bond will be produced between the CFRP and the silane coupling agent coupled with the aluminum alloy.

8.8 CASE STUDY 5: FSSW OF ALUMINUM 2024-COPPER-SILICON CARBIDE

8.8.1 Process Conditions

The FSSW is done on pure Copper C11000-Aluminum 2024 composites of 2 and 3 mm with SiC particles used as an interface layer with 50 and 250 nm [20]. The FSSW tool is prepared from H13 steel, has a shoulder of 12 mm diameter, and a pyramid-shaped pin with top and bottom diameters of 4.0 and 5.0 mm, respectively. The work piece is divided into 125 mm by 25 mm squares, with a 25 mm by 25 mm overlap. The two plates are fixed rigidly by using a mechanical fixture to complete the FSSW welding. The cube-shaped groove is made in aluminum alloy to add different sizes of SiC particles to aluminum sheets. The dwell time and rotational speed are 4 seconds and 1600 rpm, respectively, and for determining the aluminum-copper alloy interface temperature, a K-type thermocouple is employed.

8.8.2 THERMAL HISTORY

Figure 8.10 shows the variations in the temperature with respect to time and particle size. The weld temperature is observed to progressively rise with time. A lower peak temperature than the aluminum alloy's melting point can be seen close to the end of the joining process when the temperature at the entire surface in contact with the welding tool and aluminum sheet ranges from 380°C to 560°C. Since aluminum alloy has excellent heat conductivity, the copper and aluminum alloy interface temperatures rose quickly and significantly in a short amount of time.

At the commencement of the welding process, the top and bottom plate temperatures quickly rise and spread from the center to both sides. The mechanical characteristics and temperature distribution in the FSSW weld zone with different-sized nanoparticles are analyzed using the FEM. The temperature distribution is mostly focused in the stir zone, and the front of the pin is where the larger temperature area is intensive, per the FEM results. The lower sheet's temperature distribution rapidly increases and becomes concentrated in the middle at the beginning of the dwell step.

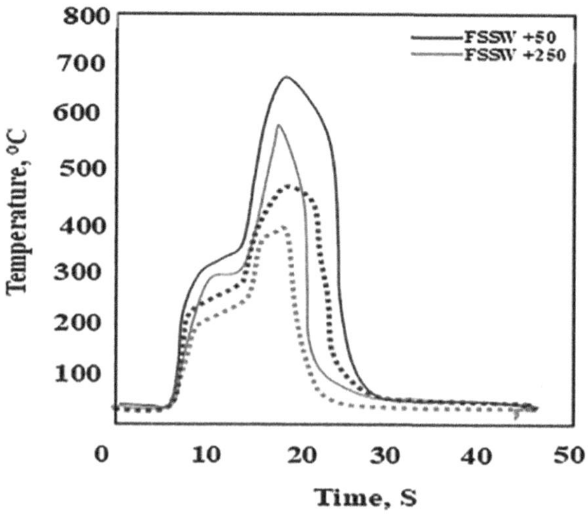

FIGURE 8.10   Temperature variation.

8.8.3 Metallurgical Characteristics

The primary factor that affects the microstructure of the material, mechanical behavior, and intermetallic compound (IMC) layer growth in copper aluminum is thermal history. Friction has occurred between the FSSW tool and the upper surface of aluminum plate; hence, a high temperature is generated around the process and the material starts to flow. When the material starts to flow, the friction is reduced, and the temperature starts to drop from its peak value. It is observed that the larger size of nanoparticles develops at a very high temperature compared with the smaller size of nanoparticles. When applying the FSSW method, two particle sizes and metals are softened, resulting in plastic flow in the weld zone. But the weld zone size enlarged as the nanoparticle size increased due to the increasing heat input during FSSW. The ever-changing recovery and re-cry process has been the primary driving force behind grain refinement in the FSSW process. Because microstructures with fine grain sizes have a large area of the grain border with significant stored energy, grain development during high-temperature deformation is naturally prone to it. Inserting reinforcing particles, such as SiC nanoparticles, slows down grain formation. Because they increase the number of nucleation spots for new recrystallized grains, these particles in the stir zone play an essential part in lowering the grain size in the stir zone. Although the grains are smaller than those for welding with bigger SiC particles, the stir zone for both sheets for FSSW demonstrates this. The stir zone grain sizes in the aluminum and copper sides decrease from 18 to 5 m and 14 to 3 m, respectively, as the reinforcing size reduces from 250 to 50 nm.

For greater clarity, increasing the size of silicon carbide from 250 to 50 nm boosts the number of particles in the matrix while offering additional nucleation sites because the capability percentage of nanoparticles remains steady. The pinning effect has a further reinforcing particle role in grain refining. This phenomenon prevents the movement of grain boundaries during grain growth. Accordingly, there is a direct correlation between the efficacy of the pinning effect and the fineness of the particle; the greater the fineness of the nanoparticle, the more effective the pinning effect. Furthermore, the breaking up of grains in the weld microstructure as a result of uneven local deformation is significantly influenced by the reinforcing particles. In contrast to one with a particle size of 250 nm, a weld specimen with a particle size of 50 nm has a higher shear strength. A few factors that influence this result are grain size, how the reinforcing particle interacts with the matrix, dislocation density, and uniform distribution of the particles, as well as residual compressive stress and dispersion hardening.

8.9 SUMMARY

Through an experimental campaign, it was determined how the FSSW process parameters affected the spreading of temperature in the welding zone, welding forces, and the mechanical characteristics of the joints. It was discovered that feed rate and rotating speed both affect the maximum temperature that may be reached during welding, with high feed rate and low rotational speed leading to lower welding forces and greater welding temperatures. These components have a direct bearing on the mechanical properties of the junctions between diverse materials and various aluminum series.

REFERENCES

1. Padhy GK, Wu CS, Gao S. Friction stir based welding and processing technologies processes, parameters, microstructures and applications: A review. *J Mater Sci Technol* 2018;1:1–38.
2. Karthikeyan R, Balasubramaian V. Optimization of electrical resistance spot welding and comparison with friction stir spot welding of AA2024-T3 aluminum alloy joints. *Mater Today Proc* 2017;4(2):1762–1771.
3. Sakano R, Murakami K, Yamashita K, Hyoe T, Fuzimoto M, Inuzuka M et al. Development of spot FSW robot system for automotive body members. In *Proceedings of the 3rd International Symposium of Friction Stir Welding*, 2001.
4. Doshi SJ, Gohil AV, Mehta ND, Vaghasiya SR. Challenges in fusion welding of Al alloy for body in white. *Mater Today Proc* 2018;5(2):6370–6375.
5. Cox CD, Aguilar JR, Ballun MC et al. The applications of a pinless tool in friction stir spot welding: An experimental and numerical study. *Proc Inst Mech Eng D J Aut* 2014;228(11):1359–1370.
6. Mei L, Yan D, Yi J, Chen G, Ge X. Comparative analysis on overlap welding properties of fiber laser and CO2 laser for body-in-white sheets. *Mater Des* 2013;9:905–912.
7. Guillaume S. *Thermal and Mechanical analysis of Friction Stir Welding Process*, vol. 1. Universidade de Coimbra, 2019.
8. Andrade DG, Leitão C, Dialami N, Chiumenti M, Rodrigues DM. Modelling torque and temperature in friction stir welding of aluminium alloys. *International Journal of Mechanical Sciences*. 2020 May;182:105725.doi: 10.1016/j.ijmecsci.2020.105725.
9. Mira-Aguiar D, Verdera, Leitão C, Rodrigues DM. Tool assisted friction welding: A FSW related technique for the linear lap welding of very thin steel plates. *J Mater Process Tech* 2016. doi:10.1016/j.jmatprotec.2016.07.006.
10. Tozaki Y, Uematsu Y, Tokaji K. A newly developed tool without probe for friction stir spot welding and its performance. *J Mater Process Technol*. 2010;210:844–851. doi:10.1016/j.jmatprotec.2010.01.015.
11. Rai R, De A, Bhadeshia HKDH, DebRoy T. Review: Friction stir welding tools. *Sci Technol Weld Join*. 2011;16(4):325–342. doi:10.1179/1362171811Y.0000000023.
12. Kan L, Liu X, Zhao Y. Research status and prospect of friction stir processing technology. 2019. doi:10.3390/coatings9020129.
13. Sun YF, Fujii H. Microstructure and mechanical properties of dissimilar Al alloy/steel joints prepared by a flat spot friction stir welding technique. no. Torque and temperature analysis in fssw of aluminium alloys. *Mater Sci*. 2013;47:350–357. doi:10.1016/j.matdes.2012.12.007.
14. FSSW in Mazda Motor Corp, https://www.assemblymag.com/articles/93337-friction-stir-spot-welding?
15. Tier MD et al., The influence of refill FSSW parameters on the microstructure and shear strength of 5042 aluminium welds. *J Mater Process Technol*. 2013;213(6):997–1005. doi:10.1016/j.jmatprotec.2012.12.009.
16. D'Urso G. Thermo-mechanical characterization of friction stir spot welded AA6060 sheets: Experimental and FEM analysis. *J Manuf Process*. 2015 Jan 1; 17 :108–119.
17. Zhang B, Chen X, Pan K, Li M, Wang J. Thermo-mechanical simulation using microstructure-based modeling of friction stir spot welded AA 6061-T6. *J Manuf Process*. 2019;37:71–81.
18. Tiwan, Ilman MN, Kusmono, Sehono. Microstructure and mechanical performance of dissimilar friction stir spot welded AA2024-O/AA6061-T6 sheets: Effects of tool rotation speed and pin geometry. *Int J Light Mater Manuf*. 2023;6:1–14. doi:10.1016/j.ijlmm.2022.07.004.

19. Ma N, Geng P, Ma Y, Shimakawa K, Choi JW, Aoki Y, Fujii H. Thermo-mechanical modeling and analysis of friction spot joining of Al alloy and carbon fiber-reinforced polymer. *J Mater Res Technol* 2021;12:1777–1793.
20. Bagheri B, Shamsipur A, Abdollahzadeh A, Mirsalehi SE. Investigation of SiC nanoparticle size and distribution effects on microstructure and mechanical properties of Al/SiC/Cu composite during the FSSW process. *Exper Simul Met Mater Int* 2022;29:1095–1112. doi:10.1007/s12540-022-01284-8

9 Wear Characterization of Friction Stir Spot Welded Joint Material

G. Prabu, N. Jeyaprakash, and Che-Hua Yang

9.1 INTRODUCTION

This chapter gives an overview of the wear characterization of the specimen, which was fabricated by friction stir spot welding (FSSW) process. Two surfaces of solid material met with each other during the execution of a specific function, as shown in Figure 9.1. The wear on the FSSW was represented as the mechanical interaction of two materials when they contact each other. Usually, this interaction occurs during the relative motions between the two bodies of the FSSW material. These relative motions were in the form of rolling, sliding, and carrying loads. Generally, stress is caused at some of the local contact points, and it leads to wear when the stress value exceeds the threshold value. It resulted in the mechanical failure of FSSW material in the local region. This mechanical failure was induced by different factors such as material property, load condition, environmental conditions, etc. Materials from FSSW joints were removed from their surface in the form of wear particles.

The major classification of the wear on the FSSW material was based on the mechanism of wear that occurred on it. The particles were removed from the FSSW joint surface during the interaction of materials, like in the machining process. This form of wear was referred to as abrasive wear and made the FSSW joint surface in the form of abrasive grooves [1]. In some cases, the material transferred from one surface of the FSSW joint to another surface due to adhesion properties between the interaction materials. This was called adhesive wear, and the FSSW joint surface lost material in the form of lumps. It made the FSSW joint surface have a lot of pits and an irregular shape. Some of the material lost its fatigue life due to continuous interaction with the material, which was denoted as fatigue wear. In these three cases, the particles at the FSSW joint were transferred from one material to another through the solid-solid or solid-fluid interaction. The solid–fluid interaction also induced wear when high stress was applied due to the high impact flow velocity. Fluid cavitation as well as the erosion of the FSSW joint was the major cause of fluid wear in the practical application. In addition, chemical wear also occurred on the component, which may induce an adverse effect or be beneficial based on the application.

Minor wear, such as diffusive wear, melting wear, impact wear, and wear, induced by electric discharge commonly occurred in the FSSW joint. However, these types of wear were important as they increased the downtime of machines. The frictional heat

DOI: 10.1201/9781003432289-9

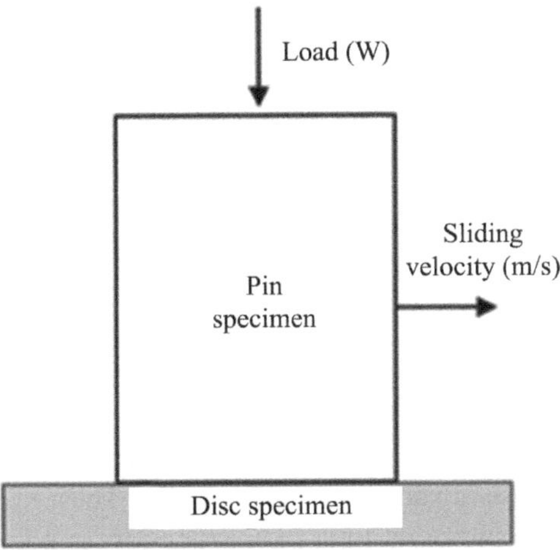

FIGURE 9.1   Schematic representation of matting of two materials.

induced the melting wear, which most commonly occurred. As a result, the coefficient of friction changed significantly. The electric discharge-induced wear occurred in most of the pentagraph assemblies and the electric motor systems. The diffusion of one material to the other occurred at high temperatures through the frictional contact that led to the diffusion wear. This type of defect occurred in the tool material due to the generation of high temperatures in the contact region. The impact wear occurred when the rotary drill bit was inserted into the rock at high hammering force.

9.2   ADHESIVE WEAR

The ability of atoms to form bonds with other atoms at the FSSW joint interacted surface was referred to as adhesion. This type of bonding occurred during the interaction of two materials [2–4]. In addition, the bonding became stronger at the interlayer of the FSSW joint when the two crystal structures interacted with each other. The influence of normal stress was less during the adhesive bonding. Usually, an oxide layer forms on the FSSW joint surface during interaction with the atmosphere. These oxide layers were broken during the interaction of materials at higher loads. The frictional force acted at the intersection region, which sheared the bond between the atoms.

The coefficient of friction was calculated by measuring the ratio of frictional force to the applied normal load. However, some of the surfaces have high adhesive strength when they have mirror-polished, cleaned FSSW joint surfaces. This surface has a high friction force even though no normal load was applied to it [5,6] Hence, the coefficient of friction was not a fundamental material property. The factor that affected the contact between the two materials significantly changed the frictional force. The real behavior of the FSSW joint surface is represented in Figure 9.2, which shows the approximate contact between the interaction surfaces. The two FSSW materials were contacted with each other at the plastic zone, and the contact area

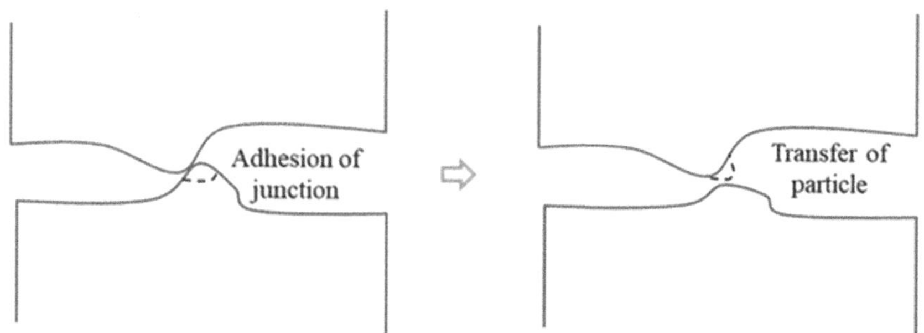

FIGURE 9.2 Model for adhesive wear.

increased with the increase in sliding among them [7]. These slidings were induced by the application of the tangential force. The contaminating layers at the intersection region established the welding between the two-contact surfaces.

However, the increase in the contamination level at the intersection region decreased the shear strength of the interface region. It resulted in a decline in the growth of bonding between the interaction layers. Hence, the COF value became finite and remained at a certain value limit. In addition, these growths in the intersection region were controlled by using a heterogeneous bearing surface rather than a homogeneous material. Moreover, the suitable FSSW joint surface finish assisted the interaction among them. The motion of the intersection line between the two-contact surfaces was interrupted frequently by finishing the surface in a particular direction [8]. Thereby, the adhesive wear was reduced based on the surface finish. This bonding between the FSSW joint materials was the initial stage of wear, and there were material losses at this stage. Furthermore, the continuous sliding of the material led to an increase in the strength of the bonding region through the work hardening mechanism. After a particular limit, the shear stress induced at the bonded region resulted in the detachment of material from one component to the surface of another component [9,10]. The bonding plan at the intersection region of the FSSW joint rotated due to the application of tangential traction. Hence, the axis of the bonding plan rotated, which interlocked the two matting surfaces. It produced the deformation bulges on each surface. The fracture at the bonding region occurred, and the material from one surface to another occurred. Then, the secondary mechanism occurred to promote the left-out particles breaking away.

This transferred material was retained on the surface as it became a single component. Further sliding led to the transfer of material from the FSSW joint to the surface of the original material, which became debris [11]. These debris consisted of groups of different particles as lumps of material that broke away. The breakage of FSSW joint material occurred when the elastic deformation energy Sf adhesive particles exceeded the surface-free energy. The fresh material appeared on the material surface after the continuous sliding process. But the fresh material did not deform immediately as the atmosphere air interacted with the fresh surface. It led to the formation of some oxide layers or the interaction of other active elements with fresh surfaces from the surrounding medium. The interaction of the surrounding medium is denoted as the healing process. The operating parameters determined these balances between the breaking of material and the healing due to atmospheric interaction. The rupturing of FSSW joint material was recognized by observing the abrupt changes in surfaces due to higher wear.

If the sliding speed increased, the time available to heal the surface would be reduced. Hence, more wear occurred on the FSSW joint at the high sliding speed. Inducting surface temperature led to an encouraging chemical reaction on the surface or restoring the bond strength of the adhesive material. Most of the wear process is initiated by the adhesive wear mechanism. However, the wear debris produced during the process led to the change of the wear mechanism to abrasive wear. Because of this change, the debris from this wear process was transformed into oxide, which became hard in nature. Hence, the abrasive wear started due to this hard debris formation. These types of situations arise when two mating surfaces of the FSSW joint are subjected to a small slip in the oscillatory movement. The wear rate may decrease as the debris reduces contact between the matting surfaces.

9.3 ABRASIVE WEAR

The FSSW joint materials were worn out from one surface through hard particles that resulted in the form of grooves. This type of wear was denoted as abrasive wear, which was similar to the machining or cutting operation [12,13]. Most of the abrasive particles possessed a negative value of rack angle, which led to the rubbing and cutting of matting FSSW joint material. The abrasive action of every piece of debris varied according to the rack angle. The abrasive particles were moved continuously between the two matting surfaces as long as the wear process continued. As a result, the particles underwent work hardening, which increased their hardness [14]. Hence, the hardness of the particles at the end of the cycle was greater than the initial stage of wear. A field emission scanning electron microscope (FESEM) image of abrasive particles (debris) formed at the Al 5356 alloy is represented in Figure 9.3. The chemical composition of the Al 5356 alloy is shown in Table 9.1. It

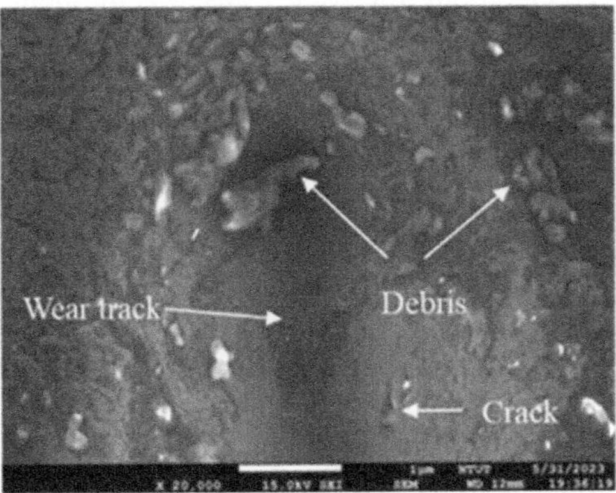

FIGURE 9.3 FESEM image of abrasive wear.

TABLE 9.1

Chemical Composition of Al 5356

Elements	Al	Mg	Fe	Cu	Mn	Others
Wt (%)	Bal.	4.2	0.4	0.1	0.05	< 0.3

led to an increase in the hardness of the abrasive particles compared to the material. So, the material removal rate increased with the increase in sliding distance. It was confirmed that the hardness of abrasive particles has some relationship with the material removal rate. The wear resistance of FSSW joint metals has a linear relationship with their hardness.

The hardness of the abrasive particles as well as the metal surface increased continuously with respect to the sliding distance. The abrasive particles blended when the hardness of the FSSW joint metal surface exceeded that of the abrasive particles [15,16]. As a result, the wear resistance of the material increased. The wear rate relationship with the hardness of the material was very useful in correlating the abrasive behavior of the material. In addition, the removal of material was changed from the abrasive debris to chip formation due to the blending of the abrasive particles [17,18]. The process occurred until the abrasive particles were retained in the plastic fatigue region. If the particles transferred to the adhesive fatigue zone, the wear characterization completely changed. This process was observed in most of the cases when the ratio of metal hardness to abrasive hardness was within the range of 0.8–1.3. Heterogeneous materials possessed different phases that contained various hardness values. Hence, the wear resistance of heterogeneous FSSW joint material was also varied accordingly. The abrasive particles were divided into fine particles in the heterogeneous material, which increased the wear resistance of the material. However, if the size of the abrasive particles increased, that would result in the adverse effect of wear resistance. The brittle debris particles behavior differed from the ductile debris particles

The presence of cracks along the wear track was an identification of abrasion action, and the corresponding wear rate was higher than the metals, which have equivalent hardness [19,20]. The behavior of fine abrasive particles became ductile, which resulted in their inability to induce wear on the brittle material. Hence, the ratio between the hardness of the base FSSW joint material and the hardness of the brittle abrasive particles less than 0.8 has not influenced the wear of the base material. However, the blunting of non-metallic brittle particles occurred at a low hardness ratio. It showed that the non-metallic brittle particles have no threshold value to induce wear on the material [21,22]. Generally, abrasive debris particles were rolled and/or slid between two wear FSSW surfaces or adhered to the opposite surface. The adhered abrasive particles induced more as they were retained at the same location and acted as the cutting edges [23]. The rolling and sliding abrasive particles induced less wear than stationary abrasive particles because the synergic effect of the rolling and sliding processes of abrasive particles led to discontinuous contact with the matting surface. Hence, the continuous abrasive actions have not been carried out. However, the abrasive particles flew along with the fluid steam, causing higher wear on the FSSW joint than the stationary [24]. This type of movement on the surface led to erosive wear on the surface due to the higher velocity of flow.

In this wear type, the impinging angle and ductility properties of the particles played an important role in causing the wear. In addition, the elasticity and brittleness of the particles influence the wear behavior of the FSSW joint. The sliding and erosive wear have the same mechanism, but the energy transferred from the fluid to the abrasive particles varies among them. In the case of erosive wear, the energy transferred to the abrasive particles by the fluid is greater than that transferred by the sliding process. As a result, the particles with higher energy impacted the surface and induced wear. Here, the kinetic energy of the particles determined the intensity of the wear. Along with the sliding process of the FSSW joint material, some of the particles were rotated due to the fluid movement. In this case, the intensity of wear was higher than that of the impingement of non-rotating particles. Because the rotating particles easily penetrated the surface compared to the non-rotating particles, the tangential force acting on the matting parts varied.

9.4 FATIGUE WEAR

The fatigue mechanism happened on the FSSW joint during sliding as well as rolling types of wear. But the stress acting on the surface differed between the sliding and rolling of abrasive particles. The rolling of abrasive particles induced more stress on the surface than the sliding surface. The contact of the abrasive particles with the matting surface behaved as elasto-hydrodynamic lubrication. Hence, adhesive interaction between the matting surface of FSSW material was reduced. Ball bearings, roller bearings, cams, and gears were practical examples of the fatigue mechanism. In this fatigue wear, pitting and spalling occurred on the matting surface during the wear process [25].

The stress acting on the rolling elements was in the range of the elastic limit. Hence, the elastic stress of the FSSW material was used to analyze the fatigue mechanism [26,27]. The shear stress along the orthogonal and unidirectional rolling elements was considered for fatigue wear. The intensity of shear stress in the orthogonal direction was higher than that in the unidirectional direction. As a result, more wear depth occurred on the surface in the orthogonal direction. Hence, the orthogonal direction was considered the critical shear stress during the analysis. However, this study was affected by different factors such as geometrical discontinuities, impurity inclusions, misalignment problems, surface flaws, pressure profiles in elastohydrodynamics, tangential traction, etc. The stress acting on the matting surface was increased in a cyclic manner [28–30]. It caused the maximum stress on the contact surface. However, the presence of a flaw in the FSSW joint material under the subsurface nucleated the crack. The crack increased with the increase in cyclic stress, which was continuously increased at higher rolling and sliding wear. It led to mislead the fatigue analysis as the crack was initiated due to the synergic effect of cyclic stress and the flaw presence at the sub-surface. It produced a large number of spalls on the matting surface and increased the surface with the increase in the number of cycles.

The movement of the rolling particles on the FSSW joint materials was subjected to tangential force, and the maximum of this force moved toward the matting surface than the maximum value of shear stress. This situation arose due to the lack of lubricant presence at the contact surface of FSSW material [31], because the asperity contact between the matting surface was not prevented by the elastohydrodynamic films. The environmental factors influence the rate of increase in surface cracks on the FSSW material more than the nucleation. Hence, the propagation of cracks increased more in severe environmental conditions. The small quantity of water present in the lubricant induced an adverse effect on crack propagation. The fluid pressure increased in the open and closed regions of the crack due to trapping.

9.5 FRETTING WEAR

In this section, the detailed view of wear induced at the interfaces due to fretting, the interconnection between fretting and fatigue, the influence of lubrication on the prevention of fretting, circumstances that induce diffusive wear, and the behavior of material when it encounters the wear can be defined as the progressive loss of substance. Fretting wear occurred on the FSSW component when it was subjected to small reciprocating and sliding movements against the contacting surface for a long period of time, as shown in Figure 9.4. As a result, the material was affected in two different ways, such as material damage due to wear and loss of fatigue life. The damage became severe to the FSSW material when the sliding of the component reached more than one million times of cyclic movement.

These types of movements were unavoidable in the interference fits where a 1 μm oscillating and sliding distance was allowed. This wear failure was usually present in all machinery and resulted in the total stoppage of some robust components. This fretting wear of FSSW joint failure problems occurred over a systematic period of time [32,33]. In some of the situations, the bearing components were fractured at a low level of stress due to this fretting wear. Hence, it was very important for designing the machine component where small amplitude movement occurred frequently with a shorter time period.

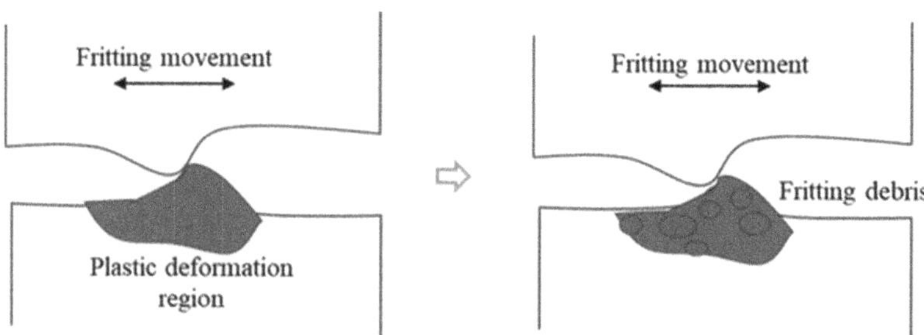

FIGURE 9.4 Model of fretting wear.

9.6 CAVITATION WEAR

The cavitation wear on the FSSW material was similar to the fluid erosive wear, but the medium and way of inducing the wear were different. In fluid erosive wear, the matting surface was damaged by fluid particles. In cases of cavitation wear, the fluid particles collapsed and released the vapor or gases in the form of bubbles. These bubbles impacted the matting surface and broke when the surface energy at the inside bubble was greater than the atmosphere [34]. The water droplet impacted the flying aircraft, which has a similar phenomenon of cavitation wear. The impact period was extremely short, which resulted in a sharp compressive force on the surface. It generated the crack in the ring form in the brittle FSSW joint material, whereas the plastic deformed in the ductile material. The surface of the ductile material was slightly depressed due to this plastic deformation [35]. If the fluid flow has not directly contacted the deformation zone and moved away from it, then shear stress has occurred on the periphery surface of FSSW material [36,37]. The repeated form of impact on the surface led to fatigue wear and pitting on the surface occurred.

The main cause of cavitation wear was the collapse of high-pressure fluid particles [38,39]. The fluid may be in the form of vapor or gas or a combination of liquid and gas. The intensity of damage on the FSSW material surface was higher in the liquid with gas combination than in the singe-phase fluid. The difference between the inside pressure and the outside surface tension from the atmosphere defined the stability of the droplet. In addition, the vapor pressure from the surrounding area induced more pressure on the periphery of the fluid droplet. The damage of the fluid particles induced by the cavitation wear on the FSSW material was measured from the surface energy [40]. However, the bulk modulus, viscosity, and liquid density of the fluid played an important role in the collapse of particles. The resilience value at the ultimate limit was used to correlate the worn-out surface damage with the property of the material. The energy dissipated from the FSSW material was determined by the tensile strength^2/elastic modulus ratio. The cavitation wear resulted in the pitting formation on the wear surface.

9.7 DELAMINATION WEAR

The lumps of FSSW material were removed from the surface during the wear process due to the adhesive process. In addition, some of the debris was in the form of blocks of irregular size. The asperities for the surface are plastically deformed and detached from it to form wear debris. Plastic deformation occurred at the topmost layer, and then the shear crack propagated, which led to the detachment of the particles from the surface (Figure 9.5). The material detachment on one surface did not depend on the adhesive forces [41,42]. The interaction between the asperities due to mechanical force was the main reason for the detachment of the asperities. The adhesive force was used to evaluate the subsequent attachment of the worn-out FSSW surface to the matting surface [43]. The formation of debris was attached to the worn-out surface due to the further sliding process. As a result, a large amount of conglomerated material was formed in the surface region. After a particular period of sliding distance, the conglomerated material detached and became wear debris. These particles were elongated and then

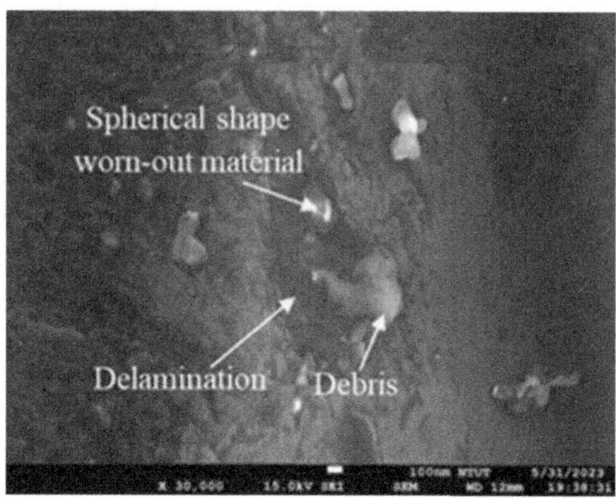

FIGURE 9.5   FESEM image of delamination wear.

flattened for further deformation in the sliding direction [44,45]. It led to the formation of plate-like debris, which was found in most of the sliding wear-tested specimens. The ratio between the thickness of flattened debris and its width was 0.3–0.5. In some of the locations in the wear track region, subsurface cracks in the FSSW material nucleated and propagated along the sliding direction.

The crack originated in the FSSW material from the plastically deformed region, which was located below the oxide and metallic mixed layer [46]. Generally, this crack nucleation was initiated from the void region due to the shearing process at the inclusion region. The nucleated creak has grown and connected with the nearer creaks. Then it propagated eventually toward the free surface of the FSSW material region and detached from the wear track region. Still, the fatigue model was studied to establish the wear behavior, but a significant result was not obtained. In the fatigue model, the high cyclic stress was considered as the propagation of cracks, whereas the low cyclic stress was assumed to be the removal of asperities [47,48]. In low and high cyclic stress, the elastic and plastic regions were considered. The crack formation during the wear process was less than that of the fatigue test because the crack induced by shear stress is different from the tensile load condition. The number of passes affected the mechanism involved in the wear test. The wear that occurred in the FSSW material depended on the chemical composition, surface film, adsorbed species on the surface, microstructure at the near surface and surface topography [49,50].

9.8   OXIDATIVE WEAR

The oxidative wear occurred in most of the material due to the presence of free oxidation energy in the surface region. It resulted in the formation of a thin layer on the FSSW material surface and covered the whole area wherever the surface contacted the atmosphere, as shown in Figure 9.6. The oxide formation increased continuously after the removal of the oxide film due to the wear process [51]. The removal occurred in the form of asperities loss from the surface. Hence, it protected the surface up to

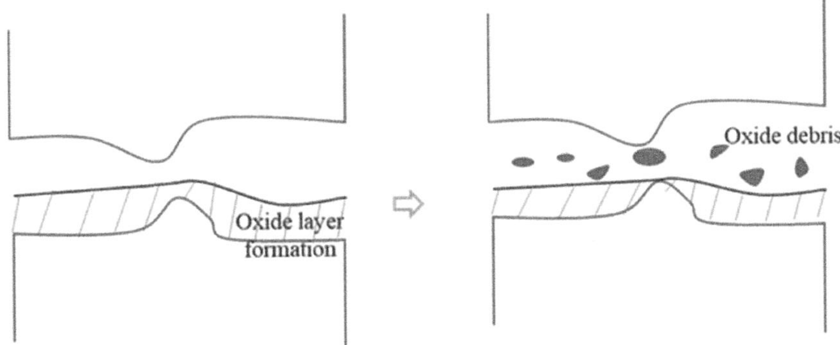

FIGURE 9.6 Model for oxidation wear.

TABLE 9.2

Chemical Composition of Steel

Elements	Fe	Mn	C	Si	Mo	Others
Wt (%)	Bal.	0.75	0.36	0.19	0.21	< 0.3

some load conditions [52,53]. The surface temperature played an important role in the formation of the oxide on the surface. The several hundred flash temperatures were generated during the sliding process. In addition, it allowed the species to diffuse in the oxide layer and promote oxide growth in it. The growth of temperature at the FSSW material surface region mainly depended on the normal load and sliding velocity during the wear test.

The moderate sliding velocity is enough to induce the high temperature to cause surface oxidation. In general, the oxide film formation suppressed the plastic delamination of the material in the wear test because it reduced the shear strength of FSSW material at the interface region. It developed the large shear strain that reduced the plastic delamination. The chemical composition of steel is listed in Table 9.2. The low surround atmosphere temperature led to oxidation at the asperity contact that was associated with the flash temperature. FeO formed at the steel surface when the surface temperature reached above 570°C. Hence, this oxide layer formed at the high temperature circumference only. The debris formation during the wear test contained the oxides with respect to the applied normal load, sliding velocity, and atmospheric temperature.

Based on the presence of temperature in the wear track region of the FSSW material, different forms of oxides were formed. The low, moderate, and high surface temperatures produced α-Fe_2O_3, Fe_3O_4, and FeO on the steel surface. The tip of the asperities was removed during the interaction with the matting parts, which was the first step in the removal of the oxide film from the worn-out surface. Then, the continuous interaction with the contour body led to the complete removal of oxide film. This process followed the order of oxidation and was then scraped. Furthermore, the increasing the surface temperature resulted in surface reoxidation. The particles from the FSSW material surface transferred to the counter body and agglomerated continuously. Further movement along the wear track led to the removal of the agglomerated particles as debris. Based on the practical example, it was found that the significant oxide film thickness built up on the worn surface. The thickness was in the range of millimeters and detached in the form of debris. The oxide film detachment occurred when the critical thickness of the film was reached and then spalled off. The homogeneous oxide film formed due to a tribo-chemical reaction on the FSSW material surface. In addition, some metallic particle-based oxide films formed after the detachment of the homogenous oxide films. The oxide particles were formed from the oxide layers and acted as abrasive particles. It induced abrasive wear as it became hard oxide debris.

9.9 DIFFERENT TYPES OF WEAR TESTING METHODS

Wear was characterized by a variety of experimental methods that were available on the market. Generally, the wear of the tested FSSW specimen mainly depended on the sliding parameters. The minor changes in the wear parameter led to a significant change in the radial direction of the lever according to the wear rate. The careful selection of testing parameters was essential to replicate the circumference of practical application. The points considered during the wear test were the surface finish of the two matting bodies, the geometry of the test FSSW specimen, the applied normal load, sliding velocity, and test environmental conditions. The behavior at the contact depended on the whole sliding system [54]. The tribometer was one of the instruments used to measure the wear rate and CoF of the specimen. In this test setup, the circular or rectangular-sized pin was pressed on the disc surface. Here, the disc was rotated at the pre-set value, whereas the pin was retained in place.

The experience gained by one material with the disc surface during the wear test was not the same as that of the other material under the same atmospheric conditions, because the asperities of these two FSSW materials were varied, which led to a difference in the points on the disc surface. Hence, asperities of one material contacted the disc surface throughout the experimentation, whereas the asperities of other FSSW material intermediately contacted the disc surfaces. As a result, the number of contact points that interacted with the disc surface varied [55]. In addition, the thermal cycles in the material changed based on the specific heat capacity of the individual material. Moreover, the chemical reaction that occurred between the wear track regions changed under different environmental conditions and the chemical composition of the material. Therefore, the wear behavior of the material with the same chemical composition was different. Hence, a number of repeated experiments were required to finalize the wear behavior of the material. The study of the wear behavior of FSSW material was similar to the pin-on-disc test, as all asperity points of the matting surface were contacted throughout the experimentation.

9.10 THE ARCHARD EQUATION FOR WEAR RATE CALCULATION

The wear on the FSSW material has happened when one surface moved on the other surface or both surfaces moved one over the other. The Archard derived the relationship between the wear rate and hardness of the material [56]. The wear coefficient was used to denote the severity of the happening in the wear region. It was initially derived for metals but was extended to be used for other materials as well. The wear initiated by the asperities of two matting surfaces came into contact [57]. Hence, the contact area was calculated by adding the area of asperities. The area of asperities contacted with the mating FSSW material surface related to the normal load acted on the material. The asperities deformed plastically when it was assumed to be metal. The shape of the asperities was assumed to be a circular cross-section. The normal load supported by this metal was calculated from Equation (9.1) as follows:

$$\text{Normal load atoneasperity} = \text{pressure act at the asperities} \times \text{cross-sectional area}$$

$$\text{Normal load atoneasperity} = \text{pressure act at the asperities}$$

$$\times \text{ radius of the asperity}^2 \tag{9.1}$$

The pressure acting on the asperities of FSSW material was considered a yield pressure as the plastic deformation started at the yield point. The asperities transferred from one metal to matting metal when the movement progressed. The transfer occurred at the contact point of the two asperities, which usually happens at their junction. This process repeated as the movement of matting metals continued for a specific period of time. The mass loss of the material depended on the fragment detachment at the junction of asperities. The volume of FSSW material loss depended on the size of the fragment detached. Hence, the removal of material volume was assumed as the hemisphere shape.

$$\text{Volume of material loss} = \frac{2}{3}\,\pi \times \text{raduis of hemisphere}^3 \tag{9.2}$$

It can be assessed that all asperities have not contributed to mass loss during the wear. Some of the asperities were only lost in the form of fragments. This portion of losses was assumed to be the wear coefficient.

$$\text{Oneasperity volume loss for unit sliding distance}$$

$$= \frac{\text{wear coefficient} \times \text{volume of material loss}}{\text{sliding distance}}$$

$$= \frac{2}{3}\,\frac{\pi \times \text{radius of hemisphere}^3}{2 \times \text{radius fo hemisphere}}$$

$$\text{Oneasperity volume loss for unit sliding distance}$$

$$= \frac{\text{wear coefficient} \times \pi \times \text{radius of hemisphere}^2}{3}$$

Total volume loss for unit sliding distance

$$= \sum \frac{\text{wear coefficient} \times \pi \times \text{radius of hemisphere}^2}{3} \tag{9.3}$$

$$\text{Total normal load} = \sum \text{pressure act at the asperities} \times \pi$$

$$\times \text{radius of the hemisphere}^2 \tag{9.4}$$

Comparing Equations (9.3) and (9.4), we can get

$$\text{Total volume loss} = \frac{\text{Wear coefficient} \times \text{total normal load}}{3 \times \text{pressure act at the asperites}}$$

The wear coefficient and the 1/3 constant were combined for convenience, and the final reduction formula was

$$\text{Total volume loss} = \frac{\text{Wear coefficient} \times \text{total normal load}}{\text{pressure act at the asperites}} \tag{9.5}$$

It was assumed that the pressure acting at the asperities was the hardness of the material during the indentation test. Hence, the volume loss equation became as

$$\text{Total volume loss} = \frac{\text{wear coefficient} \times \text{total normal load}}{\text{hardness}} \tag{9.6}$$

Equation (9.6) was called the Archard equation for wear calculation. It related the volume loss, normal load, and hardness to the asperities. The hardness was taken from the lowest hardness of the two matting FSSW material surfaces. The wear coefficient was used to refer to the severity of wear on different systems.

9.11 WORN-OUT SURFACE EXAMINATION AND WEAR DEBRIS CHANGES OF MICROSTRUCTURAL ON THE WORN-OUT SURFACE

The surface of the matting FSSW material significantly changed due to tribological action. Deformation of material occurred up to some particular distance from the top surface, as shown in Figure 9.7. However, this deformation was not the same for all materials, as it depended on the hardness, normal load, sliding distance, environmental condition, etc. Hence, the deformation depth varied from one material to another.

Grain refinement and reorientation occurred at the topmost point of the worn-out surface. Then the plastically deformed regions have grown below the grain refinement region. The undeformed region of FSSW material is below the plastically deformed zones. The plastic-deformed region possessed a lamellar structure. It was formed due to dislocation cells and sub-grains that were aligned along the

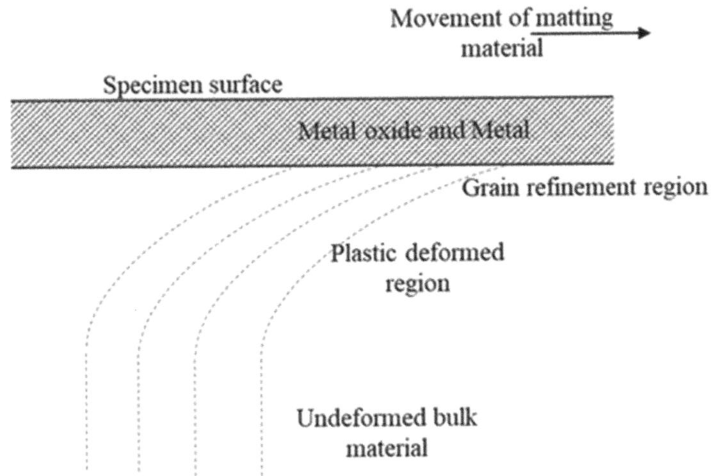

FIGURE 9.7   Schematic representation of deformation layer at the worn-out surface.

sliding direction. Strain occurred in the wear track region that led to localized deformation of the material. As a result, the fracture was initiated in the severely deformed region. It was responsible for shear the material and induced the formation of wear particles during the wear test. This strain usually occurs nearer to the worn-out surface, which results in the formation of a work-hardening region of FSSW material. Generally, the annealing process was used to remove the strain by increasing the temperature. It resulted in the dissipation of frictional heat from the strain region.

The strain of the FSSW component increased with the increase in hardness [58,59]. The microstructure refinement and recrystallization were also occurred during the annealing process. However, some of the deformed regions were preserved from the recrystallization, and the equivalent strain values were calculated from the below equation.

$$\text{Strain at the preserved region} = \frac{\tan\theta}{\sqrt{3}} \tag{9.7}$$

Two different layers were formed on the top of the worn-out FSSW material surface above the deformed region. The oxide layer was formed during the wear process due to frictional heat. The chemical reaction of oxygen with the bulk material led to the production of the oxide layers. The thickness of the oxide layer varied from 100 to 500 nm. It was used to protect the bulk FSSW material from wear, as it has not alloyed the matting surface to contact the bulk material. Hence, it was beneficial to reduce the wear, but only up to a certain normal load. Above the critical normal load, the oxide layers were broken and split into pieces as they were brittle in nature. These brittle particles acted as work-hardening elements and induced wear on the surface. The second layer was the metallic layer, which was deposited between the oxide layer and the deformed region. The chemical composition differed from the bulk FSSW material. These materials were received from the counter body during the wear process due to adhesive wear.

The wear debris from the FSSW material was generated during the wear process and deposited on the wear track region. Two types of wear debris were formed in the wear track region. The first type of wear debris was metal oxide fine particles, which were obtained from the oxide layers. The size of the oxide debris was in the range of 100 nm to 1 μm. These oxide debris grouped together and formed large agglomerate parts during the wear process. The second type of wear debris was the metallic wear debris, which has an size of up to 20 μm. These debris from the FSSW material can be viewed with the naked eye itself without using any microscopic instrument [60,61]. These debris were in the form of a plate, spherical, or needle shape based on the force acting on it.

9.12 MAPPING OF WEAR REGIME

Different forms of wear on the FSSW material dominate at a particular parameter, which was comprehensively discussed with this wear map as shown in Figure 9.8. The normalized stress acting at the contact point of two matting materials was calculated by dividing the applied normal load by the lowest hardness among the matting materials [62]. The dominant wear mode was identified with this wear map. In zone I, high contact stress occurred that led to the seizure of the matting surfaces of FSSW material. In this zone, the asperities have grown catastrophically, which has led to equalizing the contact area at the junction point

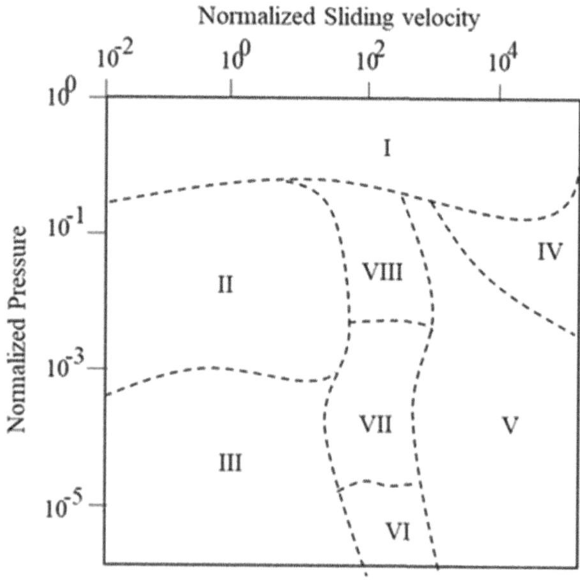

FIGURE 9.8 Schematic representation of wear map.

with the apparent area. The boundary region was calculated from the junction growth-based model [63]. In zone II, the parameters of high applied load and low sliding velocity were selected. In this zone, a thin oxide film formed and penetrated the asperity of the contact surface.

The wear occurred at this zone as high level of load was applied to the matting surface with low velocity. Hence, enough time for inducing stress on the matting material occurred. In zone III, low loads contributed to the wear on the matting parts, which led to mild wear on the oxide layers. Hence, the material was not removed from the bulk FSSW material substrate but removed from the oxide layers. The thermal effect played an important role in the formation of oxide on the FSSW material surface. Its effect was less at zone II and zone III, but it affected the wear properties at zone IV and zone V. In zone IV, the frictional heat was huge due to the high load and large sliding velocity. The frictional heat has not dissipated from the matting surface due to less thermal conductivity. Hence, the material at the interface region started to melt. It led to a rapid increase in wear, and the material was removed as molten droplets. The viscous force dissipated the heat energy from the molten layer of FSSW material. In zone V, the low load with high sliding speed was considered, in which the interface temperature was measured. It revealed that the temperature was slightly lower than the melting point. Hence, the oxidation occurred rapidly on the worn surface of the FSSW material.

In this condition, the thickness of the oxide layer increased, and some partially melted regions formed. The debris formed in the wear track region has the properties of oxide and a negligible amount of metal. In case of zone VI, zone VII, and zone VIII, mild wear occurs with low sliding velocity. In zone VI, the hotspots were formed at the tip of asperities, which increases the thermal activity around the matting region of FSSW material. As a result, the oxides have grown in patchy form in the local region. The detached oxide particles spall around the wear track region when the thickness of the oxide layers reaches about 10 µm. The metallic debris formed during the metallic contact of matting FSSW materials. These debris usually forms at the high applied load, even though the oxide patches appear. Zone VIII possessed a low wear rating due to the presence of high hardness material at the intersection region.

9.13 CONCLUSION

The different types of wear behavior that occurred on the FSSW material were studied in detail. The wear rate calculations were derived based on Archard's law. The influencing parameters on the wear rate calculation are explained. The abrasive, adhesive, oxidation, and plastic delamination wears of FSSW material were discussed well. The cause and effect of the wear process of the FSSW material on the wear debris were studied. Finally, the wear mapping was clearly studied with the different applied normal loads against the sliding velocity. Hence, this chapter would give an appropriate explanation of the wear and its type, which will be more helpful for researchers as well as industrial personnel to understand the wear characteristics of the FSSW material.

REFERENCES

1. N. Jeyaprakash, C.-H. Yang, G. Prabu, N. Radhika, Mechanism correlating microstructure and wear behaviour of Ti-6Al-4V plate produced using selective laser melting, *Metals*, 13, 2023, 575.
2. C. C. de Castro, J. Shen, A. H. Plaine, U. F. H. Suhuddin, N. G. de Alcântara, J. F. dos Santos, B. Klusemann, Tool wear mechanisms and effects on refill friction stir spot welding of AA2198-T8 sheets, *Journal of Materials Research and Technology*, 20, 2022, 857–866.
3. S. Suresh, N. Elango, K. Venkatesan, W. H. Lim, K. Palanikumar, S. Rajesh, Sustainable friction stir spot welding of 6061-T6 aluminium alloy using improved non-dominated sorting teaching learning algorithm, *Journal of Materials Research and Technology*, 9, 2020, 11650–11674.
4. P. Li, S. Chen, H. Dong, H. Ji, Y. Li, X. Guo, G. Yang, X. Zhang, X. Han, Interfacial microstructure and mechanical properties of dissimilar aluminum/steel joint fabricated via refilled friction stir spot welding, *Journal of Manufacturing Processes*, 49, 2020, 385–396.
5. N. Jeyaprakash, C.-H. Yang, M. Duraiselvam, G. Prabu, S.-P. Tseng, D. Raj Kumar, Investigation of high temperature wear performance on laser processed nodular iron using optimization technique, *Results in Physics*, 15, 2019, 102585.
6. N. Jeyaprakash, C.-H. Yang, M. Duraiselvam, G. Prabu, Microstructure and tribological evolution during laser alloying WC-12% Co and Cr_3C_2-25% NiCr powders on nodular iron surface, *Results in Physics*, 12, 2019, 1610–1620.
7. H. Zhang, X. Wu, W. Wang, Effect of fluid pressure on adhesive wear of spherical contact, Tribology International. 2023 Sep 1;187:108723. [8] A. Gerlich, P. Su, T. H. North, Tool penetration during friction stir spot welding of Al and Mg alloys, *Journal of Materials Science*, 40, 2005, 6473–6481.
9. Y. Zou et al., Refill friction stir spot welding of aluminum alloys: State-of-the-art and perspectives, *Welding in the World*, 2023 Aug;67(8):1853–85. [10] J. Wang et al., Tool wear mechanisms in friction stir welding of Ti-6Al-4V alloy, *Wear*, 321, 2014, 25–32.
11. A. Kubit et al. Failure mechanisms of refill friction stir spot welded 7075-T6 aluminium alloy single-lap joints, *The International Journal of Advanced Manufacturing Technology*, 94, 2018, 4479–4491.
12. T. Montag, J.-P. Wulfsberg, H. Hameister, R. Marschner, Influence of tool wear on quality criteria for refill friction stir spot welding (RFSSW) process, *Procedia CIRP*, 24, 2014, 108–113.
13. S. Liu, S. Zhang, R. Vaira Vignesh, O. Oladimeji Ojo, S. Mehrez, V. Mohanavel, M. Paidar, Probe-less friction stir spot processing (PFSSP) of AA5754/AA5083 joint by adding the B4C ceramics: Effect of shoulder features on mechanical properties and tribological behavior, *Vacuum*, 207, 2023, 111542.
14. G. Prabu, N. Jeyaprakash, C.-H. Yang, N. Radhika, Analysis on the transformation of melt wear to self-healing crack of wire arc additive manufactured Al 5356 alloy, *Tribology International*, 2023 Oct 1;188:108802.
15. A. Meilinger, Imre torok, The importance of friction stir welding tool, *Production Processes and Systems*, 6, 2013, 25–34.
16. Y. N. Zhang, X. Cao, S. Larose, P. Wanjara, Review of tools for friction stir welding and processing, *Canadian Journal of Metallurgy and Materials Science*, 51, 2012, 250–261.
17. A. Tiwari et al., Tool performance evaluation of friction stir welded shipbuilding grade DH36 steel butt joints, *The International Journal of Advanced Manufacturing Technology*, 103, 2019, 1989–2005.

18. T. Prater et al., A comparative evaluation of the wear resistance of various tool materials in friction stir welding of metal matrix composites, *Journal of Materials Engineering and Performance*, 22, 2013, 1807–1813.

19. D. Shaik et al., Tribological behavior of friction stir processed AA6061 aluminium alloy, *Materials Today: Proceedings* 44, 2021, 860–864.

20. J. Iwaszko et al., The effect of friction stir processing (FSP) on the microstructure and properties of AM60 magnesium alloy. *Archives of Metallurgy and Materials*, 61(3), 2016, 1555–1560.

21. G. Prabu, M. Duraiselvam, Tribological studies on AlCrFeCuCoNi high entropy alloy surface coated on Ti-6Al-4V using plasma transferred arc technique, *Archives of Metallurgy and Materials*, 67, 2022, 409–420.

22. N. Jeyaprakash, G. Prabu, C.-H. Yang, The influence of different phases on the microstructure and wear of Inconel-718 surface alloyed with AlCuNiFeCr hard particles using plasma transferred arc, *Journal of Materials Engineering and Performance* 2022 Dec;31(12):9921–34.

23. G. Madhusudhan Reddy, A. Sambasiva Rao, K. Srinivasa Rao, Friction stir processing for enhancement of wear resistance of ZM21 magnesium alloy, *Transactions of the Indian Institute of Metals*, 66, 2013, 13–24.

24. B. Yuan et al., Study on friction and wear properties and mechanism at different temperatures of friction stir lap welding joint of SiCp/ZL101 and ZL101, *Metals*, 13(1), 2022, 3.

25. P. Augusto de Paula Pacheco, C. Suque Endlich, K. Lucas Sousa Vieira, T. Reis, G. Fabiano Mendonça dos Santos, A. Antunes dos Santos Júnior, Optimization of heavy haul railway wheel profile based on rolling contact fatigue and wear performance, *Wear*, 533, 2023, 204704.

26. S. Joy-A-Ka et al., 3-Dimensional observation of the interior fatigue fracture mechanism on friction stir spot welded using 300 MPa-class automobile steel sheets, in *Proceedings of the 1st International Joint Symposium on Joining and Welding*. Woodhead Publishing, 2013.

27. B. Yuan et al., Study on friction and wear properties and mechanism at different temperatures of friction stir lap welding joint of SiCp/ZL101 and ZL101, *Metals*, 13(1), 2022, 3.

28. Castro CC. Refill Friction Stir Spot Welding: Evaluation of the welding of AA2198-T8 sheets and preliminary tool wear investigation. 2019.

29. I. Kwee, W. De Waele, K. Faes, Weldability of high-strength aluminium alloy EN AW-7475-T761 sheets for aerospace applications, using refill friction stir spot welding. *Welding in the World*, 63, 2019, 1001–1011.

30. S. F. Tebyani, K. Dehghani, Friction stir spot welding of interstial free steel with incorporating silicon carbide nanopowders, *The International Journal of Advanced Manufacturing Technology*, 79, 2015, 343–350.

31. A. Kubit et al., Failure mechanisms of refill friction stir spot welded 7075-T6 aluminium alloy single-lap joints, *The International Journal of Advanced Manufacturing Technology*, 94, 2018, 4479–4491.

32. L. Zou, D. Zeng, Y. Dong, J. Li, X. Chen, H. Zhao, L. Lu, Experimental and numerical study on the fretting wear-fatigue interaction evolution in press-fitted axles, International Journal of Fatigue. 2023 Oct 1;175:107793.

33. J. Sato, Recent trend in studies of fretting wear, *Transactions JSLE*, 30, 1985, 853–858.

34. E. Jonda, M. Szala, M. Sroka, L. Łatka, M. Walczak, Investigations of cavitation erosion and wear resistance of cermet coatings manufactured by HVOF spraying, *Applied Surface Science*, 608, 2023, 155071.

35. J. D. Escobar et al., Improvement of cavitation erosion resistance of a duplex stainless steel through friction stir processing (FSP), *Wear*, 297, 2013, 998–1005.
36. X. Jiang et al., Friction stir processing of dual certified 304/304L austenitic stainless steel for improved cavitation erosion resistance, *Applied Surface Science*, 471, 2019, 387–393.
37. S. Thapliyal, D. Kumar Dwivedi, On cavitation erosion behavior of friction stir processed surface of cast nickel aluminium bronze, *Wear*, 376, 2017, 1030–1042.
38. H. S. Grewal et al., Surface modification of hydroturbine steel using friction stir processing, *Applied Surface Science*, 268, 2013, 547–555.
39. Y. Li, Y. Lian, Y. Sun, Cavitation erosion behavior of friction stir procesed nickel aluminum bronze, *Journal of Alloys and Compounds*, 795, 2019, 233–240.
40. T. Zhao et al., Cavitation erosion/corrosion synergy and wear behaviors of nickel-based alloy coatings on 304 stainless steel prepared by cold metal transfer, *Wear*, 510, 2022, 204510.
41. A. Gerlich, P. Su, T. H. North, Tool penetration during friction stir spot welding of Al and Mg alloys, *Journal of Materials Science*, 40, 2005, 6473–6481.
42. R. Prasad et al., Microstructural, mechanical and tribological characterization of friction stir welded A7075/ZrB 2 in situ composites, *Journal of Materials Engineering and Performance*, 30, 2021, 4194–4205.
43. N. Jeyaprakash, C.-H. Yang, G. Prabu, R. Clinktan, Microstructure and tribological behaviour of Inconel-625 superalloy produced by selective laser melting, *Metals and Materials International*, 2022 Dec;28(12):2997–3015.
44. S. A. Alidokht et al., Evaluation of microstructure and wear behavior of friction stir processed cast aluminum alloy, *Materials Characterization*, 63, 2012, 90–97.
45. A. Moharami, High-temperature tribological properties of friction stir processed Al-30Mg2Si composite, *Materials at High Temperatures*, 37(5), 2020, 351–356.
46. P. Murugesan, V. Satheeshkumar, N. Jeyaprakash, G. Prabu, C.-H. Yang, Nano-level mechanical and tribological behavior of additively manufactured AlSi10Mg plates, *Archives of Civil and Mechanical Engineering*, 23, 62.
47. M. Vakili-Azghandi et al., Surface modification of pure titanium via friction stir processing: Microstructure evolution and dry sliding wear performance, *Journal of Alloys and Compounds*, 816, 2020, 152557.
48. T. Selvaraj et al., Elucidation of fatigue characteristics and fracture mechanism of friction stir spot-welded tension-shear joint steels, *Fatigue & Fracture of Engineering Materials & Structures*, 44(1), 2021, 74–84.
49. G. Prabu, M. Duraiselvam, M. Matheswaran, P. Amrish, T. Jayabalan, M. Varatharajulu, Laser surface texturing for enhancing microbial fuel cell-based electricity generation from wastewater, *Part A: Journal of Power and Energy*, 236(5), 2022, 937–948.
50. N. Jeyaprakash, G. Prabu, C.-H. Yang, Formation of different phases and their influences on the mechanical and tribological properties of surface alloyed FeCoCrNiAl particles using PTA technique, *Intermetallics* 142, 2022, 107457.
51. J. Kim, W.-Y. Lee, T. Tokoroyam, M. Murashim, N. Umehara, Tribo-chemical wear of various 3d-transition metals against DLC: Influence of tribo-oxidation and low-shear transferred layer, *Tribology International*, 177, 2023, 107038.
52. D. H. Choi et al., Frictional wear evaluation of WC-Co alloy tool in friction stir spot welding of low carbon steel plates, *International Journal of Refractory Metals and Hard Materials*, 27(6), 2009, 931–936.
53. C. de Castro Camila et al., Tool wear mechanisms and effects on refill friction stir spot welding of AA2198-T8 sheets, *Journal of Materials Research and Technology*, 20, 2022, 857–866.

54. N. Jeyaprakash, G. Prabu, C.-H. Yang, Microstructure and tribological behaviour of FeCoCrNiAlCuMn particles alloyed on Inconel-718 substrate using plasma transferred arc technique, *Intermetallics*, 139, 2021, 107360.

55. N. Jeyaprakash, S. Sivasankaran, G. Prabu, C.-H. Yang, A. Alaboodi, Enhancing the tribological properties of nodular cast iron using multi wall carbon nano-tubes (MWCNTs) as lubricant additives, *Materials Research Express*, 6, 2019, 045038.

56. A. Taghizadeh Tabrizi, H. Aghajani, H. Saghafian, F. Farhang Laleh, Correction of Archard equation for wear behavior of modified pure titanium, *Tribology International*, 155, 2021, 106772.

57. P. Trojovský, V. Dhasarathan, S. Boopathi, Experimental investigations on cryogenic friction-stir welding of similar ZE42 magnesium alloys, *Alexandria Engineering Journal*, 66, 2023, 1–14.

58. N. Jeyaprakash, G. Prabu, C.-H. Yang, Inclusion of nano silicon particle in SS316L through LPBF and its responses on corrosion behaviour, *Ceramics International*, 49(16), 2023, 26450–26468.

59. N. Jeyaprakash, G. Prabu, C.-H. Yang, M. Prasad Banda, E. Mohan, Structural and nanohardness of laser melted CM247LC nickel-based superalloy, *Transactions of the Indian Institute of Metals*, 76(2), 2022, 287–295.

60. G. Prabu, M. Duraiselvam, N. Jeyaprakash, C.-H. Yang, Microstructural evolution and wear behavior of AlCoCrCuFeNi high entropy alloy on Ti-6Al-4V through laser surface alloying, *Metals and Materials International*, 2021, 27, 2328–2340.

61. N. Jeyaprakash, C.-H. Yang, G. Prabu, K. Ganesa Balamurugan, Surface alloying of FeCoCrNiMn particles on Inconel-718 using plasma-transferred arc technique: Microstructure and wear characteristics, *RSC Advances*, 11, 2021, 28271–28285.

62. B. Zhang, X. Ma, L. Liu, H. Yu, A. Morina, X. Lu, Study on the sliding wear map of cylinder liner − piston ring based on various operating parameters, *Tribology International*, 186, 2023, 108632.

63. F. Peter Prakash, N. Jeyaprakash, M. Duraiselvam, G. Prabu, C.-H. Yang, Droplet spreading and wettability of Laser textured C-263 based Nickel superalloy, *Surface and Coatings Technology*, 397, 2020, 1260552.

10 Mechanical Properties and Temperature Distribution of Friction Stir Spot Welding Process

*Sampath Boopathi, K. Anton Savio Lewise,
J. Edwin Raja Dhas, and M. Satheesh*

10.1 INTRODUCTION

Friction stir spot welding (FSSW) is a solid-state joining technique that offers numerous advantages over conventional fusion welding methods. Unlike traditional welding, which involves the melting and solidification of base materials, FSSW creates strong and reliable joints by employing frictional heat and mechanical deformation. In FSSW, a specially designed rotating tool with a probe is plunged into the workpieces, generating heat through frictional forces. The rotating tool then traverses along the joint line, mechanically stirring and forging the material to form a solid bond. FSSW is commonly used for joining lightweight materials such as aluminum, magnesium, and their alloys, providing superior mechanical properties, reduced distortion, and enhanced fatigue performance.

FSSW is a solid-state joining process that uses a rotating tool to create a weld between two pieces of metal. It has quickly become popular in a variety of industries, such as automotive, aerospace, and shipbuilding. There have been a number of advances in FSSW in recent years, such as the development of new tool designs and new process parameters. FSSW has a number of advantages over traditional welding methods, such as no melting or remelting of the metal, precision, and versatility. The future of FSSW is bright, as it is still under development, but it has already made a significant impact on a number of industries [18].

The studies conducted on the FSSW of dissimilar Al-Cu materials encompassed various aspects, including process parameters, tool design, microstructure, and mechanical properties. The findings indicate that FSSW holds great promise as a technique for joining Al-Cu dissimilar materials, offering favorable mechanical properties, a sound microstructure, and low heat input. However, certain challenges remain, such as the formation of intermetallic compounds (IMCs) at the interface and difficulties in controlling the weld size. Despite these challenges, FSSW remains a highly promising method for effectively joining dissimilar Al-Cu materials [25].

DOI: 10.1201/9781003432289-10

A fully coupled thermo-mechanical numerical model was developed to simulate the refill FSSW process on Alclad 7075-T6 aluminum alloy sheets. The model was rigorously validated against experimental data and demonstrated excellent agreement. The results of the study revealed that the refill FSSW process is a viable technique for joining Alclad 7075-T6 aluminum alloy sheets, exhibiting desirable mechanical properties, a sound microstructure, and low heat input. The findings suggest that the refill FSSW process holds significant potential for application in various industries, including automotive, aerospace, and others. The process offers the advantages of enhanced mechanical performance, reliable microstructural integrity, and minimal thermal distortion. This makes it an attractive option for joining aluminum alloys with dissimilar properties [17].

A small-strain finite element model (FEM) was developed to simulate the FSSW process in both aluminum (Al) and magnesium (Mg) alloys. It underwent rigorous validation against experimental data and demonstrated excellent agreement. The model enables the optimization of process parameters, leading to improved weld quality. This optimization process can significantly enhance the overall quality, strength, and structural integrity of the welds. With the aid of this predictive tool, manufacturers and researchers can streamline their efforts in developing FSSW processes for Al and Mg alloys, ensuring reliable and high-quality welds [11]. A numerical model based on the finite element method (FEM) was developed to simulate the thermo-mechanical behavior of the FSSW process. The results showed that the Coulomb friction model was the most accurate for predicting heat generation in FSSW. This model provides valuable insights into optimizing FSSW processes, enabling manufacturers and researchers to improve the weld quality and overall performance of welded components [17]. An FEM of the three stages of the FSSW process—plunging, dwelling, and withdrawal—are presented. The model was validated against experimental results and was found to be in good agreement. The results of the study showed that the FEM model is a viable tool for predicting the weld quality in FSSW. It can be used to optimize the process parameters and improve the quality of the welds [4]. Water jet spot welding (WJSW) and FSSW are two solid-state joining processes used to join two pieces of metal. A numerical model of the WJSW and FSSW processes was validated against experimental results and found to be in good agreement. The results showed that the WJSW process produces similar weld quality but is more energy-efficient and produces less heat input, which can be beneficial for some applications [2].

An FEM of the FSSW process, based on the commercial software DEFORM 3D, was investigated. The model was validated against experimental results and was found to be in good agreement. The results showed that the FEM model is a viable tool for predicting the weld quality in FSSW and can be used to optimize the process parameters and improve the quality of the welds [8]. A numerical simulation of the FSSW process for aluminum alloys based on the FEM has been presented. It was validated against experimental results and found to be in good agreement. The results showed that the numerical simulation is a viable tool for predicting the weld quality in FSSW and can be used to optimize the process parameters and improve the quality of the welds [16]. The FEM is a valuable tool for understanding the friction stir welding (FSW) process and predicting the weld quality. It is used to study the effect of various process parameters on the weld quality, predict the microstructure and mechanical properties of the weld, and optimize the process parameters. The results show that the weld microstructure is characterized by a fine-grained zone and a heat-affected zone (HAZ), and the mechanical properties are comparable to the base metal properties [15].

An FEM of the FSW process for 6061 aluminum alloy is used to investigate the effects of tool rotation and translation rates on the temperature and strain distributions in the weld zone. The results show that the peak welding temperature can vary from 360°C to 500°C, while the cumulative effective strain can range from 12 to 45. The secondary deformation due to the shoulder edge is primarily concentrated in the near-surface layer. The model can be modified to account for material properties, but this could be a significant limitation for some applications [21]. The experimental and numerical work involved the fabrication of FSSW lap joints using a CNC machine tool. The process parameters were varied to investigate their effects on the temperature distribution, welding forces, and mechanical properties of the joints. The results showed that the peak welding temperature, welding forces, and mechanical properties of the joints were all affected by the process parameters. The optimal process parameters for obtaining high-quality joints were found to be rotational speed: 1200 rpm; axial feed rate: 0.1 mm/s; plunge depth: 0.5 mm; and dwell time: 0.5 s. The FEM model was able to predict the temperature distribution, welding forces, and mechanical properties of the joints with good accuracy. The findings of the study are valuable for understanding the FSSW process and optimizing the welding parameters for AA6060 sheets [7].

An FEM of the FSW process for 6061 aluminum alloy was reported. The results show that the peak welding temperature can vary from 360°C to 500°C, while the cumulative effective strain can range from 12 to 45. The secondary deformation due to the shoulder edge is primarily concentrated in the near-surface layer. The results of the study are valuable for understanding the FSW process and optimizing the welding parameters. However, the model does not account for the effects of material properties on the temperature and strain distributions, which could be a significant limitation for some applications. Overall, the paper is a valuable contribution to the field of FSW [26]. The thermomechanical model for FSW is based on the FEM and takes into account the heat generation due to friction, plastic deformation, and stirring. It is validated against experimental results for the FSW of AA2024-T351 aluminum alloy and predicts the temperature distribution, strain distribution, and flow behavior during the FSW. The results show that the weld quality is affected by the tool rotation speed, tool traverse speed, and tool offset. The optimum process parameters are found to be a tool rotation speed of 1200 rpm, a tool traverse speed of 100 mm/min, and a tool offset of 2 mm. The model is a valuable tool for understanding the FSW process and predicting weld quality [10]. The model is based on the FEM and takes into account the heat generation due to friction, plastic deformation, and stirring. It is validated against experimental results for the underwater friction stir welding (UFSW) of low carbon steel. The experiments showed that the welds made underwater were of better quality than the welds made in the air. The results of the simulation showed that the heat generation in the underwater weld was lower than the heat generation in the air weld due to the cooling effect of the water. The findings suggest that UFSW is a viable alternative to conventional FSW for welding steel in underwater environments. The underwater welds have better quality than the air welds and are therefore more resistant to corrosion and fatigue [20].

A thermomechanical model for FSSW based on the FEM was presented. The model is validated against experimental results for the FSSW of AA 6061-T6 aluminum alloy. The experiments showed that the welds had a good surface finish and a sound microstructure, with a tensile strength of 120 MPa. The simulation showed that the heat generation in the FSSW process was concentrated in the stir zone (SZ), with a peak temperature of 600°C and a high temperature gradient. The findings suggest that FSSW is a viable alternative to conventional welding processes for joining aluminum alloys, producing welds with good mechanical properties and a sound microstructure [29,13]. An experimental study and thermo-mechanical modeling of UFSW of polycarbonate (PC) was conducted on 3 mm thick PC plates. The results showed that the welds made underwater were of better quality than the welds made in air, with higher tensile strength, lower hardness, and a finer microstructure than the air welds. The heat generation in the underwater weld was lower than the heat generation in the air weld due to the cooling effect of the water. The findings suggest that UFSW is a viable alternative to conventional FSW for welding PC in underwater environments, as the underwater welds have better quality and are more resistant to corrosion and fatigue [6].

This chapter aims to provide a comprehensive understanding of FSSW by examining its principles, applications, and benefits. It explores the mechanical properties and temperature distribution analysis of FSSW joints, utilizing computer-aided design (CAD) modeling and ANSYS analysis. The chapter further discusses the optimization of process parameters to achieve high-quality welds. By delving into the fundamental aspects of FSSW, this chapter lays the foundation for subsequent sections that delve into the mechanical properties, temperature distribution analysis, CAD modeling, and ANSYS analysis. The insights gained from this chapter will contribute to enhancing the understanding and utilization of FSSW in various industries and promoting its application in manufacturing processes requiring robust and reliable joints.

10.2 FRICTION STIR SPOT WELDING

The research objectives of this chapter are as follows:

- To provide a comprehensive overview of the mechanical properties of FSSW joints, including joint strength, ductility, and toughness. This involves understanding the factors influencing these properties and exploring experimental methods and numerical simulations for their analysis.
- To investigate the temperature distribution during the FSSW process and its impact on the HAZ and potential phase transformations. This involves utilizing CAD modeling techniques to simulate the FSSW process and ANSYS analysis for temperature distribution simulation.

- To explore the use of CAD modeling and ANSYS analysis for process optimization, with the aim of enhancing weld quality, reliability, and performance. This involves analyzing the effects of various process parameters on mechanical properties and temperature distribution and identifying optimal settings.
- By addressing these research objectives, this chapter aims to provide valuable insights into FSSW, enabling researchers, engineers, and practitioners to optimize the welding process, understand the mechanical behavior of FSSW joints, and improve the overall performance and reliability of welded components.

10.2.1 PRINCIPLES AND PROCESS DESCRIPTION

FSSW is a solid-state joining process that creates high-strength welds by utilizing frictional heat and mechanical deformation. The pictorial representation of FSSW is shown in Figure 10.1. The process involves the following principles and steps [18]:

- Tool design: FSSW employs a specially designed rotating tool with a small, non-consumable pin or probe at the end. The tool typically consists of a shoulder and a threaded pin, which facilitate material movement during the welding process.
- Preparation: The workpieces to be joined are prepared by aligning their surfaces and clamping them securely in place. The workpieces can be sheets, plates, or other forms of similar geometry.
- Tool placement: The rotating tool is positioned on the surface of the workpieces, with the pin in contact with the joint area. The tool's shoulder rests on the top surface of the workpieces, providing downward force and preventing material from escaping.
- Plunge and rotation: The tool is then plunged into the workpieces, creating frictional heat due to the contact between the pin and the materials. Simultaneously, the tool rotates, generating additional heat and facilitating material movement.

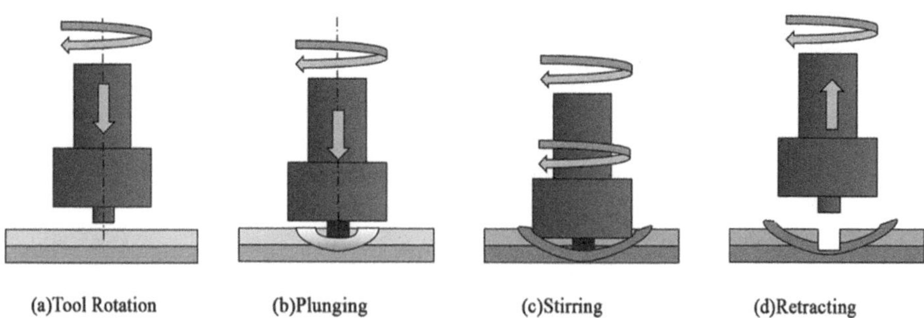

FIGURE 10.1 Schematic of FSSW process.

- Welding zone formation: As the tool rotates and traverses along the joint line, the material in the joint area experiences plastic deformation due to the combined action of heat and mechanical pressure. The tool's probe generates a mixing action, mechanically stirring and forging the material to form a solid bond.
- Cooling and solidification: After the tool passes through the joint area, the heat generated during the welding process rapidly dissipates, allowing the material to cool and solidify. The result is a solid-state joint with metallurgical bonding between the workpieces.

In the temperature analysis experimental setup, a specially fabricated fixture will be utilized to hold the entire specimen, including a k-type thermocouple (2, 3, or 4 channel) instrument. On each sample, four holes in the breadth direction, each with a diameter of 1mm, were made, each spaced from the specimen center by a particular amount. The depth of these holes, which were cut to a depth equal to half the breadth of the specimens, was carried out at the same height as the samples that overlapped. During the FSSW trials, four thermocouples with a 1mm diameter were placed within the holes. The specimens were blocked with a special clamping mechanism, and the thermocouples' entry was made possible by a number of tiny, rectangular slots that were machined. Figure 10.2 shows a four-channel k-type thermocouple installation setup along with the FSSW specimen.

The key advantage of FSSW lies in its solid-state nature, as it avoids the problems associated with melting and solidification in conventional fusion welding techniques. The absence of a molten phase eliminates issues such as solidification cracking, porosity, and alloy segregation, leading to improved joint strength and integrity. FSSW is particularly well-suited for joining lightweight materials, such as aluminum and its alloys, as well as other non-ferrous metals. It offers several benefits, including high joint strength, excellent fatigue performance, reduced distortion, and a minimized HAZ. The principles and process description provided here form the foundation for understanding the subsequent discussions on the mechanical properties, temperature distribution analysis, and optimization of FSSW through CAD modeling and ANSYS analysis.

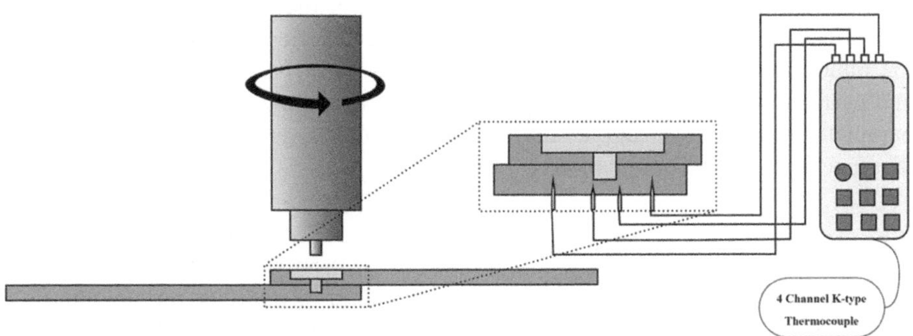

FIGURE 10.2   Schematic of thermocouple placement with FSSW setup.

10.3 MECHANICAL PROPERTIES OF FSSW JOINTS

10.3.1 JOINT STRENGTH ANALYSIS

The joint strength analysis of FSSW joints is a crucial aspect of evaluating the mechanical performance and reliability of the welded joints. The strength of FSSW joints is influenced by various factors, including process parameters, material properties, and joint geometry. To analyze the joint strength of FSSW, several testing methods can be employed [19]:

- Tensile testing: Tensile testing is commonly used to evaluate the tensile strength of FSSW joints. In this test, specimens with FSSW joints are subjected to an axial load until failure. The load-displacement curve is recorded to determine the ultimate tensile strength, yield strength, and elongation of the joint.
- Shear testing: Shear testing is performed to assess the shear strength of FSSW joints. Specimens with FSSW joints are subjected to a shear force until failure. Shear testing provides information about the joint's resistance to shear loading and is particularly relevant for applications where the joint is subjected to transverse loads.
- Peel testing: Peel testing is employed to evaluate the peel strength of FSSW joints. In this test, a peel force is applied to separate the FSSW joint interface. Peel testing provides insights into the joint's resistance to peel or delamination forces.
- Cross-sectional analysis: Cross-sectional analysis involves cutting and preparing FSSW joints for microscopic examination. This analysis enables the evaluation of the weld nugget, a HAZ, and any potential defects or discontinuities within the joint. The joint strength can be correlated with the weld microstructure and the presence of defects.

The joint strength of FSSW can be influenced by various parameters, including rotational speed, plunge depth, tool design, tool material, and material properties of the workpieces. Optimization of these parameters can lead to improved joint strength and mechanical properties. Furthermore, numerical simulations and finite element analysis (FEA) can be utilized to predict and analyze the joint strength of FSSW joints. FEA can simulate the mechanical behavior of the joint and provide insights into stress distribution, load transfer, and potential failure modes. Understanding the joint strength of FSSW joints is essential for ensuring the structural integrity and performance of welded components. It aids in optimizing the welding process parameters, material selection, and joint design to achieve robust and reliable joints that meet the specific application requirements.

10.3.2 DUCTILITY AND TOUGHNESS EVALUATION

In addition to joint strength analysis, the evaluation of ductility and toughness is important for assessing the mechanical properties of FSSW joints. Ductility refers to the ability of a material to deform plastically without fracturing, while toughness is the material's ability to absorb energy and resist fracture. To evaluate the ductility and toughness of FSSW joints, the following methods can be employed:

- Charpy impact testing: Charpy impact testing is commonly used to measure the toughness of materials, including FSSW joints. In this test, a notched specimen is subjected to an impact load, and the energy absorbed during fracture is measured. This test provides information about the resistance of the joint to sudden impact loads.
- Bend testing: Bend testing involves subjecting FSSW specimens to a bending load to assess their ductility. The specimens are bent to a specified angle, and any cracking or fracture behavior is examined. Bend testing helps evaluate the joint's ability to withstand deformation and absorb energy before failure.
- Hardness testing: Hardness testing is widely employed to evaluate the mechanical properties of FSSW joints, including their ductility and toughness. Hardness measurements provide an indirect indication of a material's strength and deformation behavior. Hardness tests, such as Vickers or Rockwell hardness test, can be performed on FSSW joints to assess their hardness values and infer their mechanical properties.

10.3.3 Factors Affecting Mechanical Properties

Several factors can influence the mechanical properties of FSSW joints, including the following:

- Process parameters: Parameters such as rotational speed, plunge depth, dwell time, and tool geometry have a significant impact on the mechanical properties of FSSW joints. Optimizing these parameters helps achieve desirable joint strength, ductility, and toughness.
- Material properties: The mechanical properties of the base materials being joined, such as their strength, ductility, and toughness, directly affect the mechanical properties of the FSSW joint. Material selection plays a critical role in achieving the desired mechanical performance.
- Microstructure: The microstructure of FSSW joints, including grain size, phase distribution, and precipitates, can influence their mechanical properties. Proper control of heat input and cooling rates during the welding process helps in obtaining a desirable microstructure.
- Defects and discontinuities: The presence of defects, such as voids, inclusions, or insufficient bonding, can significantly affect the mechanical properties of FSSW joints. Proper process control and inspection techniques are essential to minimize or eliminate these defects.
- Joint geometry: The geometry of the FSSW joint, including the overlap length, nugget diameter, and shoulder diameter, can impact the mechanical properties. Optimization of joint design parameters helps in achieving the desired mechanical performance.

Understanding the factors affecting the mechanical properties of FSSW joints enables the optimization of process parameters, material selection, and joint design to achieve joints with desirable strength, ductility, and toughness. This knowledge assists in ensuring the reliability and performance of welded components in various applications.

10.3.4 Experimental Methods for Property Analysis

To analyze the mechanical properties of FSSW joints, various experimental methods can be employed. These methods provide valuable insights into the material behavior, strength, ductility, toughness, and other mechanical properties of the welded joints. Some commonly used experimental methods for property analysis are illustrated in Figure 10.3.

- Tensile testing: Tensile testing is a widely used method to evaluate the tensile strength, yield strength, elongation, and other mechanical properties of FSSW joints. Specimens with FSSW joints are subjected to an axial load until failure, and the load-displacement curve is recorded. This test provides information on the joint's ability to withstand tensile forces.
- Shear testing: Shear testing is conducted to assess the shear strength and shear behavior of FSSW joints. In this test, specimens with FSSW joints are subjected to a shear force until failure, and the shear load-displacement curve is recorded. Shear testing helps evaluate the joint's resistance to shear forces and provides insights into its shear behavior.
- Bend testing: Bend testing is used to evaluate the ductility and fracture behavior of FSSW joints. The specimens are subjected to a bending load, and any cracking, deformation, or fracture behavior is examined. Bend testing helps assess the joint's ability to withstand deformation and provides insights into its ductility and toughness.

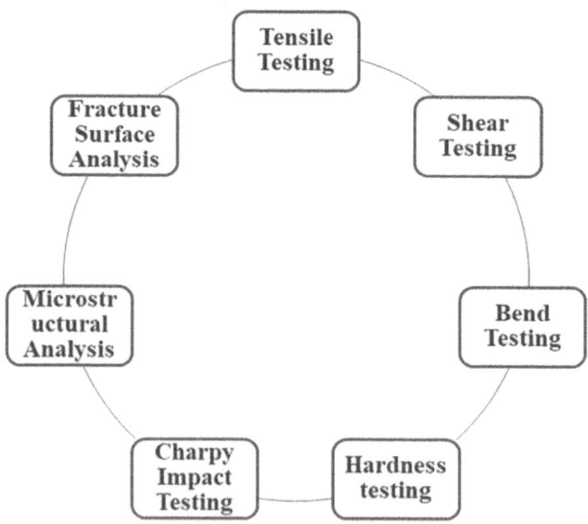

FIGURE 10.3 Various mechanical testing for FSSW.

- Hardness testing: Hardness testing is a non-destructive method used to evaluate the hardness and mechanical properties of FSSW joints. Various hardness testing methods, such as Vickers or Rockwell hardness tests, can be employed to measure the hardness values of the joint. Hardness measurements provide an indication of the joint's strength and deformation behavior.
- Charpy impact testing: Charpy impact testing is commonly used to assess the toughness and impact resistance of FSSW joints. Notched specimens are subjected to an impact load, and the energy absorbed during fracture is measured. Charpy impact testing helps evaluate the joint's ability to resist sudden impact loads and provides insights into its toughness.
- Microstructural analysis: Microstructural analysis involves examining the microstructure of FSSW joints using optical microscopy, scanning electron microscopy (SEM), or transmission electron microscopy (TEM). This analysis helps evaluate the grain structure, phase distribution, weld nugget characteristics, and any defects or discontinuities present in the joint.
- Fracture surface analysis: Fracture surface analysis is performed to understand the failure mechanisms and fracture behavior of FSSW joints. The fractured surfaces of failed specimens are examined using microscopy techniques to identify the fracture modes, such as ductile, brittle, or mixed-mode fractures. Fracture surface analysis aids in understanding the joint's failure characteristics.

These experimental methods provide valuable data for assessing the mechanical properties of FSSW joints, understanding their behavior under different loading conditions, and optimizing the welding parameters to achieve the desired mechanical performance. Proper experimental planning, specimen preparation, testing techniques, and data analysis are essential for accurate property analysis.

10.3.5 NUMERICAL SIMULATIONS FOR PERFORMANCE OPTIMIZATION

Numerical simulations play a crucial role in optimizing the performance of FSSW joints. They provide valuable insights into the welding process, heat distribution, material behavior, and mechanical properties of the joints. By conducting virtual experiments through numerical simulations, researchers and engineers can optimize the FSSW process parameters, joint design, and material selection to achieve the desired performance characteristics [28]. Some common numerical simulation techniques used for performance optimization in FSSW are shown in Figure 10.4.

- Finite element analysis (FEA): FEA is a widely used numerical method for simulating the FSSW process. It involves dividing the joint region into small elements and solving the governing equations to determine the temperature distribution, stress-strain behavior, and deformation characteristics during welding. FEA allows for studying various process parameters, such as rotational speed, plunge depth, dwell time, and tool geometry, to optimize the joint quality and mechanical properties.
- Thermal analysis: Thermal analysis simulations focus on predicting the temperature distribution, SZ, HAZ, and thermo-mechanically affected zone

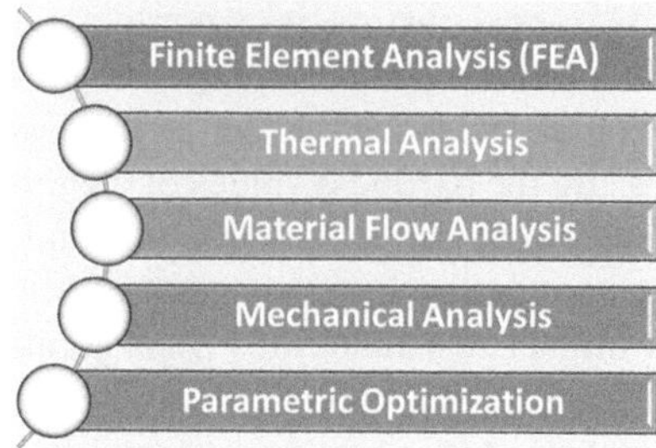

FIGURE 10.4   Common numerical simulation techniques for optimization of FSSW.

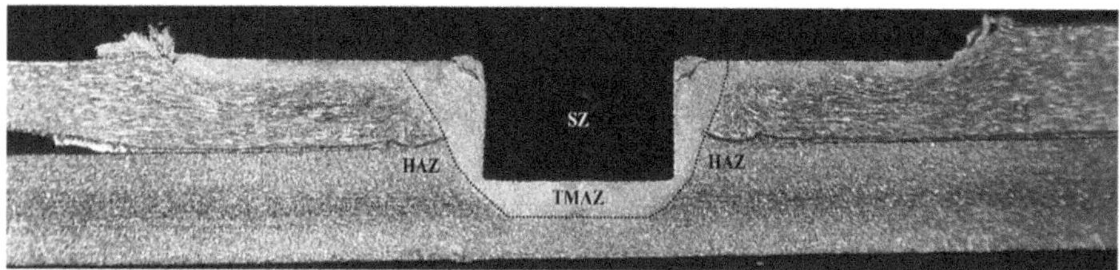

FIGURE 10.5   Different zones of FSSW process.

(TMAZ) during FSSW, which are clearly highlighted in Figure 10.5. By modeling the heat generation, heat transfer, and cooling rates, thermal simulations help optimize the welding parameters to minimize the HAZ size, reduce thermal distortion, and control the microstructure development in the joint region.

- Material flow analysis: Material flow analysis simulations enable the visualization and prediction of material flow patterns during FSSW. They provide insights into the material mixing, deformation, and recrystallization processes occurring within the weld nugget. By optimizing the tool geometry, plunge depth, and rotational speed, material flow simulations help achieve uniform mixing, refine the microstructure, and enhance the mechanical properties of the joint.

- Mechanical analysis: Mechanical analysis simulations focus on predicting the stress distribution, deformation behavior, and mechanical properties of FSSW joints. They consider factors such as material properties, joint geometry, and loading conditions to evaluate the joint strength, ductility, toughness, and fatigue performance. By varying the process parameters and analyzing the mechanical response, these simulations aid in optimizing the welding conditions for enhanced joint performance.

- Parametric optimization: Numerical simulations can be coupled with optimization algorithms to perform parametric optimization studies. By defining objective functions and constraints, these studies aim to find the optimal combination of process parameters and joint design variables to maximize joint strength, ductility, or any other desired performance metric. Parametric optimization simulations facilitate systematic exploration of the design space and can significantly improve the performance of FSSW joints.

Numerical simulations allow for virtual testing and evaluation of various scenarios, reducing the need for costly and time-consuming experimental trials. They provide insights into the complex phenomena occurring during FSSW and offer valuable guidance for process optimization and performance improvement. However, it is essential to validate the simulation results through experimental verification to ensure their accuracy and reliability.

10.4 TEMPERATURE DISTRIBUTION ANALYSIS IN FSSW

10.4.1 THERMAL BEHAVIOR DURING FSSW PROCESS

Understanding the temperature distribution and thermal behavior during the FSSW process is essential for optimizing the welding parameters, controlling the HAZ, and achieving the desired joint quality. The temperature distribution in FSSW is influenced by various factors, including process parameters, tool design, material properties, and heat transfer mechanisms. Analyzing the thermal behavior provides insights into the heat generation, heat transfer, and cooling rates, which directly affect the weld microstructure and mechanical properties [22]. During the FSSW process, the following stages contribute to the thermal behavior:

- Heat generation: Frictional heat is generated by the interaction between the rotating tool and the workpiece materials. The mechanical work and frictional forces result in localized heating at the tool-workpiece interface. Heat generation is influenced by factors such as rotational speed, plunge depth, dwell time, and tool geometry.
- Heat transfer: Heat generated during FSSW is transferred through the workpiece material and the tool. Heat is conducted radially from the weld zone into the surrounding regions, including the shoulder, retreating side, and advancing side. The heat transfer process is influenced by the thermal conductivity of the materials involved, as well as their thermal properties, such as specific heat capacity and thermal diffusivity.
- Cooling: After the welding process, cooling occurs as heat is dissipated from the weld zone to the surrounding areas and the environment. Cooling rates affect the microstructure development, grain growth, and the resulting mechanical properties of the joint. The cooling rate is influenced by factors such as the cooling medium, tool geometry, and ambient conditions.

Numerical simulations, such as FEA, can be employed to analyze the temperature distribution and thermal behavior during FSSW. FEA models consider factors such as heat generation, heat transfer, and cooling to predict the temperature evolution in the weld zone and surrounding regions. The simulations provide a visual representation of the temperature distribution, enabling researchers and engineers to analyze the thermal behavior and optimize the process parameters.

Experimental techniques, such as infrared thermography, can also be utilized to measure and analyze the temperature distribution during FSSW. Infrared cameras capture the surface temperature of the welded joint, allowing for real-time monitoring and analysis. These experimental measurements validate the simulation results and provide a better understanding of the thermal behavior during the welding process. By analyzing the temperature distribution and thermal behavior during FSSW, researchers and engineers can optimize the welding parameters to achieve desirable weld quality, control the HAZ size, minimize thermal distortion, and ensure the desired microstructure and mechanical properties of the joint. This understanding aids in the development of robust and reliable FSSW processes for various applications.

10.4.2 CAD Modeling Techniques for Heat Generation and Transfer

CAD modeling techniques can be utilized to simulate and analyze the heat generation and transfer during FSSW. CAD models provide a virtual representation of the welding process, allowing for the visualization and prediction of the temperature distribution in the weld zone and surrounding regions [5]. Several CAD modeling techniques can be employed to simulate heat generation and transfer in FSSW:

- Solid modeling: Solid modeling involves creating a three-dimensional (3D) digital representation of the FSSW joint and the surrounding components. The CAD model includes the tool geometry, workpiece materials, and other relevant features. Solid modeling allows for accurate visualization of the joint geometry and facilitates the analysis of heat generation and transfer.
- Finite element analysis (FEA): FEA is a numerical simulation technique that can be integrated with CAD models to analyze the heat generation and transfer during FSSW. FEA divides the CAD model into smaller, finite elements and solves the governing equations for heat conduction, convection, and radiation. FEA simulations provide detailed information on the temperature distribution and heat flow within the weld zone and surrounding regions. Figure 10.6 shows the numerical simulation of the FSSW process executed in ABAQUS software. The different colors show the temperature distribution present in various locations of the welded zone.
- Material property assignment: CAD models allow for assigning material properties to different regions within the joint. Material properties, such as thermal conductivity, specific heat capacity, and thermal diffusivity, play a vital role in heat generation and transfer. By assigning appropriate material properties to different regions, CAD models enable accurate analysis of temperature distribution and thermal behavior.

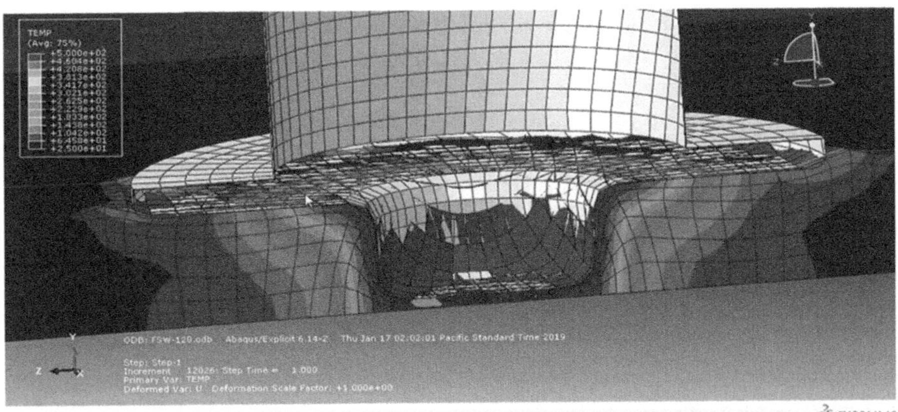

FIGURE 10.6   Numerical simulation of FSSW process using ABAQUS.

- Boundary conditions: CAD models can incorporate boundary conditions to simulate the heat transfer environment during FSSW. Boundary conditions include factors such as ambient temperature, convective heat transfer coefficients, and contact conditions between the tool and workpiece. By defining these boundary conditions, CAD models enable the simulation of realistic heat transfer scenarios.
- Coupling with thermal analysis software: CAD models can be coupled with specialized thermal analysis software, such as ANSYS or COMSOL, to perform detailed heat generation and transfer simulations. These software packages provide advanced numerical algorithms and solvers to accurately predict the temperature distribution and thermal behavior during FSSW. The CAD models serve as the input geometry for the thermal analysis software.

By employing CAD modeling techniques for heat generation and transfer analysis in FSSW, researchers and engineers can gain valuable insights into the temperature distribution, heat flow patterns, and thermal behavior of the welded joint. These insights aid in optimizing the welding parameters, tool design, and material selection to achieve desirable weld quality, control the HAZ size, and ensure the appropriate microstructure and mechanical properties of the joint.

10.4.3   General Procedure of ANSYS Analysis for Temperature Distribution Simulation

ANSYS is a widely used FEA software that offers advanced capabilities for simulating temperature distribution during FSSW. ANSYS allows for accurate and detailed analysis of the thermal behavior of FSSW joints, providing insights into the temperature distribution, heat generation, heat transfer, and cooling rates [14,28]. Here are the key aspects of using ANSYS for temperature distribution simulation in FSSW:

1. Geometry import: ANSYS provides tools to import CAD models of FSSW joints into the software. The CAD geometry can be imported in various formats, such as STEP, IGES, or native CAD formats. ANSYS allows for easy manipulation and refinement of the geometry to prepare it for the thermal analysis.
2. Mesh generation: ANSYS offers meshing capabilities to divide the FSSW joint geometry into smaller, finite elements. The meshing process determines the resolution of the analysis and affects the accuracy of the temperature distribution results. ANSYS provides different meshing methods, such as tetrahedral, hexahedral, or a combination of both, to generate a suitable mesh for the analysis.
3. Material property assignment: ANSYS allows for assigning material properties to different regions of the FSSW joint. Material properties, such as thermal conductivity, specific heat capacity, and thermal diffusivity, are critical for accurate temperature distribution simulation. ANSYS provides an extensive material library or allows for user-defined material properties to be assigned.
4. Boundary conditions: ANSYS enables the definition of boundary conditions for the FSSW thermal analysis. Boundary conditions include factors such as initial temperature, heat fluxes, convective heat transfer coefficients, and contact conditions between the tool and workpiece. Defining realistic boundary conditions is essential to accurately simulating the temperature distribution during FSSW.
5. Solver configuration: ANSYS provides various solvers, such as steady-state or transient heat transfer solvers, to simulate the temperature distribution in FSSW. The choice of solver depends on the specific requirements of the analysis. ANSYS allows for customization of solver settings to achieve desired accuracy and computational efficiency.
6. Results visualization: Once the analysis is complete, ANSYS provides visualization tools to interpret and analyze the temperature distribution results. Temperature contour plots, temperature profiles, and animations can be generated to visualize the thermal behavior during FSSW. ANSYS also allows for post-processing and extraction of quantitative data, such as maximum temperature, cooling rates, and HAZ size.

By utilizing ANSYS for temperature distribution simulation in FSSW, researchers and engineers can gain insights into the thermal behavior of the welded joint, optimize the welding parameters, control the HAZ size, and ensure the desired microstructure and mechanical properties. It is important to validate the ANSYS simulation results through experimental measurements to ensure accuracy and reliability.

10.5   CAD MODELING OF FSSW JOINTS

Geometric representation of FSSW joints: In CAD modeling of FSSW joints, the geometric representation involves creating a digital model of the joint geometry. This includes accurately capturing the shape, size, and features of the FSSW joint, such as the tool geometry, workpiece materials, and any other relevant components. Geometric representation is crucial for simulating the welding process, analyzing the heat distribution, and predicting the mechanical behavior of the joint [19,22].

Material and boundary conditions: In CAD modeling of FSSW joints, material properties and boundary conditions need to be assigned to accurately simulate the welding process. Material properties, such as thermal conductivity, specific heat capacity, and mechanical properties, are assigned to the different regions of the joint. These properties influence heat transfer and mechanical behavior during FSSW. Boundary conditions, including initial temperature, heat fluxes, convective heat transfer coefficients, and contact conditions, are defined to simulate the heat transfer environment and interaction between the tool and workpiece.

Modeling techniques and considerations: CAD modeling techniques and considerations play a significant role in accurately representing FSSW joints. Some key modeling techniques and considerations include (Figure 10.7):

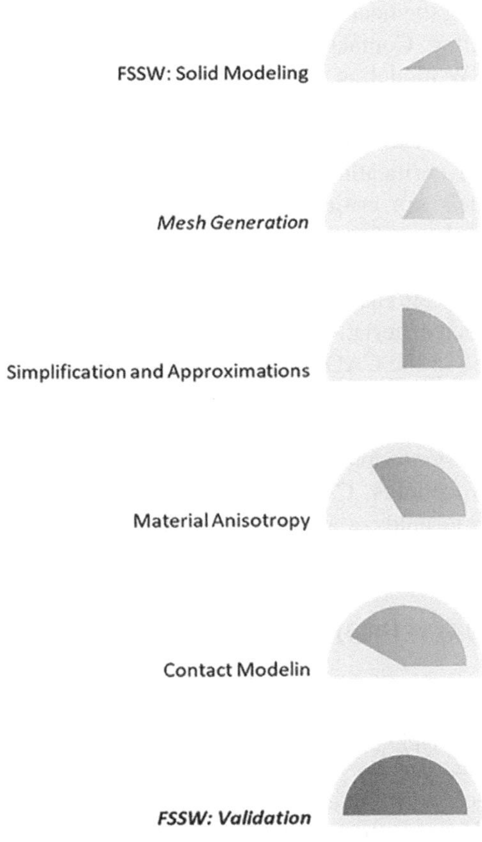

FIGURE 10.7   Stages of CAD modeling of FSSW for analysis.

- Solid modeling: Solid modeling involves creating a 3D representation of the FSSW joint geometry using CAD software. This allows for accurate visualization and manipulation of the joint geometry, enabling precise analysis of the welding process.
- Mesh generation: Meshing is the process of dividing the CAD model into smaller, finite elements to solve the governing equations for heat transfer and mechanical behavior. Proper mesh generation is essential for capturing the details of the joint geometry and obtaining accurate simulation results.
- Simplification and approximations: CAD models often involve simplifications and approximations to reduce computational complexity while maintaining the desired level of accuracy. This may include simplifying complex joint features, assuming axisymmetric conditions, or neglecting certain minor geometrical details.
- Material anisotropy: In some cases, FSSW joints may involve materials with anisotropic properties, meaning their properties vary with direction. Proper modeling techniques should consider material anisotropy, especially when simulating the heat transfer and mechanical behavior of the joint.
- Contact modeling: Contact between the tool and workpiece is an essential aspect of FSSW modeling. Proper contact modeling techniques should be employed to accurately represent the interaction between the tool and workpiece during the welding process.
- Validation and verification of CAD models: Validation and verification of CAD models are critical to ensuring their accuracy and reliability. Validation involves comparing the simulation results with experimental data or established analytical solutions to assess the model's accuracy. Verification involves ensuring that the CAD model behaves as expected and meets predefined criteria. Validation and verification processes help establish confidence in the CAD model's capability to predict the heat distribution, material behavior, and mechanical properties of FSSW joints. Through careful geometric representation, assignment of material properties and boundary conditions, appropriate modeling techniques, and validation and verification procedures, CAD models of FSSW joints enable researchers and engineers to analyze and optimize the welding process, predict joint performance, and enhance the design and manufacturing of FSSW joints.

10.6 ANSYS ANALYSIS OF FSSW JOINTS

10.6.1 FEA Approach

ANSYS offers a FEA approach to simulate and analyze FSSW joints. FEA is a numerical method that divides the CAD model of the joint into smaller, finite elements and solves the governing equations to predict the behavior of the joint under various conditions. Here are the key aspects of using ANSYS for FSSW joint analysis (Figure 10.8):

- Mesh generation: ANSYS provides tools for mesh generation that divide the CAD model into smaller, finite elements. The meshing process determines the resolution of the analysis and affects the accuracy of the results. ANSYS offers different meshing techniques, such as tetrahedral, hexahedral, or a combination of both, to generate an appropriate mesh for FSSW joint analysis.
- Material property assignment: ANSYS allows for the assignment of material properties to different regions of the FSSW joint. Material properties, such as elastic modulus, yield strength, and thermal properties, play a crucial role in accurately predicting the mechanical and thermal behavior of the joint. ANSYS provides an extensive material library or allows for user-defined material properties to be assigned.

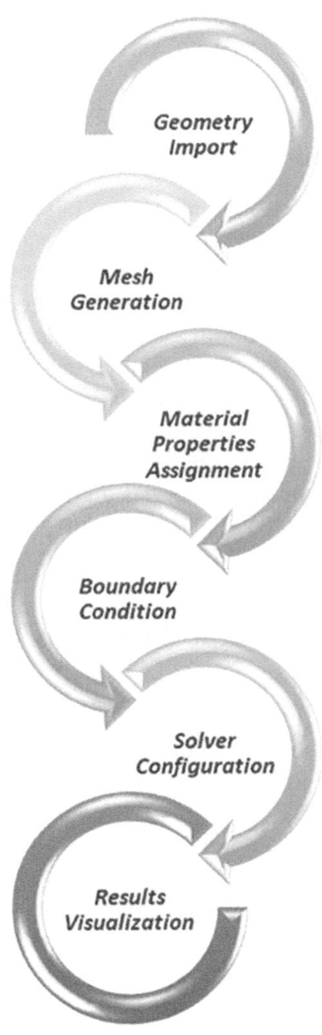

FIGURE 10.8   ANSYS analysis for temperature distribution simulation.

- Boundary conditions: ANSYS enables the definition of boundary conditions for FSSW joint analysis. Boundary conditions include factors such as applied loads, contact conditions, temperature, and constraints. Properly defining the boundary conditions is essential to simulating the welding process accurately.
- Solver configuration: ANSYS provides a range of solvers to solve the governing equations for FSSW joint analysis. The choice of solver depends on the specific requirements of the analysis, such as static or dynamic analysis, heat transfer analysis, or coupled field analysis. ANSYS allows for customization of solver settings to achieve the desired accuracy and computational efficiency.
- Results visualization: Once the analysis is complete, ANSYS offers visualization tools to interpret and analyze the results. This includes generating stress and strain contour plots, displacement profiles, temperature distributions, and other relevant results. ANSYS also allows for the post-processing and extraction of quantitative data, such as maximum stresses, distortions, and temperature gradients.

10.6.2 Modeling Welding Process Parameters

ANSYS enables the modeling of welding process parameters in FSSW joint analysis. Process parameters such as tool rotational speed, plunge depth, dwell time, and axial force significantly affect the heat generation, temperature distribution, and mechanical behavior of the joint. By modeling these parameters in ANSYS, researchers and engineers can evaluate their impact on joint performance and optimize the welding process for desired outcomes [19,22]. By employing ANSYS for FSSW joint analysis, researchers and engineers can gain insights into the mechanical behavior, thermal behavior, and overall performance of the joint. ANSYS enables the simulation and optimization of welding process parameters, helps in predicting joint strength, fatigue life, and distortion, and allows for design improvements and optimization. It is important to validate the ANSYS analysis results through experimental measurements to ensure accuracy and reliability.

10.6.3 Simulation of Temperature Distribution

In the ANSYS analysis of FSSW joints, the simulation of temperature distribution is a crucial aspect. ANSYS provides tools and capabilities to accurately predict and visualize the temperature distribution during the welding process [1,24]. Here are the key aspects of simulating temperature distribution using ANSYS:

- Heat generation: ANSYS allows for modeling the heat generation in FSSW joints. This involves considering the frictional heat generated at the interface between the tool and workpiece, as well as the plastic deformation heat caused by the stirring action. ANSYS considers these heat sources and calculates the temperature distribution within the joint.
- Heat transfer: ANSYS simulates the heat transfer within the FSSW joint, taking into account factors such as conduction, convection, and radiation. Heat transfer analysis helps predict how the generated heat is distributed

throughout the joint and the surrounding regions. It provides insights into the HAZ size, cooling rates, and thermal gradients.

- Material properties: Accurate modeling of material properties is essential for temperature distribution simulation. ANSYS allows for assigning appropriate material properties, such as thermal conductivity and specific heat capacity, to accurately capture the heat transfer characteristics of the materials involved in the joint.
- Boundary conditions: ANSYS allows for defining boundary conditions that affect the temperature distribution. This includes specifying the initial temperature, applying heat fluxes, considering convective heat transfer from the surrounding environment, and defining contact conditions between the tool and workpiece.
- Results visualization: ANSYS provides visualization tools to analyze and interpret the temperature distribution results. This includes generating temperature contour plots, temperature profiles, and animations to visualize the thermal behavior during the FSSW process. ANSYS also allows for post-processing and extraction of quantitative data, such as peak temperatures, temperature gradients, and cooling rates.

10.6.4 ANALYSIS OF RESIDUAL STRESSES AND DISTORTIONS

In addition to temperature distribution, ANSYS analysis of FSSW joints allows for the analysis of residual stresses and distortions [1]. Residual stresses and distortions are important factors that can affect the mechanical properties and performance of the joint. Here's how ANSYS enables the analysis of residual stresses and distortions:

- Material behavior: ANSYS considers the material behavior, such as plasticity and elasticity, during the welding process. This allows for the simulation of the mechanical response of the joint, including the development of residual stresses and distortions.
- Structural analysis: ANSYS performs structural analysis to predict the distribution of residual stresses and distortions in the FSSW joint. The structural analysis accounts for factors such as material properties, applied loads, and boundary conditions to simulate the mechanical behavior of the joint.
- Results visualization: ANSYS provides visualization tools to analyze and visualize the distribution of residual stresses and distortions. This includes generating stress contour plots, displacement profiles, and animations to understand the mechanical response of the joint after the welding process.

By simulating the temperature distribution, residual stresses, and distortions using ANSYS, researchers and engineers can gain insights into the thermal and mechanical behavior of FSSW joints. This information aids in optimizing welding parameters, understanding joint performance, and developing strategies to minimize distortion and control residual stresses for improved joint quality.

10.6.5 Future Extension

In the ANSYS analysis of FSSW joints, the optimization of process parameters plays a vital role in achieving enhanced weld quality. By adjusting the welding process parameters, such as tool rotational speed, plunge depth, dwell time, and axial force, researchers and engineers can optimize the FSSW process to improve weld quality and performance. Here's how ANSYS facilitates the optimization of process parameters:

- Design of experiments (DoE): ANSYS allows for the setup and execution of DoE to systematically explore different combinations of process parameters. By defining a set of parameter combinations and running simulations in ANSYS, researchers can evaluate the effects of various process parameter settings on the weld quality metrics, such as temperature distribution, residual stresses, distortions, and mechanical properties [3].
- Response surface methodology (RSM): ANSYS supports the application of RSM to model and optimize the relationship between process parameters and weld quality metrics. RSM involves fitting mathematical models to the simulation results obtained from ANSYS to predict the weld quality response based on the process parameter settings. These models can be further utilized to optimize the process parameters and find the optimal combination for enhanced weld quality [27,12].
- Parametric studies: ANSYS allows for parametric studies where individual process parameters can be varied while keeping others constant. By systematically varying the process parameters in ANSYS simulations, researchers can observe the effects of each parameter on the weld quality metrics. This helps in understanding the sensitivity of the process parameters and identifying the optimal range for achieving enhanced weld quality [9,23].
- Optimization algorithms: ANSYS provides optimization algorithms that can be employed to find the optimal process parameter settings automatically. These algorithms use mathematical optimization techniques to search for the best parameter combination that maximizes or minimizes the defined objective function, such as minimizing residual stresses or maximizing joint strength. ANSYS allows for incorporating these optimization algorithms to streamline the search for the optimal process parameters [23,27].
- Sensitivity analysis: ANSYS enables sensitivity analysis, which helps in identifying the most influential process parameters on the weld quality metrics. By systematically varying individual parameters and analyzing the corresponding changes in the weld quality, researchers can prioritize the process parameters for optimization efforts. This ensures that the most critical parameters are targeted for achieving enhanced weld quality.

Through the optimization of process parameters using ANSYS, researchers and engineers can improve weld quality, reduce defects, enhance mechanical properties, and optimize the performance of FSSW joints. The iterative process of simulating different process parameter settings, analyzing the results, and optimizing the parameters can lead to significant improvements in weld quality and process efficiency.

10.6.6 SUMMARY

This chapter provides a comprehensive overview of FSSW joints, focusing on mechanical properties analysis, temperature distribution analysis, CAD modeling techniques, and ANSYS analysis. The findings of this research have practical implications for industries that utilize FSSW as a welding technique. Understanding the mechanical properties of FSSW joints helps in predicting their strength and reliability, optimizing welding parameters, reducing defects, and minimizing thermal distortion. The CAD modeling techniques and ANSYS analysis presented in this chapter offer valuable insights into the behavior of FSSW joints, enabling optimization efforts for enhanced weld quality. This research contributes to the advancement of FSSW as a reliable and efficient welding technique, facilitating its adoption in industries such as automotive, aerospace, and manufacturing.

APPENDIX: LIST OF SYMBOLS AND ABBREVIATIONS

ANSYS: Analysis System
CAD: Computer-Aided Design
CFD: Computational Fluid Dynamics
DoE: Design of Experiments
FEA: Finite Element Analysis
FSSW: Friction Stir Spot Welding
HAZ: Heat-Affected Zone
RSM: Response Surface Methodology

REFERENCES

1. M. Awang, *Simulation of Friction Stir Spot Welding (FSSW) Process: Study of Friction Phenomena*, West Virginia University, 2007.
2. B. Bagheri, M. Abbasi, A. Abdollahzadeh, A.O. Moghaddam, Numerical modeling and experimental analysis of water jet spot welding and friction stir spot welding: a comparative study, *Journal of Materials Engineering and Performance*. 30 (2021) 1454–1471.
3. S. Boopathi, Experimental investigation and multi-objective optimization of cryogenic Friction-stir-welding of AA2014 and AZ31B alloys using MOORA technique, *Materials Today Communications*. 33 (2022) 104937.
4. G.E. Carr, N. Biocca, G.A. Lombera, S.A. Urquiza, FEM modelling of the three stages of friction stir spot welding, *Proceedings of the Institution of Mechanical Engineers, Part C: Journal of Mechanical Engineering Science*. 2023 Aug;237(15):3483–92.
5. C.D. Cox, B.T. Gibson, D.R. Delapp, A.M. Strauss, G.E. Cook, A method for double-sided friction stir spot welding, *Journal of Manufacturing Processes*. 16 (2014) 241–247.
6. H.A. Derazkola, E. Garcia, M. Elyasi, Underwater friction stir welding of PC: Experimental study and thermo-mechanical modelling, *Journal of Manufacturing Processes*. 65 (2021) 161–173.
7. G. D'urso, Thermo-mechanical characterization of friction stir spot welded AA6060 sheets: Experimental and FEM analysis, *Journal of Manufacturing Processes*. 17 (2015) 108–119.
8. G. D'urso, C. Giardini, FEM model for the thermo-mechanical characterization of friction stir spot welded joints, *International Journal of Material Forming*. 9 (2016) 149–160.

9. V. Haribalaji, S. Boopathi, M.M. Asif, Optimization of friction stir welding process to join dissimilar AA2014 and AA7075 aluminum alloys, *Materials Today: Proceedings.* 50 (2021) 2227–2234.

10. X.C. He, Thermo-mechanical modelling of friction stir welding process, Advanced Materials Research. (Volumes 774–776) (2013) 1155–1159.

11. P. Jedrasiak, H. Shercliff, Small strain finite element modelling of friction stir spot welding of Al and Mg alloys, *Journal of Materials Processing Technology.* 263 (2019) 207–222.

12. N. Jeyaprakash, M. Duraiselvam, S.V. Aditya, Numerical modeling of Wc-12% Co laser alloyed cast iron in high temperature sliding wear condition using response surface methodology, *Surface Review and Letters.* 26 (2018) 1–19.

13. N. Jeyaprakash, C.H. Yang, M. Duraiselvam, G. Prabu, Microstructure and tribological evolution during laser alloying WC-12%Co and Cr_3C_2-25%NiCr powders on nodular iron surface, *Results in Physics.* 12 (2019) 1610–1620.

14. A.W.H. Khuder, M.A. Muhammed, H.K. Ibrahim, Numerical and experimental study of temperature distribution in friction stir spot welding of AA2024-T3 aluminum alloy, *International Journal of Innovative Science Engineering and Technology (IJISET).* 6 (2017) 1111–1121.

15. D. Kim, H. Badarinarayan, I. Ryu, J. Hoon Kim, C. Kim, K. Okamoto, R. Wagoner, K. Chung, Numerical simulation of friction stir welding process, *International Journal of Material Forming.* 2 (2009) 383–386.

16. D. Kim, H. Badarinarayan, I. Ryu, J.H. Kim, C. Kim, K. Okamoto, R. Wagoner, K. Chung, Numerical simulation of friction stir spot welding process for aluminum alloys, *Metals and Materials International.* 16 (2010) 323–332.

17. A. Kubit, T. Trzepiecinski, A fully coupled thermo-mechanical numerical modelling of the refill friction stir spot welding process in Alclad 7075-T6 aluminium alloy sheets, *Archives of Civil and Mechanical Engineering.* 20 (2020) 117.

18. M. Li, C. Zhang, D. Wang, L. Zhou, D. Wellmann, Y. Tian, Friction stir spot welding of aluminum and copper: A review, *Materials.* 13 (2019) 156.

19. W. Li, J. Li, Z. Zhang, D. Gao, W. Wang, C. Dong, Improving mechanical properties of pinless friction stir spot welded joints by eliminating hook defect, *Materials & Design.* (1980-2015). 62 (2014) 247–254.

20. S. Memon, J. Tomków, H.A. Derazkola, Thermo-mechanical simulation of underwater friction stir welding of low carbon steel, *Materials.* 14 (2021) 4953.

21. V. Mishin, I. Shishov, A. Kalinenko, I. Vysotskii, I. Zuiko, S. Malopheyev, S. Mironov, R. Kaibyshev, Numerical simulation of the thermo-mechanical behavior of 6061 aluminum alloy during friction-stir welding, *Journal of Manufacturing and Materials Processing.* 6 (2022) 68.

22. S. Patil, F. Baratzadeh, H. Lankarani, Preliminary study on modeling of the deformation and thermal behavior of FSSW using SPH approach, in: *Proceedings of 11th European-DYNA Conference Salzbg,* Austria, 2017, pp. 9–11.

23. B. Sampath, V. Haribalaji, Influences of welding parameters on friction stir welding of aluminum and magnesium: A review, *Materials Research Proceedings.* 19 (2021) 222–230.

24. P. Sathiya, N. Siva Shanmugam, T. Ramesh, R. Murugavel, Temperature distribution modeling of friction stir spot welding of AA 6061-T6 using finite element technique, *Multidiscipline Modeling in Materials and Structures.* 4 (2008) 1–14.

25. Z. Shen, Y. Ding, A.P. Gerlich, Advances in friction stir spot welding, *Critical Reviews in Solid State and Materials Sciences.* 45 (2020) 457–534.

26. V. Soundararajan, S. Zekovic, R. Kovacevic, Thermo-mechanical model with adaptive boundary conditions for friction stir welding of Al 6061, *International Journal of Machine Tools and Manufacture.* 45 (2005) 1577–1587.

27. P. Trojovský, V. Dhasarathan, S. Boopathi, Experimental investigations on cryogenic friction-stir welding of similar ZE42 magnesium alloys, *Alexandria Engineering Journal*. 66 (2023) 1–14.

28. S. Witte, R.T. Zinkstok, W. Hogervorst, K. Eikema, Numerical simulations for performance optimization of a few-cycle terawatt NOPCPA system, *Applied Physics B*. 87 (2007) 677–684.

29. B. Zhang, X. Chen, K. Pan, M. Li, J. Wang, Thermo-mechanical simulation using microstructure-based modeling of friction stir spot welded AA 6061-T6, *Journal of Manufacturing Processes*. 37 (2019) 71–81.

11 Residual Stress in FSSW Process

Durairaj Raj Kumar, Karuppiah Madhan
Muthu Ganesh, Kannaian Vijayan, and
Paramasivam Thamizhvalavan

11.1 INTRODUCTION

Residual stress (RS) presents in any processed material without any external load or force. It is a lack of strain in the processed component, which has plastic deformation and distortion. RS can be classified as undesirable and desirable form. For example, the turbine blade has an RS developed through laser peening. The glass display in smartphones has an RS, which is used to protect the glass from crack-scratch through resistance. RS is generated, in which the component (welded component, in particular) is stressed above the elastic limit as a result of plastic deformation. Thermal gradient, phase transformation and mechanical processing are reasons for RS formation in the components [1]. In thermal gradient, after welding, the component is cooled at a high temperature. There is a difference in the cooling rate of components. The cooling rate differences found on the outside surface and inside the component surface were due to the localized differences in thermal contraction; as a result, non-uniform stress developed. The high rate of cooling is used on components, resulting in compressive stress developing in the heated component at the center. Consequently, in a similar process, the inside component has a tensile RS and the outside component has a compressive RS. In phase transformation, the volume difference is formed between the new phase and the surrounding phase, resulting in stress formation because of the expansion or contraction in the material. In mechanical processing, it occurs in components with non-uniform plastic deformation by rolling, extrusion, drawing and bending. There are two types of deformation: plastic and elastic. The elastic deformation is converted into plastic deformation after load removal. RS is affecting the mechanical properties of thermal expansion coefficient, melting point, kind of groove to be prepared for weld (U, V, W, J), welding sequence (instead of continuous welding, selecting the forward skip, backward skip method), preheating and post-welding heat treatment (stress-relieving, normalizing and annealing). These are methods that are used to reduce the magnitude of RS. In friction stir welding (FSW), both tensile and compressive RSs are presented. The RS value is varying from positive to negative. The positive RS is found in components due to the compressive stress on components (laser peening). The benefit of RS is used to strengthen the thin section or toughen the brittle surface. Similarly, the negative RS is found due to the tensile stress on components (bending, drawing and machining) and used to strengthen the thick section and improve the surface integrity [2].

DOI: 10.1201/9781003432289-11

The measurement of RS is performed by destructive (contour and slitting methods), semi-destructive (deep hole and center hole drilling) and non-destructive (neutron diffraction, synchrotron X-ray diffraction (XRD) and X-ray diffraction). Besides, the RS can be reduced by various techniques such as post-weld heat treatment, selection of welding parameters, shot peening, cold rolling and stretching.

11.2 RS IN WELDING

The welding RS and its deformation affect the mechanical properties of the weldment, such as compressive stability, stiffness and fatigue performance. Actually, the welding deformation is unacceptable due to the fabrication accuracy and dimensional stability. Before welding, preheating the base metal (BM) reduces the thermal gradient. After welding, the cooling rate with holding time reduces the thermal gradient [3]. Mechanical methods include external force or vibration in weldment, while thermal methods include localized cooling or heating. These methods improve corrosion resistance, fatigue resistance, toughness and hardness and affect the RS. The benefit of reducing or eliminating RS in welded parts is to improve the quality, reliability, increasing performance, durability, rising efficiency and productivity, which have been used to reduce rework and defects, increase the service life, decrease maintenance and repair costs and time. Among the various methods, severe plastic deformation (SPD) such as equal channel angular rolling (ECAR), constrained groove pressing (CGP), repetitive corrugation and straightening (RCS), tube channel pressing (TCP), cyclic extrusion compression (CEC), repetitive tube expansion and shrinking (RTES), cryorolling (CR), accumulative roll-bonding (ARB), high pressure torsion (HPT) and equal channel angular pressing (ECAP) were used to improve the mechanical properties and microstructure [4].

11.3 THE PROCESSES OF FSSW

In 1993, the process was developed by Mazda Motor Corporation, and it is an extension of the FSW process. They found that in the welding of aluminum alloys, the process takes less electricity, there is no coolant involved and no compressed air is used compared with conventional FSW. The general sequence of friction stir spot welding (FSSW) is shown in Figure 11.1.

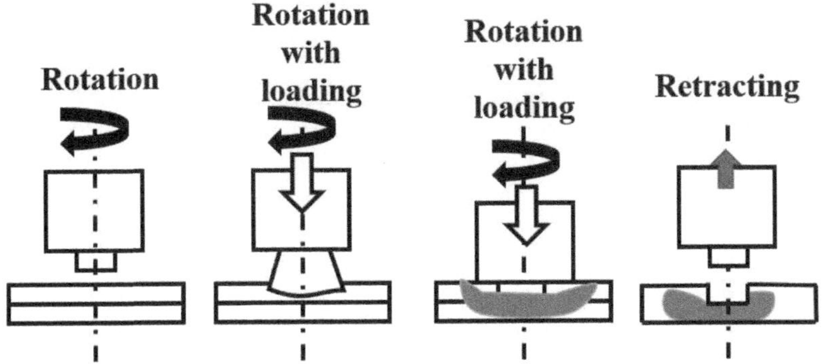

FIGURE 11.1 Schematic diagram of sequence of process on FSSW.

The important features of FSSWs are high strength, high fatigue life, low distortion, small RS and reduced corrosion resistance [5]. The process includes plunging, stirring and retracting. The work begins with the tool rotating at an angular speed. The tool is forced on the work material up to the tool shoulder contacts with the work material to be weld spot. The plunger movement of the rotational tool causes the softening of work materials. In the stirring stage, the rotational tool reaches a calculated depth when it rotates constantly. The heat is developed between the plunger and stirrer; as a result, the tool-heated work material softens when a joint is formed in this stirring stage due to the bonding. The tool retracted away from the work material. The keyhole is formed in this process due to the plunge tool shape, and it affects the strength of joints. So, in order to avoid keyhole formation on the work material, swing FSSW, pinless FSSW and refill FSSW are used. In refill FSSW, the pin and shoulder are equally placed on the upper sheet surface and rotated to generate heat and plunge. Both pin and shoulder are entered in the work material, and the pin retracts. Then the shoulder is retracted and the pin is plunged; as a result, a void is formed by the shoulder. In swing FSSW, the sliding cam is used to swing the tool in an arc motion. In pinless FSSW, no pin is used for plunging.

11.4 RS IN ALUMINUM ALLOY

The low-density material aluminum alloys are applied in the automotive and aerospace industries to decrease fuel consumption. In order to join the low-density similar and dissimilar material grades for assembly/repair purpose, there are many methods used, such as riveting, laser spot welding and resistance spot welding. The weight of components increased when using rivets on two or more work materials. Thereby, the riveting process is used in the application of components such as cranes, bridges and building frames that have a thickness of more than 6 mm, called a plate. The thickness of the material is less than 6 mm, called a sheet. The porosity defects are found in the laser spot welding on work materials. Besides, tool consumption is found in conventional resistance spot welding. To overcome all the above problems or alternative techniques required in it, in 1991, FSW was developed by TWI owing to its green welding, high efficiency, uniform microstructure, good mechanical properties and low thermal deformation. The researchers are consistently working on a developed FSW process for welding aluminum alloys, magnesium alloys, copper alloys, steel and titanium alloys, superalloys and composites. The FSSW is an alternative process to resistance welding and riveting and is widely applied to aerospace, aviation and automobile applications. Changing welding settings for a regulated quantity of generated heat, material flow and RSs has a significant impact on the desired weld quality of FSW joints in terms of the strength that may be achieved and joint soundness [6]. The refill friction stir spot welding (RFSSW) method is used to reduce the RS because solid-state joints replace conventional welding in space applications. The quality of the new welding process is directly influenced by the welding parameters selected. The finite element analysis is used to study the complexity of the welding process and validate it using controlling parameters [7]. The FEA-coupled CEL numerical model simulates the plunge, welding steps and dwellings in FSW and finds the RS in the aluminum alloy AA 6082-T6 weld joint. It was found that a higher RS is found in the longitudinal direction than in the transverse direction, and the M-shaped profile was found [8]. The RS was studied in different heat treatments with three different weld configurations. The heat treatment, non-precipitation hardening, precipitation hardening and annealing were carried out on weld AA2519. It was found that the precipitation hardening of AA2519 produced a low RS [9]. By varying the transverse speed and tool, the RS of aluminum alloy AA5083 was studied. It was found that the RS analysis showed a weld zone that was in tension in both the longitudinal and transverse directions. The peak longitudinal stresses increased as the traverse speed increased. This increase was probably due to sharper thermal gradients in welding and reduced time for stress relaxation to occur [10]. Different tool effects, fillet + scroll tool T_{FS}, fillet + cavity tool T_{FC} and fillet tool T_F on RS of 1.5 mm thick AA 6082 using FSW. It was found that the same result for RS was found using T_{FC} and T_F. The variation was found in the RS value due to the varying heat inputs. Hence, T_{FC} showed a different RS value [11]. The RS effect on fatigue cracks in FSW 2024-T351 joints was studied in different directions. It was found that RS was relieved mechanically, and the influence of crack propagation was found. FSW joints were dominated by RS; microstructure and hardness changes had a few influences [12]. The rotation speed and tool traverse on the RS are evaluated in between the non-age-hardening AA5083 and the age-hardening AA6082. The weld line had a tensile RS, whereas the parent materials had a compressive stress. It was observed that three times more longitudinal RS stress was developed compared with transverse direction [13]. The RS states in AA2024 weld were based on mechanical tensioning. It was found that the mechanical tensioning produced a large compressive stress in the weld. The RS in the untensioned sheets had more tensile stresses in the longitudinal direction, with peak values of 130 MPa [14]. The longitudinal tensile RS of the AA2024 aluminum alloy was studied by varying the roller tensioning. The roller tensioning and post-weld roller tensioning were used in FS weld lines. On increasing the loading, a greater effect was observed due to the reduction in the longitudinal tensile RS. The reverse sign of weld line RS was found using 20 kN [15]. When varying the welding parameters in FSW, the RS was studied. The experimental result was that the RS was tensile in the weld region and compressive in the parent plate balancing. The width in the tensile stress region with maximum tensile RS showed a linear correlation with heat input calculated. It was also found that more heat input gave a high RS [16]. The FSW model for Al6061-T6 was performed using ANSYS. The RS study found that the tensile load was highly affected by the RS [17]. By varying the welding speed and then the rotational speed of the tool, the RS was studied using a genetic algorithm. In this case, RS was considered as minimization. Results found that the high RS was found using a high welding speed with a fixed rotational speed. The high RS was found in the tension zone, and the wide tension zone led to a high RS tensile force [18]. The comparative study was carried out between conventional and underwater welding to study the RS formation in the AA6061 joint. The effect of underwater FSW with varying process parameters on RS was studied. It was found that the high RS was found using underwater welding compared with conventional welding due to the cooling effect [19]. The RS formation was studied with varying process parameters and fixed nanoparticles added to dissimilar 6061-T6 and 7075-T6 aluminum alloys. It was found that the RS was reduced by increasing the rotational speed and transverse speed and adding SiO_2 nanoparticles to the FSW joint. The reduction in RS was found due to the addition of nanoparticles, resulting in heat generation [20]. The RS was formed with stress relief annealing, heat treatment and CGP in copper sheet. The RS was changed from compressive to tensile stress with varying thickness. By increasing the constrained groove passes, the RS decreased with stress relief annealing rises and increased microstructure heterogeneity [21]. The ECAR process was used to evaluate the RS by varying the identical direction and the alike direction by 180 degrees. One pass gave a high RS compared with four passes, owing to the increasing temperature of the strips after ECAR passes. The same direction produced a high RS that was less than the identical direction [22]. By varying the FSW process parameters, the RS was studied in the cast aluminum alloy AlSi9Mg. It was found that by varying the tool geometry, number of passes and rotational speed, low rotational speed, retreating side and low number of passes gave low RS compared with conventional FSW [23]. A hybrid model was used to analyze the RS of the as-cast AlSi9Mg aluminum alloy. It was found that an asymmetric temperature distribution was found, and a low processing temperature on the retreating side produced a low tensile RS [24]. The effect of rotational speed, tool type and process parameters on RS in the cast aluminum alloy AlSi9Mg was studied. Results found the bead profile generated tensile RS and compressive stresses in the base. The low RS was found on the retreating side compared to the advancing side [25]. The ultrasonic vibration was used to study the RS in AA601 T6 joints. By altering the vibration frequency and process factors, the RS was investigated. It was inferred that the low ultrasonic vibration reduced tensile longitudinal RS. This method was also used to reduce the defect-free weld and tunnel defects [26]. The bending mode ultrasonic-assisted friction stir welding (BM-UAFSW) was used to study the RS. It was found that BM-UAFSW decreased the maximum longitudinal RS compared with conventional FSW. The ultrasonic method with amplitude gave the best performance in welding [27]. A two-dimensional FEA model was formed in FSW of an Al alloy to study the RS. The maximum longitudinal RS was found in the welding line. The region of tensile RS was correlated with plastic strain. The RS in longitudinal was directly related to pin velocity [28]. The RS was developed between the AA 6061 and A365 dissimilar friction-stir-weld. It was observed that the RS was developed due to the use of rotational direction, traverse direction and tool geometry. A multi-pass weld was used to alter the RS due to the increasing temperature [29]. The RS in dissimilar Al/Mg was studied using the normal method and the ultrasonic method. It was observed that the Mg and Al sides had tensile and compressive stresses, respectively [30]. The RS was investigated by joining the two dissimilar materials, 5052 AA and dual-phase steel. It was found that with increasing rotation speed, RS in the zone of weld increased due to the welding temperature and cooling rate. It was also found that the M-shape for Al alloy and the W-shape for steel were found near the interfacial RSs due to the various thermal expansions [31]. The RS was reduced by FSW carried out between Mg alloys and Al alloys with varying post-weld treatment and age. It was found that the aging period and post-weld heat treatment were used to reduce the RS and internal defects. The high post-weld treatment temperature was used to reduce RS more due to the refinement [32]. The post-weld heat treatment was carried out on aluminum alloy 2024 under two different aging conditions in FSW. It was found that high-temperature aging conditions produced a better result. The PWHT produced improved strength and fatigue life resistance owing to the uniform distribution of hardness in the weld zone and low RS as compared to the as-welded joint [33]. Besides, the FSW shows that the temperature rises from 400°C to 550°C in the nugget zone. The coarsen precipitation was found at high temperature, and it depended on the alloy type and maximum temperature. To improve the mechanical properties of joints, heat treatment was carried out on the 7xxx/2xxx series to study the mechanical properties. The hardness of the sample was highly related to RS. Hence, the hardness of FSW alloys fabricated between 2024-T3 and 7075-T6. Analysis found that more hardness was found on the advancing side and reduced on the retreating side of the weld sample [34]. The heat input and PWHT affect the hardness of the stir zone in the FSW 2024 aluminum alloy. The analysis found that while welding speed improved, the hardness in the stir zone decreased after PWHT owing to over-aging [35].

More work is performed on heat treatment after the welded sample and focused on heat treatment temperature/holding time, which are used as a single variable only and not as combined variables. To evaluate the combined heat treatment parameters, heat treatment temperature, holding time and cooling rate effect on RS in AA7075-T6/AA2024-T4 are studied using FSSW. Orthogonal experiments are used to plan the work. Variance in analysis is used to study the most effective factor. XRD is used to find the particles in the stir zone. In order to evaluate the RS at AA7075-T6 and AA2024-T4, the cast studies have been performed using post-weld heat treatment.

11.5 EXPERIMENTAL DETAILS

The AA7075-T6 and AA2024-T4 were considered as experimental materials. The FSSW process, schematic diagram of work materials and tool profile are shown in Figure 11.2. The microstructure of the base materials, namely AA 7075-T6 and AA 2024-T3, is shown in Figure 11.3. Each work material size was 200 mm × 100 mm × 6 mm, and the lab joint was made in it. Table 11.1 shows the specific composition of the BM and its mechanical properties. Based on the actual experimental parameters, welding operations were carried out manually. A set of welding factors was planned on the basis of 100 A current, 23 V voltage and 2.6 mm/s speed [36]. The post-weld heat treatment was performed after welding, as shown in Table 11.2. To decrease the joint RS in AA 7075-T6 and AA 2024-T3, heat treatment was used. The values of heat treatment temperature, holding time and brine cooling rate were selected. Based on the number of levels and factors, the nine orthogonal arrays were selected, and spot welding was performed on the AA 7075-T6 and AA 2024-T3 joints. The design of the experiment and its responses are shown in Table 11.3. The RS in AA 7075-T6 and AA 2024-T3 was measured using the XRD method.

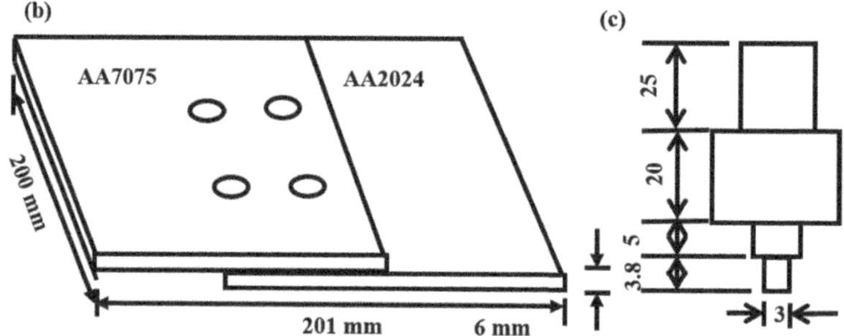

FIGURE 11.2 (a) FSSW process, (b) schematic diagram of FSSW and (c) tool profile.

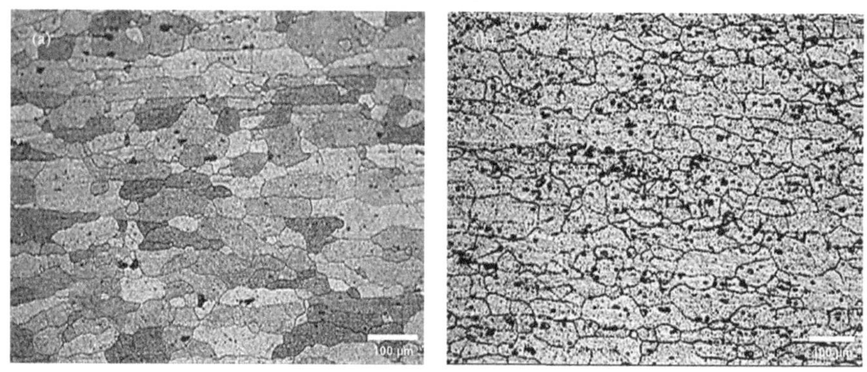

FIGURE 11.3 Base material microstructure: (a) AA 7075-T6 and (b) AA 2024-T3.

TABLE 11.1

Chemical Composition of Alloy (Mass Fraction, %) and Mechanical Properties

Alloy	Al	Fe	Mn	Ti	Cr	V	Cu	Mg	Zn
AA7075-T6	Bal.	0.163	0.165	0.116	0.391	0.162	1.58	2.62	5.21
AA2024-T4	Bal.	0.123	0.425	0.039	-	0.01	4.089	1.443	-

Mechanical properties		
Alloys	AA7075-T6	AA2024-T4
Yield strength (MPa)	460	345
Ultimate strength (MPa)	524	485
Hardness (HV)	175	125

TABLE 11.2

Heat Treatment Parameters with Values

Levels	Heat Treatment (°C)	Holding Time (minutes)	Cooling Rate (°C/h)
1	120	30	25
2	150	60	50
3	180	90	75

TABLE 11.3

Experimental Cases with Its Responses

Ex. No.	Heat Treatment Temperature (°C)	Holding Time (minutes)	Cooling Rate (°C/h)	Residual Stress at AA7075	Residual Stress at AA2024
1	120	30	25	61	15
2	120	60	50	44	12.8
3	120	90	75	76	12.1
4	150	30	50	40	15
5	150	60	75	72	11.1
6	150	90	25	55	10
7	180	30	75	68	8
8	180	60	25	51	10.9
9	180	90	50	34	7

11.6 RS MEASUREMENT

The RS in the FSSW joint was studied through XRD [37]. Based on Bragg's law, the XRD was used, as shown in Equation (11.1):

$$2d \sin\theta = n\lambda \qquad n = 1, 2,\ldots. \tag{11.1}$$

Here, d is the distance between the surface and diffraction crystal plane, λ is the wavelength of the X-ray beam and θ is the diffraction angle. The μ-X360n X-ray instrument is used for measurement.

11.7 HEAT TREATMENT TEMPERATURE

The mean RS is calculated based on the level of process parameters for each parameter. The delta was calculated by the difference between the high and low values of each response. The delta was used to find the most influential factor among the selected factors. Generally, a high delta is denoted as the first rank. The rank was decreased by decreasing the delta value, as shown in Table 11.4. Based on the mean RS, Figure 11.4 is drawn. The heat treatment temperature was an important factor used to control the physical properties of weld and base materials. Hence, the

TABLE 11.4

Mean Residual Stress Calculations

Levels	Heat Treatment Temperature (°C)	Holding Time (minutes)	Cooling Rate (°C/h)
Residual Stress at AA7075			
1	51.07	55.07	39.07
2	55.87	55.73	55.87
3	60.4	56.53	72.4
Delta	9.33	1.47	33.33
Rank	2	3	1
Residual Stress at AA2024			
1	60.33	56.33	55.67
2	55.67	55.67	39.33
3	51	55	72
Delta	9.33	1.33	32.67
Rank	2	3	1

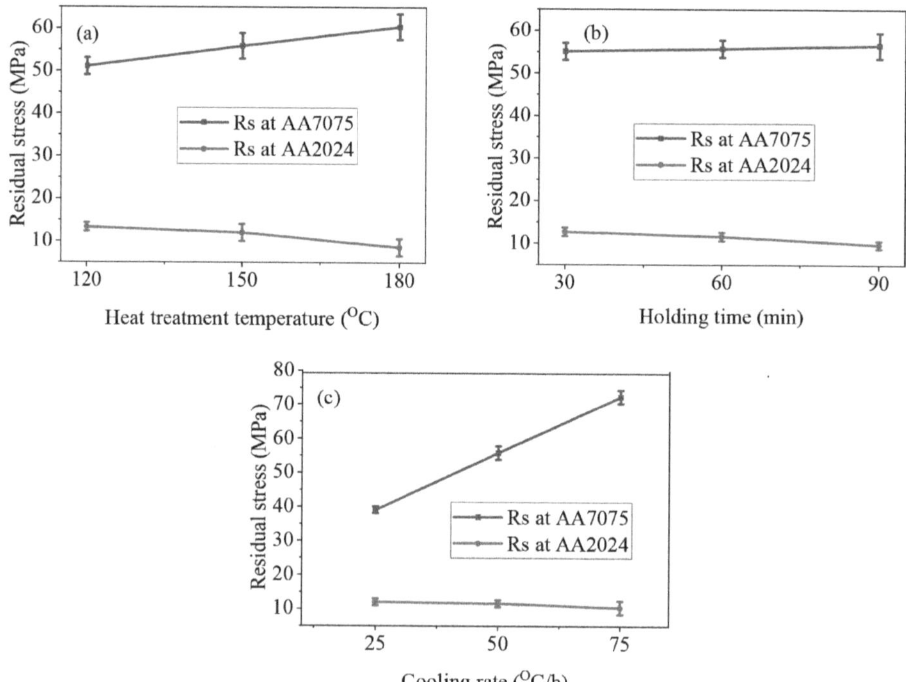

FIGURE 11.4 Effect of residual stress over: (a) heat treatment, (b) holding time and (c) cooling rate.

selection of heat treatment temperature was an important task after welding; as a result, heat generation, distribution and material flow were decided. The heat was generated owing to friction heat between the work material and tool. The rate of frictional heat generation was directly related to plastic deformation [38]. The value of RS formation was related to welding parameters such as title angle, heat input, current, voltage and speed based on the material thickness. The compressive RS distribution was consistent throughout the whole experiment; moreover, the RS was varied. Obviously, the RS was higher prior to heat treatment; the RS from experimental number one to nine was changed, which shows that the RS decreased/increased after heat treatment, and the reduction/addition was not much larger. The high RS was found on the AA7075, showing that heat treatment was more effective on the AA7075. The possible reason was more plastic deformation with the release of the RS. After spot welding, the AA7075 releases more RS. Comparing the RS made within the experiments in AA7075 shows that the RS was increased from the experimental number one to nine; the increment of RS becomes more obvious. Similarly, comparing RS within experiments in AA2024 shows that the RS was decreased from the experimental number one to nine; the reduction of RS becomes more obvious.

The RS in AA2024 was reduced significantly and shows the delamination phenomenon happened due to the RS in the contact point small varied from AA2024 to weld; as a result, the RS was most concentrated [39]. The RS reduction in experimental numbers one to three was small; in the experimental number four to six, the RS was decreased; in the experimental number seven to nine, the RS was largest. When the temperature increased, the stress release rate in AA2024 was higher. The effect of heat treatment temperature on RS was small compared with holding time and cooling rate. Therefore, it was observed that the heat treatment temperature was less. In Figure 11.5a, on increasing the heat treatment from 120°C to 180°C, the RS in AA2024 was reduced. Similarly, the RS in AA7075 was increased at the same heat treatment temperature. The gap was high between the physical properties of AA2024 and weld. The plastic deformation of AA2024 was greater than that of the weld. The high physical properties and plastic deformation combined effect on RS after welding.

11.8 HOLDING TIME

The holding time of heat treatment temperature was processed, which affected the microstructure, heat transfer rate and distribution rate and was used to control the physical properties of weld and base materials. Hence, the selection of heat treatment temperature time/holding time was an important task after welding; as a result, heat generation, high elongation, high load carrying, distribution and material flow were decided. Holding time was limited due to the possibilities of melting of the BM and weldment. The high holding time was providing high mechanical properties change in both base materials and spot welds [40]. Similarly, the shorter the holding time the less it affected the properties and RS of the base material and spot weld. The RS was affected by holding time. In this case, the holding time varied from 30 to 90 minutes. Hence, the selection of holding time was an important criterion to reduce/increase RS of base materials and spot welds.

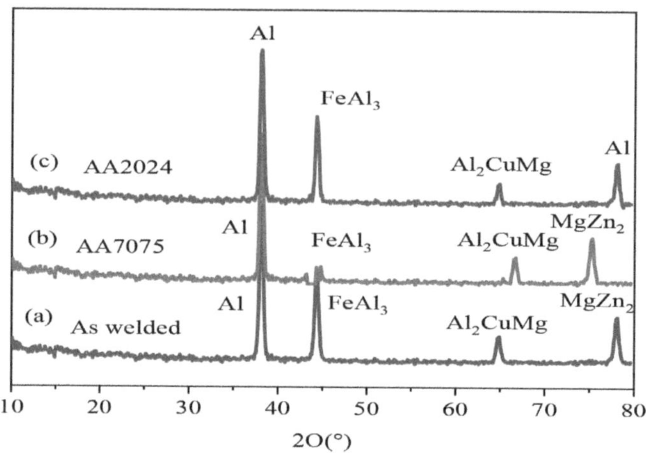

FIGURE 11.5   XRD patterns of nugget zone of welds.

The high RS was found on the AA7075, showing that heat treatment time was more effective on the AA7075. The reason was more plastic deformation, which releases the RS compared to base materials. After holding time, the AA7075 releases more RS. A comparison of RS was made within experiments in AA7075; it shows that the RS was slightly increased from experimental number one to nine; the increment of RS was very obvious. Similarly, comparisons were made within experiments in AA2024, which show that the RS was decreased from experimental number one to nine; the reduction of RS was very obvious. The holding time effect was less on the AA2024 compared with the AA7075. The possible reason was that less delamination occurred due to the RS in the contact point's small variance from AA2024 to weld; as a result, the RS was concentrated. In this high temperature holding time, the AA2024 has a low plastic deformation rate owing to its low plasticity and low stress release rate. RS reduction in experimental numbers one to three was small; in experimental numbers four to six, the RS was decreased; in experimental numbers seven to nine, the RS was small. Moreover, the difference between each group level was small. When the holding time increased, the possible reason for the stress release rate in AA2024 was less. In Figure 11.4b, by increasing the holding time from 30 to 90 minutes, the RS in AA2024 was reduced. Similarly, the RS in AA7075 was increased at the same holding time. The variation of RS between the AA2024 and AA7075 was found due to its holding time effect. The temperature holding time was not much affected by the RS because of the low stress release rate itself and poor plasticity [41]. Compared to the heat treatment on RS in both AA2024 and AA7075, the influence of holding time on RS in both AA2024 and AA7075 was less. Finally, it was found that the selected holding time range did not greatly affect the RS formation in the spot weld.

11.9   COOLING RATE

The cooling rate in the welding had a significant effect on the metallurgical properties and mechanical behavior of the weld. The hardness, grain growth and microstructure were dependent on the cooling rate. The combined effect of cooling rate and heat treatment temperature was also affecting the RS of the weld [42]. In this case, the cooling rate was varied from 25°C/h to 75°C/h with 25°C/h increments, and the RS was changed. In Figure 11.4c, the graph was drawn between the cooling rate and RS in AA2024 and AA7075. It was observed that the RS in AA7075 was higher than that of AA2024. The higher RS was due to less stress relief itself and poor plasticity. Moreover, the coefficient of thermal expansion of AA7075 (23.4/°C) was higher than that of AA2024 (23.2/°C). It was found that 0.2/°C deviation was found between the two alloys. Thereby, the coefficient of thermal expansion was not much affecting the RS. But the RS variations were high between the two alloys due to their cooling effect. The cooling medium had a high cooling capacity and specific heat and was reasoned for effective cooling on RS in spot weld. The low RS in the AA2024 was found owing to the large plastic deformation, better plasticity and stress rate [43]. The cooling effect was highly effective in reducing the RS in AA2024 compared with AA7075. On using a short cooling rate of 25°C/h, it was found that the RS of AA7075 was higher than that of AA2024. In other words, using a fast cooling rate of 75°C/h, it was found that the RS of AA7075 was higher than that of AA2024. On increasing the cooling rate of AA7075, it was found that the RS was gradually increasing. Similarly, by increasing the cooling rate of AA2024, it was observed that the RS decreased. In the AA7075, the short cooling rate was to reduce the RS. The high cooling rate was to reduce the RS in the AA2024. The change in microstructure of the stir zones of AA7075 and AA2024 was one of the main reasons for reducing the RS, and the cooling effect also affected the stress. The hardness of the weld zone increased owing to its cooling rate effect; as a result, the RS was affected. Comparing the RS over the heat treatment

TABLE 11.5

Analysis of Variance for AA7075 and AA2024

Source	DF	Seq SS	Adj SS	Adj MS	F	P
			Residual stress at AA7075			
Heat treatment	2	130.67	130.67	65.33	**	–
Holding time	2	2.67	2.67	1.33	**	–
Cooling rate	2	1600.67	1600.67	800.33	**	–
Error	2	0	0	0		
Total	8	1734				

** Denominator of F-test is zero or undefined.

$S = 2.820467\text{E}{-}15$; $R^2 = 100.00\%$; R^2 (adj) $= 100.00\%$.

Source	DF	Seq SS	Adj SS	Adj MS	F	P
			Residual stress at AA2024			
Heat treatment	2	34.942	34.942	17.471	3.88	0.205
Holding time	2	13.549	13.549	6.774	1.5	0.4
Cooling rate	2	4.029	4.029	2.014	0.45	0.691
Error	2	9.016	9.016	4.508		
Total	8	61.536				

$S = 2.12315$; $R^2 = 85.35\%$; R^2 (adj) $= 41.40\%$.

temperature, cooling rate and holding time, it was found that the cooling rate was highly contributing to reducing the RS in both alloys.

11.10   ANALYSIS OF VARIANCE

ANOVA was carried out to find the most influential factor, and it was developed from the MINITAB statistical software [44]. Based on the input parameters such as heat treatment temperature, cooling rate, holding time and RS in AA7075 and AA2024, the values were formed, as presented in Table 11.5. The most influential factor was found [45]. In this AA7075, it was observed that F and P values were not found because the denominator of the F-test was zero [46]. The model strength of AA7075 was good because of its low S value, high R^2 and high R^2 (Adj) value [47]. In this AA2024, it was found that heat treatment, cooling rate and holding time were not influenced by reducing the RS due to the p value greater than 0.5. The ANOVA model was also good because of its low S value and high R^2 value. But the R^2(Adj) value was less than 40% due to the more parameters needed to be considered for improving the RS of spot weld. Based on the mean response calculations [48], it was found that the cooling rate was highly affecting the RS in both alloys.

11.11   PHASE CHARACTERIZATION

Figure 11.5 shows XRD and found the precipitates Al_2CuMg and $MgZn_2$ present as-welded and particles Fe ($FeAl_3$ compound) present the structure. The precipitates, Al_2CuMg and $MgZn_2$, nucleate after post-weld. On applying AA2024 heat treatment, Al_2CuMg particles exhibited peaks and also showed the formation of $MgZn_2$ after treatment. Similarly, after using the AA7075 treatment, $MgZn_2$ was diagnosed after heat treatment. The $MgZn_2$ and Al_2CuMg precipitates were known as strengthening precipitates.

11.12   CONCLUSION

The post-heat treatment effect on dissimilar welds 7075-T6/2024-T4 aluminum alloys AA is investigated in FSSW. The effect of heat treatment temperature, cooling rate and holding time on RS in AA7075/AA2024 is studied. The phase characterization is also studied.

- The ninth (heat treatment at 180°C, holding time of 90 minutes and cooling speed of 50°C/h) has been found to have low RS due to the good plastic deformation and plasticity.
- The low RS in the joint has been limited, and the reduction of the RS on the AA2024 has been significantly low compared with the weld and AA7075 after treatment.
- The cooling rate is the most influential factor that affects the RS, and the low holding time and cooling rate affect the RS.
- Microstructure observation showed the precipitations dissolved in the stir region, and particles, Fe and Mn, remain in the stir region after treatment.

ACKNOWLEDGMENT

We sincerely acknowledge the editors of this book, who provided valuable suggestions for improving the book chapter.

REFERENCES

1. Jeyaprakash, N., Yang, C.H. and Sivasankaran, S., 2020. Laser cladding process of Cobalt and Nickel based hard-micron-layers on 316L-stainless-steel-substrate. *Materials and Manufacturing Processes*, 35(2), pp. 142–151.
2. Raj Kumar, D., Jeyaprakash, N., Yang, C.H. and Ramkumar, K.R., 2020. Investigation on drilling behavior of CFRP composites using optimization technique. *Arabian Journal for Science and Engineering*, 45, pp. 8999–9014.
3. Jeyaprakash, N., Yang, C.H. and Raj Kumar, D., 2020. Machinability study on CFRP composite using Taguchi based grey relational analysis. *Materials Today: Proceedings*, 21, pp. 1425–1431.
4. Raj Kumar, D., Jeyaprakash, N., Yang, C.H. and Sivasankaran, S., 2021. Optimization of drilling process on carbon-fiber reinforced plastics using genetic algorithm. *Surface Review and Letters*, 28(3), p. 2050056.
5. Jeyaprakash, N., Yang, C.H. and Raj Kumar, D., 2020. Minimum cutting thickness and surface roughness achieving during micromachining of aluminium 19000 using CNC machine. *Materials Today: Proceedings*, 21, pp. 755–761.
6. Shi, L., Wu, C.S. and Fu, L., 2020. Effects of tool shoulder size on the thermal process and material flow behaviors in ultrasonic vibration enhanced friction stir welding. *Journal of Manufacturing Processes*, 53, pp. 69–83.
7. Bîrsan, D.C., Păunoiu, V. and Teodor, V.G., 2023. Neural networks applied for predictive parameters analysis of the refill friction stir spot welding process of 6061-T6 aluminum alloy plates. *Materials*, 16(13), p. 4519.
8. Salih, O.S., Ou, H. and Sun, W., 2023. Heat generation, plastic deformation and residual stresses in friction stir welding of aluminium alloy. *International Journal of Mechanical Sciences*, 238, p. 107827.
9. Śnieżek, L., Kosturek, R., Wachowski, M. and Kania, B., 2020. Microstructure and residual stresses of AA2519 friction stir welded joints under different heat treatment conditions. *Materials*, 13(4), p. 834.
10. Peel, M., Steuwer, A., Preuss, M. and Withers, P.J., 2003. Microstructure, mechanical properties and residual stresses as a function of welding speed in aluminium AA5083 friction stir welds. *Acta materialia*, 51(16), pp. 4791–4801.
11. De Giorgi, M., Scialpi, A., Panella, F.W. and De Filippis, L.A.C., 2009. Effect of shoulder geometry on residual stress and fatigue properties of AA6082 FSW joints. *Journal of Mechanical Science and Technology*, 23, pp. 26–35.
12. Bussu, G. and Irving, P.E., 2003. The role of residual stress and heat affected zone properties on fatigue crack propagation in friction stir welded 2024-T351 aluminium joints. *International Journal of Fatigue*, 25(1), pp. 77–88.
13. Steuwer, A., Peel, M.J. and Withers, P.J., 2006. Dissimilar friction stir welds in AA5083-AA6082: The effect of process parameters on residual stress. *Materials Science and Engineering: A*, 441(1–2), pp. 187–196.
14. Staron, P., Kocak, M., Williams, S. and Wescott, A., 2004. Residual stress in friction stir-welded Al sheets. *Physica B: Condensed Matter*, 350(1–3), pp. E491–E493.
15. Altenkirch, J., Steuwer, A., Withers, P.J., Williams, S.W., Poad, M. and Wen, S.W., 2009. Residual stress engineering in friction stir welds by roller tensioning. *Science and Technology of Welding and Joining*, 14(2), pp. 185–192.

16. Lombard, H., Hattingh, D.G., Steuwer, A. and James, M.N., 2009. Effect of process parameters on the residual stresses in AA5083-H321 friction stir welds. *Materials Science and Engineering: A*, 501(1–2), pp. 119–124.

17. Feng, Z., Wang, X.L., David, S.A. and Sklad, P.S., 2007. Modelling of residual stresses and property distributions in friction stir welds of aluminium alloy 6061-T6. *Science and Technology of Welding and Joining*, 12(4), pp. 348–356.

18. Tutum, C.C. and Hattel, J.H., 2010. Optimisation of process parameters in friction stir welding based on residual stress analysis: A feasibility study. *Science and Technology of Welding and Joining*, 15(5), pp. 369–377.

19. Fathi, J., Ebrahimzadeh, P., Farasati, R. and Teimouri, R., 2019. Friction stir welding of aluminum 6061-T6 in presence of watercooling: Analyzing mechanical properties and residual stress distribution. *International Journal of Lightweight Materials and Manufacture*, 2(2), pp. 107–115.

20. Jafari, H., Mansouri, H. and Honarpisheh, M., 2019. Investigation of residual stress distribution of dissimilar Al-7075-T6 and Al-6061-T6 in the friction stir welding process strengthened with SiO_2 nanoparticles. *Journal of Manufacturing Processes*, 43, pp. 145–153.

21. Nazari, F., Honarpisheh, M. and Zhao, H., 2019. Effect of stress relief annealing on microstructure, mechanical properties, and residual stress of a copper sheet in the constrained groove pressing process. *The International Journal of Advanced Manufacturing Technology*, 102, pp. 4361–4370.

22. Honarpisheh, M., Haghighat, E. and Kotobi, M., 2018. Investigation of residual stress and mechanical properties of equal channel angular rolled St12 strips. *Proceedings of the Institution of Mechanical Engineers, Part L: Journal of Materials: Design and Applications*, 232(10), pp. 841–851.

23. Węglowski, M.S., Sedek, P. and Hamilton, C., 2016. Experimental analysis of residual stress in friction stir processed cast AlSi9Mg aluminium alloy. *Key Engineering Materials*, 682, pp. 18–23.

24. Hamilton, C., Węglowski, M.S., Dymek, S. and Sedek, P., 2015. Using a coupled thermal/material flow model to predict residual stress in friction stir processed AlMg9Si. *Journal of Materials Engineering and Performance*, 24, pp. 1305–1312.

25. Węglowski, M.S., Sędek, P. and Hamilton, C., 2016. The effect of process parameters on residual stress in a friction stir processed cast aluminium alloy AlSi9Mg. *Engineering Transactions*, 64(3), pp. 301–309.

26. Alinaghian, I., Amini, S. and Honarpisheh, M., 2018. Residual stress, tensile strength, and macrostructure investigations on ultrasonic assisted friction stir welding of AA 6061-T6. *The Journal of Strain Analysis for Engineering Design*, 53(7), pp. 494–503.

27. Alinaghian, I., Honarpisheh, M. and Amini, S., 2018. The influence of bending mode ultrasonic-assisted friction stir welding of Al-6061-T6 alloy on residual stress, welding force and macrostructure. *The International Journal of Advanced Manufacturing Technology*, 95, pp. 2757–2766.

28. Zhang, H.W., Zhang, Z. and Chen, J.T., 2005. The finite element simulation of the friction stir welding process. *Materials Science and Engineering: A*, 403(1–2), pp. 340–348.

29. Sabry, N., Stroh, J. and Sediako, D., 2023. Characterization of microstructure and residual stress following the friction stir welding of dissimilar aluminum alloys. *CIRP Journal of Manufacturing Science and Technology*, 41, pp. 365–379.

30. Muhammad, N.A., Geng, P., Wu, C. and Ma, N., 2023. Unravelling the ultrasonic effect on residual stress and microstructure in dissimilar ultrasonic-assisted friction stir welding of Al/Mg alloys. *International Journal of Machine Tools and Manufacture*, 186, p. 104004.

31. Geng, P., Morimura, M., Wu, S., Liu, Y., Ma, Y., Ma, N., Aoki, Y., Fujii, H., Ma, H. and Qin, G., 2022. Prediction of residual stresses within dissimilar Al/steel friction stir lap welds using an Eulerian-based modeling approach. *Journal of Manufacturing Processes*, 79, pp. 340–355.

32. Muruganandam, D., 2018. Influence of post weld heat treatment in friction stir welding of AA6061 and AZ61 alloy. *Russian Journal of Nondestructive Testing*, 54(4), pp. 294–301.

33. Yadav, V.K., Gaur, V. and Singh, I.V., 2020. Effect of post-weld heat treatment on mechanical properties and fatigue crack growth rate in welded AA-2024. *Materials Science and Engineering: A*, 779, p. 139116.

34. Zadpoor, A.A., Sinke, J., Benedictus, R. and Pieters, R., 2008. Mechanical properties and microstructure of friction stir welded tailor-made blanks. *Materials Science and Engineering: A*, 494(1–2), pp. 281–290.

35. CHEN Yu, DING Hua, LI Ji-zhong, ZHAO Jing-wei, FU Ming-jie, LI Xiao-hua. Effect of welding heat input and post-welded heat treatment on hardness of stir zone for friction stir-welded 2024-T3 aluminum alloy. *Transactions of Nonferrous Metals Society of China*, 2015, 25, pp. 2524–2532.

36. Huang, B., Liu, J., Zhang, S., Chen, Q. and Chen, L., 2020. Effect of post-weld heat treatment on the residual stress and deformation of 20/0Cr18Ni9 dissimilar metal welded joint by experiments and simulations. *Journal of Materials Research and Technology*, 9(3), pp. 6186–6200.

37. Boucherit, A., Abdi, S., Aissani, M., Mehdi, B., Abib, K. and Badji, R., 2020. Weldability, microstructure, and residual stress in Al/Cu and Cu/Al friction stir spot weld joints with Zn interlayer. *The International Journal of Advanced Manufacturing Technology*, 111, pp. 1553–1569.

38. Maalekian, M., Kozeschnik, E., Brantner, H.P. and Cerjak, H., 2008. Comparative analysis of heat generation in friction welding of steel bars. *Acta Materialia*, 56(12), pp. 2843–2855.

39. Ni, Y., Fu, L. and Chen, H.Y., 2019. Effects of travel speed on mechanical properties of AA7075-T6 ultra-thin sheet joints fabricated by high rotational speed micro pinless friction stir welding. *Journal of Materials Processing Technology*, 265, pp. 63–70.

40. Yin, Y.H., Sun, N., North, T.H. and Hu, S.S., 2010. Hook formation and mechanical properties in AZ31 friction stir spot welds. *Journal of Materials Processing Technology*, 210(14), pp. 2062–2070.

41. Luo, C., Li, X., Song, D., Zhou, N., Li, Y. and Qi, W., 2016. Microstructure evolution and mechanical properties of friction stir welded dissimilar joints of Mg-Zn-Gd and Mg-Al-Zn alloys. *Materials Science and Engineering: A*, 664, pp. 103–113.

42. Liu, X.C., Li, W.T., Zhou, Y.Q., Li, Y.Z., Pei, X.J., Shen, Z.K. and Wang, Q.H., 2023. Multiple effects of forced cooling on joint quality in coolant-assisted friction stir welding. *Journal of Materials Research and Technology*, 25, pp. 4264–4276.

43. Zhou, S., Wu, K., Yang, G., Wu, B., Qin, L., Guo, X. and Wang, X., 2023. Friction stir welding of wire arc additively manufactured 205A aluminum alloy: Microstructure and mechanical properties. *Materials Science and Engineering: A*, 876, p. 145154.

44. Kannan, V.S., Lenin, K., Srinivasan, D. and Raj Kumar, D., 2023. Analysis of microstructural, mechanical and surface properties of aluminium hybrid composites obtained through stir casting. *Journal of the Institution of Engineers (India): Series D*, pp. 1–12.

45. Prabu, G., Jeyaprakash, N., Yang, C.H. and Radhika, N., 2023. Analysis on the transformation of melt wear to self-healing crack of wire arc additive manufactured Al 5356 alloy. *Tribology International*, 2023 Oct 1;188:108802..

46. Ragunath, S., Radhika, N., Krishna, S.A. and Jeyaprakash, N., 2023. Enhancing microstructural, mechanical and tribological behaviour of AlSiBeTiV high entropy alloy reinforced SS410 through friction stir processing. *Tribology International*, 2023 Oct 1;188:108840..
47. Jeyaprakash, N., Yang, C.H., Susila, P. and Karuppasamy, S.S., 2023. Laser cladding of NiCrMoFeNbTa particles on Inconel 625 alloy: Microstructure and corrosion resistance. *Transactions of the Indian Institute of Metals*, 76(2), pp. 599–612.
48. Kumar, M.S., Yang, C.H., Ishfaq, K., Jeyaprakash, N. and Rehman, M., 2023. Evaluating the effects of vortex generation and the settling time of reinforcing particles on the mechanical characteristics of the PRMMC: A real-time simulation and experimental study. *Journal of Manufacturing Processes*, 98, pp. 80–94.

12 Corrosion in Friction Stir-Welded Joints

Ishita Koley, Arindam Dhar, and Avinash Kumar

12.1 INTRODUCTION

Corrosion resistance is a fundamental characteristic relating to how a material stops degradation in a specific environment. All metals have a propensity to steadily stabilize again. This natural tendency leads to the classification of metals according to increasing nobleness, which then results in the classification of reducing activity and rising potential. Degradation of material is a rising concern for the industry that deals with metal and structures for potential financial loss and catastrophic calamity. As far as industrial structures and components are concerned, welding on them is a regular manufacturing process. However, welded joints have a tendency to react in harsh environments and develop severe corrosion. The welding process parameters sometimes play a pivotal role in lowering corrosion. However, regular welding techniques, such as shielded metal arc welding, gas metal arc welding, gas tungsten arc welding, etc., are fusion welding techniques that are susceptible to corrosion in harsh environments. However, with the advancement of welding techniques, new solid-state welding methods are now in consideration. These methods are not only environmentally friendly but also reduce the degradation of materials in engineering components. FSW stands for friction stir welding, which is a solid-state welding technique where a non-consumable rotating tool is used to weld together two metal parts without melting the metal during the process of welding (Figure 12.1). It has become a popular method of joining metals, particularly in the aerospace and

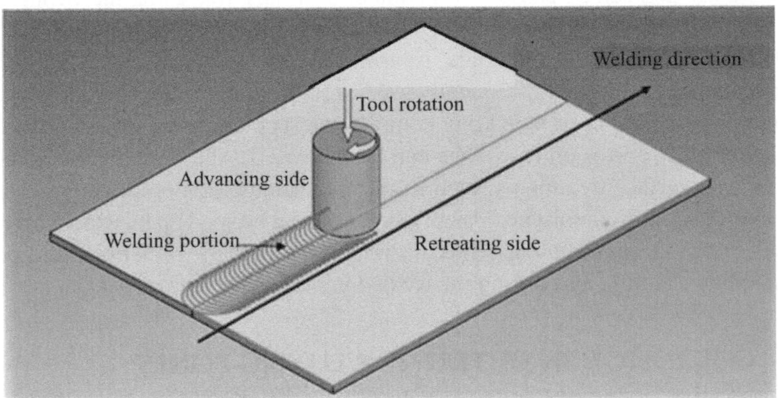

FIGURE 12.1   Schematic diagram of friction stir welding.

DOI: 10.1201/9781003432289-12

automotive industries. A rotating tool is inserted into the interface between the two metal components. The material is then heated and deformed plastically as the tool is moved along the joint. The metal becomes malleable as a result of the heat and deformation, which enables the tool to mix and interlock the metal from both sides of the joint to create a solid-state bond. The material undergoes extensive plastic deformation and thermal exposure during the FSW process, resulting in the development of multiple different zones in the welded connection. These zones are:

- Stir zone (SZ) is the area where the rotating tool meets and stirs the material, creating significant plastic deformation and mixing of the material from both sides of the joint. An equiaxed and fine grain microstructure is formed in this zone, replacing the original grain structure.
- Thermo-mechanically affected zone (TMAZ): It is the area around the SZ where there is significant plastic deformation and heat exposure, but the material is not completely melted. The microstructure of the TMAZ is partially altered, with a mixture of recrystallized and deformed grains.
- Heat-affected zone (HAZ): This is the region surrounding the TMAZ where the material is heated but not deformed. The heat from the welding process affects the microstructure of the HAZ, but not as much as it does the TMAZ.
- The base material (BM) is the original material that was used for welding and has not been subjected to considerable plastic deformation or thermal exposure.

Depending on the welding settings and material qualities, the attributes of each of these zones can vary dramatically. The welded joint's microstructure and mechanical properties are important to its overall performance, and careful management of the FSW process is required to obtain the correct attributes in each zone. In FSW joints, however, the metal might experience plastic deformation throughout the procedure, which could alter its microstructure and result in the production of intermetallic compounds (if the joint is not between similar materials) that are more prone to corrosion. In FSWed material, corrosion can be caused by a number of things, like using the wrong materials, being in severe conditions, and not doing enough post-weld processing.

Selection of the right materials is essential for reducing corrosion in FSW. Aluminium and its alloys are frequently used in FSW, and by choosing an alloy with a high aluminium content or by adding components such as chromium, copper, or zinc to the alloy, their corrosion resistance can be increased. Additionally, the application of coatings or surface treatments, such as chromate conversion or anodizing, can offer additional safeguards against corrosion. Corrosion can be sped up by exposure to moisture, salt, or acidic environments; thus, it's important to do the right post-weld processing, including cleaning and drying the welded joint, to stop corrosion from happening.

12.2 CORROSION IN DIFFERENT WELDING ZONES

Three distinct welding zones (Figure 12.2) show different affinities for corrosion. Equiaxed microstructure in the SZ of FSW is thought to be less vulnerable to corrosion than the HAZ and TMAZ. However, various factors can affect the SZ's corrosion

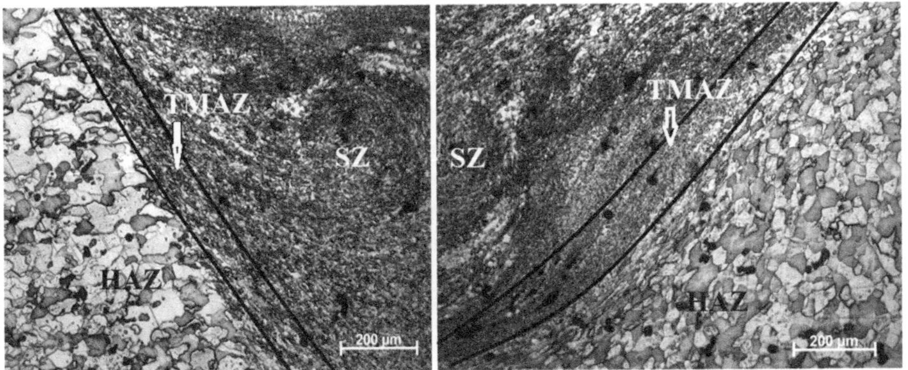

FIGURE 12.2   SZ, TMAZ, and HAZ in friction stir welding joints.

resistance, including welding parameters, the material that is being welded, and the post-weld treatment. The development of intermetallic compounds, which can be more corrosive than the BM, is one potential concern in the SZ for dissimilar welding conditions. Intermetallic compounds can arise as a result of the significant shear and deformation that occur in the SZ during the welding process. Intermetallic compounds can be reduced by selecting proper welding parameters such as tool rotation speed and travel speed, as well as materials with a low sensitivity to intermetallic production. The presence of residual stress is another element that can affect the corrosion resistance of the SZ. Thermal and mechanical strains present during the welding process might cause residual stress. There can be localized high stress areas as a result of these strains, which increases corrosion susceptibility. A post-weld stress reduction treatment can reduce residual tensions and improve corrosion resistance in the SZ. In general, corrosion resistance of the SZ can be improved by carefully controlling welding settings, utilizing appropriate materials, and using post-weld treatments to decrease residual stresses and reinforce the welded joint.

The TMAZ of an FSW is often believed to be more susceptible to corrosion than the SZ and the BM because of its modified microstructure and residual stresses present in this area. During the welding process, the TMAZ experiences substantial temperature and mechanical changes that could affect how resistant it is to corrosion. The development of a heterogeneous microstructure in the TMAZ, which can have different levels of sensitivity to corrosion, is a cause for concern. A variety of recrystallized and deformed grains may be present in the TMAZ, and these grains may have various chemical make-ups and crystallographic orientations, which can affect how corrosion occurs. The TMAZ's enhanced sensitivity to corrosion may also be a result of intermetallic compound production there. The residual tensions that remain after welding can also have an impact on the TMAZ's ability to resist corrosion. The TMAZ may experience high residual thermal and mechanical stresses as a result of the FSW process, which could make it more vulnerable to stress corrosion cracking (SCC). In order to increase the corrosion resistance of the TMAZ, these residual stresses can be decreased by applying the proper post-weld procedures, such as stress relief annealing. By cautiously tracking the welding settings, choosing suitable materials, and applying post-weld treatments to lessen residual stresses and prevent the formation of intermetallic compounds, the corrosion resistance of the TMAZ can be increased.

Due to its altered microstructure and residual stresses, the HAZ of an FSW is typically thought to be more prone to corrosion than the SZ and the TMAZ. The HAZ is exposed to high temperatures through the welding process, which could change its capacity to withstand corrosion. The creation of a coarse-grained microstructure in the HAZ, which may have less corrosion resistance than the BM and the SZ, is a problem. A loss of strength and corrosion resistance may result from the growth of grains in the HAZ brought on by the heat input during the welding process. Additionally, the HAZ may be more susceptible to sensitization, a process that can result in the development of chromium carbides and lower the material's corrosion resistance. Also, the residual stress that remains after welding can have an impact on the HAZ's ability to resist corrosion. The residual thermal and mechanical stresses created by the FSW process may have a considerable impact on the HAZ, making the material more susceptible to SCC and less resistant to corrosion. Several steps can be taken to increase the corrosion resistance of the HAZ, including lowering the heat input during welding, choosing materials with low sensitization susceptibilities, and applying post-weld treatments, such as stress relief annealing or solution heat treatment, to improve the microstructure and lower residual stresses. These actions can support the integrity of the welded junction and increase the HAZ's corrosion resistance.

12.3 CORROSION IN ALUMINIUM ALLOY

During the early stages of FSW, the techniques were mostly applied to aluminium (Al) and its various alloys. The aluminium-copper alloy AA2219-T87 was joined by FSW, and the corrosion properties of the joints were analysed in a 3.5% NaCl solution, where the pH of all tests was raised to 10 by adding potassium hydroxide. It was found that the initiation sites for pitting corrosion are the intermetallics in Al–Cu alloys. The local matrix dissolution that results from galvanic interaction between intermetallics and the surrounding matrix is what leads to pitting. Due to the higher heat generated during FSW and the lack of second-phase particles, the corrosion behaviour of the SZ is nobler than that of the base metal [1]. Another Al alloy, AA 2050, was welded by FSW. The corrosion reactions at the SZ were studied using conventional 3-day immersion tests and also with a three-electrode electrochemical cell, which was performed in a 0.7 M NaCl solution. Heterogeneous corrosion was observed between the top and bottom of the nugget. Two metallurgical states, NHT (as-welded joint) and PWHT (post-welding heat treatment), were taken into consideration. Investigations demonstrated that the PWHT dramatically altered the weld nugget's corrosion behaviour. While the PWHT nugget had intragranular as well as intergranular corrosion (IGC) characteristics, the nugget zone of NHT was more susceptible to IGC due to galvanic coupling phenomena [2]. A nugget with finer grains has a higher propensity for cathodic corrosion than one with coarser grains [3].

Yong Chen et al. reported the interactions between the different zones of the 7050-T76 aluminium alloy FSW joint's corrosion response. The FSW tool parameter was set at 800 rpm and the traverse speed at 200 mm/min. Corrosion tests were conducted in a 3.5% NaCl solution. The increased difference in solute content between the GB (GB) and grain interior induces aggravated corrosion along GBs in the nugget zone; visible coarsening of precipitates in HAZ leads to Cu enrichment in precipitates but depletion in the matrix; for that, the HAZ boundary region next to NZ experiences the strongest negative current and has the greatest potential for corrosion [4]. SCC is an inevitable part of the FSW joints. Shuaihao Gian studied FSW joints of 2219 aluminium alloy and their corrosion resistance properties. Salt spray corrosion tests were performed with a 5% NaCl-mixed solution with a probable pH value of 3 as the acidic corrosive medium. The initial salt spray was 1–2 mL/80 cm²h, and the chamber temperature was maintained at 50°C ± 0.5°C. After ultrasonic impact treatment (UIT), the SCC sensitivity index of friction stir-welded joints decreased in a 3.5% NaCl solution, and a significant decrease in crack initiation was observed during the SCC process [5]. Many slip systems in FCC structures are often found in Al alloy, which affect its crack propagation route [6]. Under the influence of UIT, precipitates were dissolved, their density and size were reduced, and the remaining precipitates were spread in grains. The chance of IGC naturally decreased as a result [5].

The AA6061-T6 aluminium alloy was FSWed, and the corrosion rates of the AA6061-T6 alloy were estimated through weight loss measurements in a 3.5% NaCl solution for 24 hours. Environments with NaCl cause the development of aluminium chlorides, which reduces the ability of the oxide layer to prevent corrosion and thus increases the rate of degradation of the surface [7]. Aviation aluminium alloy or simply 7075 Al alloy had been underwater FSWed and 0.97 mol NaCl + 0.3 mol H_2O_2 + 1 L distilled water solution was used as an electrolyte to analyse the electrochemical behaviour joints. The FSWed joints exhibit IGC with certain corrosion pits but no reticular pattern. Intermetallics such as $MgZn_2$ clustered on the GB, so the GB became the main corrosion pathway. The vulnerability to corrosion was decreased by the high potential phase $MgCu_2$. However, the clustering of $MgCu_2$ also produced a few precipitation-free zones (PFZs), which may be quickly consumed, after which grains tend to exfoliate. The rate of exfoliation corrosion increased as these PFZs continued to peel off [8]. Gupta et al. reported the corrosion property of 7475 Al alloy FSWed butt joints in a 3.5% NaCl electrolyte solution and found that SZ is more corrosion resistant than TMAZ, and its resistivity is higher in lower rpm compared to higher rpm. They also mentioned galvanic corrosion due to the presence of halide ions [9]. The corrosion resistance of FSWed butt joints of AA 2024-T3 in a 3.5 wt.% NaCl aqueous solution was found to have a pitting tendency on the surface of the BM, whereas the weld zone shows a passive nature and, compared to the advancing and retreating sides, later shows improved corrosion behaviour [10]. Galvanic corrosion occurs mainly when there is a large difference in electromotive force (EMF) between two metals. C. Zhang et al. experimented with the corrosion properties of the dissimilar FSWed AA2024/7075 joints, and galvanic corrosion occurred in that material. It was further discovered that the NZ second phases and grain microstructure have an impact on the dissimilar FSW joints' corrosion resistance. By FSW at appropriate RPM, precipitates and dispersoids are produced to effectively prevent corrosion attack [11]. For crevice corrosion (Figure 12.3) in Al under the presence of Cl^- ion, it transformed into aluminium hydroxide.

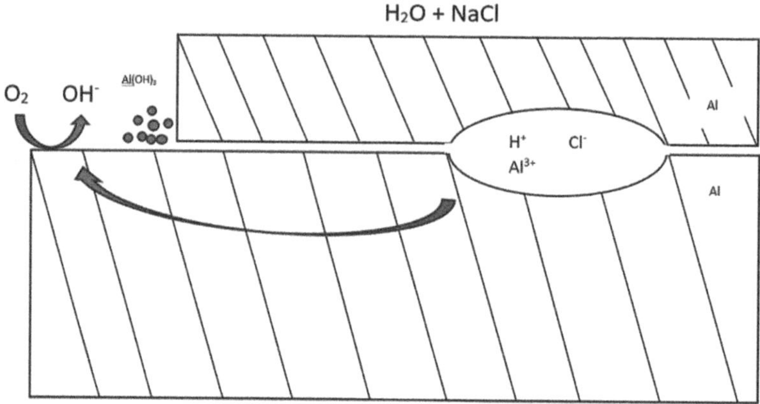

FIGURE 12.3   Schematic diagram of crevice corrosion in Al.

12.4   CORROSION IN MAGNESIUM ALLOY

Magnesium (Mg) is also a low melting metal, like Al. Compared to regular fusion welding techniques, it is easy to join through FSW. Corrosion is an unavoidable situation in FSWed joints under various environments. Different types of corrosion occur in those joints depending on the materials used for FSW. Strass et al. investigated an Al/Mg dissimilar joint by ultrasound supported friction stir welding (US-FSW) and the joint corrosion characteristics analysed in a 1.0 M NaCl solution. On the Mg side, the material was greatly dissolved, while the Al region appeared to be less corroded. This is a representation of galvanic corrosion, which protects one material (in this case, Al) at the expense of another, less noble substance (Mg). Also, the α-Mg matrix, $Mg_{17}Al_{12}$ (β-phase), and precipitates of Al_8Mn_5 (Fe) and Mg_2Si have been detected, which cause the corrosion phenomenon [12]. Chen et al. prepared a protective ceramic-coated FSWed AZ31B magnesium alloy and examined it in a 3.5% NaCl solution. In this experiment, HAZ is severely affected due to pitting corrosion occurred whereas SZ is showing little corrosion due to presence of β phase particles ($Mg_{17}Al_{12}$), which act as a corrosion barrier.

Plasma electrolytic oxidation (PEO) coating improved the corrosion resistance property of the FSWed AZ31B magnesium alloy [13]. AZ31B magnesium alloy and titanium butt joints immersed in a 3.5% NaCl solution showed that PEO-treated samples had less corrosion, whereas without coating, AZ31B alloy was severely corroded and the degradation of the titanium was comparatively negligible [14]. Salt spray testing of AZ61A magnesium alloy FSWed butt joints resulted in an effect of pH value, spraying time, and chloride ion concentration. The greater the chloride ion concentration, the greater the effect of corrosion [15]. FSWed LAZ933 magnesium-lithium (Mg-Li) alloy in 3.5 wt.% NaCl solution shows that weld zones exhibit better corrosion-resistant properties than base metal, and microgalvanic corrosion occurred in regions where $Mg(OH)_2$, Li_2O_2, and $Al(OH)_3$ are the corrosion-triggering components [16]. Liu et al. investigated how a duplex coating on FSWed AZ31 metal corrodes in a solution made of 3.5 wt% NaCl and concluded that the corrosion resistance

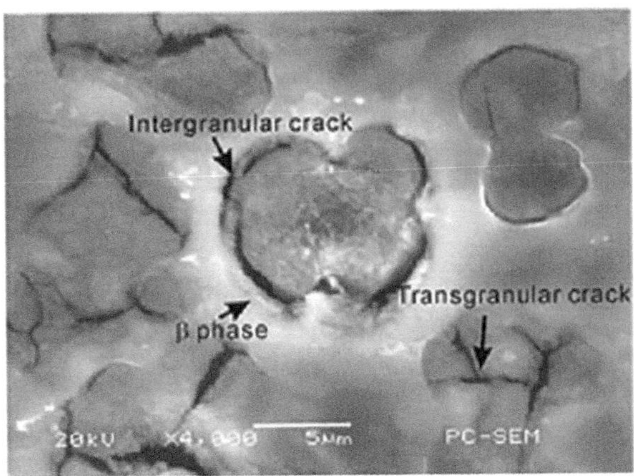

FIGURE 12.4 Intergranular corrosion in base material after salt spray corrosion test [19].

in the HAZ was lowered by galvanic coupling between the coarse Al-Mn-Fe and α-Mg phases. Combined MAO/EN plating technology helps reduce the corrosion rate in the welded zone [17]. Corrosion behaviour of new Al–Zn–Mg friction stir weld joints in standard exfoliation corrosion solution GB/T 22639-2008 through immersion testing indicates that TMAZ on the retreating side has the highest corrosion due to coarse Si and Fe impurity intermetallics and wide PFZs, whereas base metal has pitting corrosion and the nugget zone is affected by IGC. $Al_3Sc_xZr_{1-x}$ nanoparticles help reduce the corrosion rate [18].

Zeng et al. investigated FSWed magnesium alloy AM50 through 5% NaCl salt spray and conventional polarization tests in a 3.5% NaCl solution. In the salt fog test, the weld with thicker oxide layers appears to have more corrosion resistance than the substrate. The welds exhibited pitting corrosion, intergranular cracks, and transgranular cracks in the BM (Figure 12.4). Polarization test results showed that TMAZ is the most corrosion-prone region [19]. Lingampalli et al reported that the corrosion behaviour of FSWedFSW but joints of Mg-Zn-Mn alloy (ZM21) in 3.5 wt.% NaCl and concluded that corrosion resistance is observed to be adversely affected by compressive stresses in the SZ for some samples. Also, because of the large grain refinement, it was found that the corrosion resistance in the weld zone was higher than it was for base metal [20].

12.5 CORROSION IN STEEL

Conventional welding in steel resulted in more joint defects such as distortion, pin holes, spatter, etc. FSW of steel is a remedy to minimize weld defects in joints. Gang et al. investigated underwater FSW between 16Mn steel and X65 steel. The corrosion behaviour of the welded joints was simulated in seawater (approximately similar to a 3.5 wt.% NaCl solution). Variation in Ni coating exhibits different corrosion rates; coating with a 100% duty ratio shows the best impedance compared to others; and corrosion products such as iron oxides such as Fe_2O_3, Fe_3O_4, and FeOOH

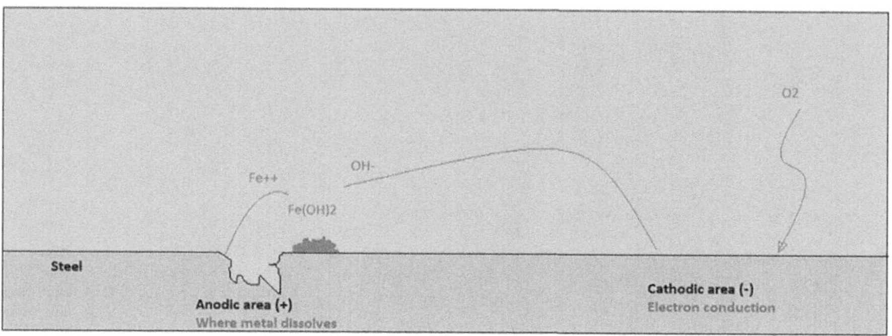

FIGURE 12.5 Schematic diagram of steel corrosion.

were generated on the surfaces of the corroded samples [21]. The common chemical reaction for corrosion in steel is shown in Figure 12.5. Zhang et al. examined the behaviour of high-nitrogen steel FSWed joints' pitting corrosion and concluded that the splitting up of coarse inclusions and the grain refining during FSW are primarily responsible for the NZ's superior pitting corrosion resistance [22]. SCC in steel often comes up with hydrogen embrittlement (HE). SCC behaviour of FSWed maraging steel in air and a 3.5 wt.% NaCl solution was analysed. FSWed samples showed better SCC resistance properties than base metal and GTAW samples, and it appears that cracks were propagated along the lath martensitic sites. Segregation of alloying elements and retained austenite, grain size, changes in GB density, and residual stress control the SCC in the joint. Higher compressive residual stresses, which are present in the FSWed sample, exhibit better SCC resistance [23]. Mohan et al. implemented induction heated friction stir welding (IH-FSW) to join martensitic AISI 410 stainless steel and examined its corrosion property in a 0.5 M H_2SO_4 solution through the weight loss method. The homogenous grains that developed in the weld zone led to an enhancement in corrosion resistance compared to base metal. Due to the uniformly sized, refined grains that formed in the weld zone, the two-layer oxidation film created was less damaging to the IH-FSW specimens [24]. Li et al. welded high-nitrogen, nickel-free austenitic stainless steel by FSW. It was found that the formation of a Cr-depleted zone, which is susceptible to corrosion, would result from the precipitation of secondary phases Cr_2N and $Cr_{23}C_6$, and pitting and IGC create less resistance in the SZ compared to base metal [25]. Another SCC was analysed by Rajasekaran et al. in which defect-free 316L grade stainless steel joints were FSWed and the SCC resistance of the weld was calculated in boiling $MgCl_2$.

TMAZ of the retreating side of the joint showed high crack propagation phenomena and higher chloride SCC. Base metal and FSW exhibited transgranular SCC in brittle mode [26]. FSWed nickel titanium-shaped memory alloy with stainless steel in a 3.5% (w/v) NaCl aqueous solution showed that welded joints had less corrosion resistance than base metals. NiTi and SS made a good galvanic pair, and a highly protective passive film was formed, and as a result of that, minimum galvanic corrosion was detected [27]. Electrochemical properties of FSWed shipbuilding steels (DH36, S690QL, and 80HLES) in a 0.01 M NaCl solution showed that among these three steels, the 80HLES steel sample showed major differences in the surface reactivity between the nugget and the other vicinity zones due to galvanic corrosion [28]. Matlan et al. welded stainless steel (316SS) and low-carbon steel together using a double-sided butt joint using FSW. The SCC test was conducted using a solution of sodium chloride (NaCl) at a concentration of 30,000 parts per million (ppm). Pitting initiation was observed on the low-carbon steel side, where the joints were not affected by chloride cracking [29]. Koley et al. reported the electrochemical behaviour of FSWed ultra-low-carbon steel plates in a 3.5% NaCl solution. It was found that SZ had the least corrosion resistance property compared to other zones due to grain size reduction in this zone. With the reduction in grain size, the number of grains was higher in the SZ compared to TMAZ, HAZ, and BM, which indicates an increase in GB density in the SZ. As grain boundaries are initiation sites for pitting corrosion, the corrosion resistance in the SZ was found to be less [30].

12.6 CORROSION PHENOMENON IN FRICTION STIR SPOT WELDING JOINT

Friction stir spot welding (FSSW) is a recent advancement of the FSW process where the tool doesn't traverse along the interface of workpieces, resulting in localized spot welding. The FSSW process consists of three stages: plunging, stirring, and retracting. Similar to FSW, FSSW also generates the weld nugget zone, TMAZ, and HAZ. James et al. investigated the corrosion behaviour of the AZ31 alloy through the scanning electrode reference technique (SERT) and an immersion test. It was found that the spot region appeared to have galvanic corrosion. Out of three, in the SZ, higher corrosion potential can be observed and inherited pitting attacks on the circumference. Reduction of the second-phase precipitate of $Mg_{17}Al_{12}$ in the mechanically mixed region causes a reduction in the microgalvanic cell, and due to severe mechanical mixing in the SZ and TMAZ, the number of twins and dislocations was reduced. This in turn resulted in less corrosion in the mechanically mixed zone [31]. Kazeem et al. investigated the corrosion behaviour of FSSWed, commercially pure copper. It was found that the tool rotational speed had an effect on the corrosion rate in the welded material. Samples processed at a lower rotational speed showed a higher corrosion rate compared to samples processed at a higher rotational speed. In lower rotational speed samples, the inclusion of more particles from the tool was found as a result of high tool wear. This, in turn, increased the corrosion rate at a lower rotational speed for the processed sample [32]. Savguira et al. studied the corrosion behaviour of FSSWed AZ31B. It was found that with an increase in dwell time, the size of the SZ increased. In the corrosion behaviour experiments, it was found that samples with more SZ-occupied areas are resistant to corrosion attack. It was concluded that with a larger SZ, more cathodic zones were observed in the galvanic cell experiment [33]. Zhang et al. observed exfoliation and pitting corrosion on the aluminium alloy at the galvanized steel FSSW joint. At the point wherein the interface and the interfacial layer of the aluminium alloy met, it existed with a continuous corrosion hole, whereas the steel surface showed a minimum corrosion effect. In this dissimilar FSSW welding, galvanic cells were formed during corrosion. Elements of silicon and ferrous, which have a high electrode potential, and magnesium and aluminium, which have a low electrode potential, created the galvanic

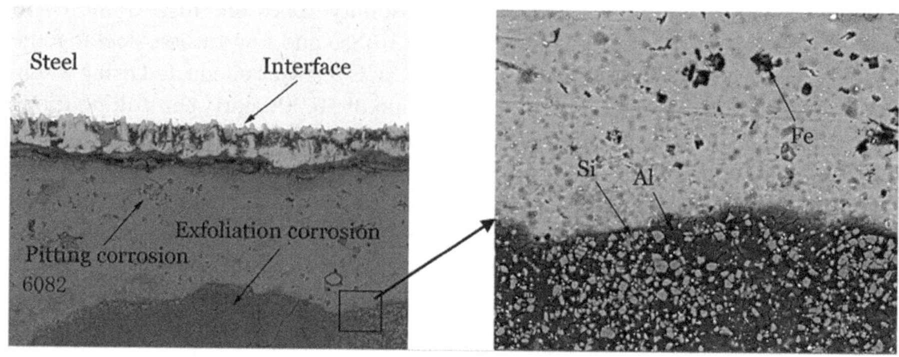

FIGURE 12.6 Microstructural representation of different corrosion phenomena during FSSW [34].

corrosion couplings. Around silicon elements, magnesium and aluminium phases of Mg_2Si and Si-containing solid-solution phase α (Al) tended to create anodic dissolution and exfoliation corrosion (Figure 12.6). And on the steel side (cathodic region), the Fe-rich phase (Al_3Fe) corroded the magnesium and created pitting corrosion surrounding the Mg-rich phase (Al_3Mg_2), which served as the anode [34].

Li et al. studied the corrosion behaviour of FSSWed AA-galvanized steel. In this experiment, a refilled FSWed joint between aluminium alloy and galvanized steel sleeve-affected zone (SAZ) and pin-affected zone (PAZ) showed severe corrosion attack rather than other welded zones such as TMAZ, HAZ and BM. Due to excessive corrosion in SAZ and PAZ, these two regions became the weakest of the entire welded section. Different types of corrosion phenomena took place, such as pitting corrosion and IGC. Pitting formed through the microcrack in the Zn-enriched layer, while another type of corrosion, i.e., IGC, was caused by the accumulation of Zn and Mg elements with lower potential in the Zn-dominant layer's GB. Also, the galvanic cell between steel and Al alloy caused severe degradation of the aluminium alloy side [35]. To control the effect of corrosion, PEO coatings are a real-time solution for magnesium alloy (AZ31), joined by FSSW. Different intermetallics, such as β-$Mg_{17}Al_{12}$ and Al-Mn, generate microgalvanic cells and induce the corrosion rate in the welding zone. The welded joint's corrosion resistance increased when the coating thickness increased and decreased with the rising coating roughness [36]. Another refilled fiction stir spot welding of Aleris Superlite 200ST aluminium alloy with ST06Z galvanized steel and ST16 uncoated steel sheets generates various corrosion regions in the welding zones. Various sheets create galvanic coupling and enhance corrosion. The surface ratio between anodic and cathodic sites plays a very important role in increasing it; if the surface ratio is smaller in count, then the galvanic corrosion rate will be higher in amount [37,38]. It has been observed that the mixed layer exfoliation increases the corrosion current density, which increases the overall corrosion rate. The passivation layer on the steel sheet determines the effect of corrosion on the welded surface [39].

FSSWed aluminium 5083-H116 alloy showed optimized tool rotational speed and mixing of the samples, which will produce good joint control of the electrochemical characteristics. The development of grain refinement and crystallographic changes in the welded region have an impact on the corrosion resistance of the FSSW samples. The sample's rate of corrosion is heavily influenced by the size of the grains in the welded sample; more refined grains boost the material's stress hardening, which facilitates the formation of a passive layer and decreases the material's susceptibility to corrosion. For the FSSW samples, the corrosion loss only takes place gradually over time. However, coarse granules also help to generate a less passive layer that delays corrosion failure and ensures failure to take place gradually [40]. FSSW of AA1050-O/AA6061-T4 dissimilar joints corrosion behaviour exhibits that the tool rotation speed, dwell time, and plunging depth have a significant impact on it. Increase in the rotation speed of the tool used during welding reduces the corrosion rate. It has been found that with higher rotation speeds, the corrosion resistance increases, while increasing the dwell time tends to reduce the corrosion rate. This suggests that prolonged contact between the tool and workpiece may have a protective effect against corrosion. Plunging depth also tends to reduce the corrosion rate up to a certain value; deeper penetration may improve corrosion resistance. During welding, a balance needs to be struck because within specified parameters, increasing tool rotation speed, dwell duration, and plunging depth can help lower corrosion rates, but going above those parameters might have negative impacts, especially in the agitated zones of welded joints [41]. The area where the two metals AZ31 and AZ80 connected through FSSW seems to be the site of the worst onslaught of degradation. It stipulates a path of corrosion attack along the welding direction. Corrosion on the bottom surface was confined to the area where the two metals touched, and the AZ31's dissolution was accelerated. Similar to the corrosion attack on the bottom surface of the joint, it was present on the top surface as well, but it was displaced away from the weld centreline and towards the AZ80 side. In this scenario, AZ80 showed the best corrosion resistance compared to AZ31 and joint area, respectively. Instead of microstructural changes brought on by the welding process, as is the case with similar junctions, the corrosion of the weld was controlled by the galvanic coupling of the dissimilar metals [42].

The presence of salts in the corrosive oil solution likely enhances the corrosive effects because salts can act as electrolytes and accelerate the corrosion process. Additionally, the different types of corrosion observed in various welding areas suggest that the susceptibility to corrosion can vary depending on factors such as the welding process, the type of metal used, and the specific environmental conditions. Projection friction stir spot welded (PFSSW) AA2024-T3 aluminium sheet exhibited that TMAZ is more corrosion-prone compared to other welded zones. Here, the corrosion type was pitting in nature. The SZ of the weld was where most of the corrosion products were present. Controlling the type and amount of precipitation present in the welded region is a crucial aspect of ensuring its long-term durability. This often requires a comprehensive understanding of the alloy's metallurgical behaviour and careful process control. Galvanic corrosion was seen in some places as a result of the potential differential between precipitates and particles containing the base metal.

The first melting of copper-rich particles like Al_2CuMg may have contributed to the rise in sensitivity to IGC at the PFSSW process [43]. Controlling the plunge depth during the FSSW process is critical for achieving good corrosion resistance in the welded joints. Changes in penetration depth during FSSW into the material during the welding process have a significant influence on the corrosion resistance of the joints. Deeper or shallower plunge depths can result in different corrosion behaviours. In contrast to plunge depth, variations in tool rotational speed had a comparatively lesser effect on the corrosion aspects of the joints. Similarly, variations in dwell time were found to have a lesser impact on corrosion compared to plunge depth. Deeper or shallower plunges can lead to variations in microstructure and the presence of cracks, which, in turn, affect the corrosion behaviour of the joints, i.e., areas with cracks or defects in the welded joint are particularly vulnerable to corrosion [44].

It has been observed that careful process control and optimization of plunge depth are essential for producing FSSW joints with improved corrosion resistance and mechanical integrity. The occurrence of a keyhole in the middle of the weld is not only unsightly, but it also reduces the bond area, weakens the weld, and leaves a possible space for the initiation of crevice corrosion. The keyhole can create a potential cavity or crevice within the weld [45]. Often, Al alloy layer cladding on the sheet surface reduces the chances of corrosion on the surface, which leads to a longer work life for the joints. In FSSW, the presence of a pinhole at the centre of a weld nugget, as described when using a conventional tool, can indeed create a potential vulnerability in terms of corrosion because pinholes can act as corrosion initiation sites as they provide a confined space where moisture, such as rainwater, can accumulate and remain in contact with the metal. Moisture and air are key uncontrollable variables in promoting corrosion, and the presence of a pinhole can create an environment conducive to corrosion. The bottom of the pinhole is often difficult to reach with protective coatings or body paint. This lack of coverage leaves the exposed metal at the bottom of the pinhole more susceptible to corrosion because it is not adequately shielded from the external environment, and a combination of trapped moisture and the absence of protective coatings can lead to localized corrosion at the pinhole. This type of corrosion is typically more aggressive and can cause significant damage in a concentrated area. Pit formation in this area enhances its degradation. In comparison between different welding techniques, gas pocket-assisted friction stir spot welded samples exhibit better corrosion properties than underwater FSSW and basic FSSW on AA5083-H112. Due to the effectiveness of the GA-FSSW technology, further innovations have been made to lessen or completely prevent the penetration of corrosive molecules like chlorine ions from seawater within the welding zone. Changes in microstructure, grain refinement, and the presence of fine precipitations resulted in GA FSSW being a better welding method to control the corrosion property than other techniques [46]. To determine optimized process parameters that have the best impact on the strength, mechanical characteristics, and corrosion behaviour of the aluminium alloy (commercial aluminium alloy 6063 sheet), it would be easier to understand how the process parameters affect the alloy's changing properties.

Both the tool rotation rate and dwell time can influence the corrosion rate of welds, primarily through their effects on the quality and microstructure of the weld. Increasing the tool rotation rate can lead to higher heat input and greater mixing of the material. This might result in improved metallurgical bonding and potentially reduced porosity in the weld, which can enhance corrosion resistance. Conversely, a lower rotation rate may reduce heat input and mixing, which could affect the quality of the weld. In some cases, this might result in a higher susceptibility to corrosion, particularly if the welding process leaves behind defects or inclusions that can promote corrosion. Since an anodic reaction is known to aid in the corrosion of a material, it can be seen that as the tool rotation rate increases, more anodic reactions take place. Increasing the dwell time allows for more prolonged heating and greater interaction between the welding tool and the material. This can lead to better consolidation of the weld and potentially improved corrosion resistance. A shorter dwell time may limit the heating and interaction, which could result in incomplete bonding or less uniform mixing. In such cases, there may be areas of the weld that are more susceptible to corrosion [47]. In general, the corrosion rate of welds, installations, machineries, and devices has a significant impact on their durability and lifetime. When exposed to the environment, welded connections, particularly those with different types of welds, are more vulnerable to corrosion.

12.7 CONCLUSION

Generally, in FSSWed, dissimilar materials pose galvanic corrosion due to the EMF difference between the two metals. This galvanic corrosion tends to cause the more anodic metals, which have higher chances of corroding to go through quicker degradation, while the other cathodic metals get protected due to that. Weld defects play a vital role in corrosion in FSSWed joints. Formation of hook defects is one of the active sites for corrosion. The defects act as a corrosion initiation site, allowing congeal electrolyte to penetrate the join and speed up the corrosion sensations. Thermal expansion is also responsible for generating corrosion. For dissimilar welding joints with different thermal expansion coefficients, a thermal gradient is generated during welding, which in turn affects the metallurgy of the joining materials. Also, during the heat generation process or stirring actions of the tool, metals with a higher thermal expansion coefficient will cause stress generation, which will cause cracking. The presence of weld cracks due to improper process parameters and uncontrollable variables will endorse the corrosion. In the case of the joining of dissimilar metals, intermetallic phases are inevitable. The presence of intermetallics in welded structures is one of the crucial reasons for welding corrosion. The presence of intermetallics makes them active sites for corrosion. Intermetallic sites are more susceptible to Cl^--ion attack. However, for similar materials, corrosion depends mainly on the variation in grain size and the GB density. Also, the chances of crevice corrosion and pitting corrosion are quite high in similar metal FSSW joints. On the other hand, residual stresses that are generated during FSSW can be the reason for SCC. Residual stresses are mainly generated inside the joint section due to the presence of a thermal gradient, improper material flow, and excessive strain energy inside the material. Therefore, proper post-processing methods are required inside to mitigate residual stresses and avoid the chances of SCC. Appropriate material selection, better surface preparation, application of coatings or inhibitors and other corrosion protection techniques will reduce the chances of corrosion in welded joints.

REFERENCES

1. R. Gundla, D.B. Naik, C.H.V. Rao, K.S. Rao, G.M. Reddy. Optimization of friction stir welding parameters for improved corrosion resistance of AA2219 aluminum alloy joints. *Defence Technology* 11, no. 4 (2015) 330–337.
2. P. Vincent, J. Alexis, E. Andrieu, J. Delfosse, M.C. Lafont, C. Blanc. Characterisation and understanding of the corrosion behaviour of the nugget in a 2050 aluminium alloy Friction Stir Welding joint. *Corrosion Science* 73 (2013) 130–142.
3. T.S. Mahmoud. Effect of friction stir processing on electrical conductivity and corrosion resistance of AA6063-T6 Al alloy. *Proceedings of the Institution of Mechanical Engineers, Part C: Journal of Mechanical Engineering Science* 222, no. 7 (2008) 1117–1123.
4. Y. Chen, Y. Wang, G. Meng, B. Liu, J. Wang, Y. Shao, J. Jiang. Macro-galvanic effect and its influence on corrosion behaviors of friction stir welding joint of 7050-T76 Al alloy. *Corrosion Science* 164 (2020) 108360.
5. S. Qian, Shuaihao, T. Zhang, Y. Chen, J. Xie, Y. Chen, T. Lin, H. Li. Effect of ultrasonic impact treatment on microstructure and corrosion behavior of friction stir welding joints of 2219 aluminum alloy. *Journal of Materials Research and Technology* 18 (2022) 1631–1642.
6. L. Dandan, L. Zhang, Y. Du, H. Xu, S. Liu, L. Liu. Assessment of atomic mobilities of Al and Cu in fcc Al-Cu alloys. *Calphad* 33, no. 4 (2009) 761–768.
7. S. Rajakumar, C. Muralidharan, V. Balasubramanian. Predicting tensile strength, hardness and corrosion rate of friction stir welded AA6061-T6 aluminium alloy joints. *Materials & Design* 32, no. 5 (2011) 2878–2890.
8. Q. Wang, Y. Zhao, K. Yan, S. Lu. Corrosion behavior of spray formed 7055 aluminum alloy joint welded by underwater friction stir welding. *Materials & Design* 68 (2015) 97–103.
9. R.K. Gupta, H. Das, T.K. Pal. Influence of processing parameters on induced energy, mechanical and corrosion properties of FSW butt joint of 7475 AA. *Journal of Materials Engineering and Performance* 21 (2012) 1645–1654.
10. A. Squillace, A. De Fenzo, G. Giorleo, F. Bellucci. A comparison between FSW and TIG welding techniques: Modifications of microstructure and pitting corrosion resistance in AA 2024-T3 butt joints. *Journal of Materials Processing Technology* 152, no. 1 (2004) 97–105.
11. C. Zhang, Y. Cao, G. Huang, Q. Zeng, Y. Zhu, X. Huang, N. Li, Q. Liu. Influence of tool rotational speed on local microstructure, mechanical and corrosion behavior of dissimilar AA2024/7075 joints fabricated by friction stir welding. *Journal of Manufacturing Processes* 49 (2020) 214–226.
12. B. Strass, G. Wagner, C. Conrad, B. Wolter, S. Benfer, W. Fürbeth. Realization of Al/Mg-hybrid-joints by ultrasound supported friction stir welding-mechanical properties, microstructure and corrosion behavior. *Advanced Materials Research* 966 (2014) 521–535.
13. T. Chen, W. Xue, Y. Li, X. Liu, J. Du. Corrosion behavior of friction stir welded AZ31B magnesium alloy with plasma electrolytic oxidation coating formed in silicate electrolyte. *Materials Chemistry and Physics* 144, no. 3 (2014) 462–469.
14. S. Aliasghari, A. Rogov, P. Skeldon, X. Zhou, A. Yerokhin, A. Aliabadi, M. Ghorbani. Plasma electrolytic oxidation and corrosion protection of friction stir welded AZ31B magnesium alloy-titanium joints. *Surface and Coatings Technology* 393 (2020) 125838.
15. A. Dhanapal, S.R. Boopathy, V. Balasubramanian. Developing an empirical relationship to predict the corrosion rate of friction stir welded AZ61A magnesium alloy under salt fog environment. *Materials & Design* 32, no. 10 (2011) 5066–5072.

16. S. Gao, H. Zhao, G. Li, G. Sun, L. Zhou, Y. Zhao. Strengthening mechanism and corrosion behavior of friction stir welded LAZ933 magnesium-lithium alloy. *Journal of Magnesium and Alloys* (2023).

17. J. Liu, S. Li, Z. Han, R. Cao. Improved corrosion resistance of friction stir welded magnesium alloy with micro-arc oxidation/electroless plating duplex coating. *Materials Chemistry and Physics* 257 (2021) 123753.

18. Y. Deng, R. Ye, G. Xu, J. Yang, Q. Pan, B. Peng, X. Cao. Corrosion behaviour and mechanism of new aerospace Al-Zn-Mg alloy friction stir welded joints and the effects of secondary Al3ScxZr1-x nanoparticles. *Corrosion Science* 90 (2015) 359–374.

19. R.C. Zeng, J. Chen, W. Dietzel, R. Zettler, J.F. dos Santos, M.L. Nascimento, K.U. Kainer. Corrosion of friction stir welded magnesium alloy AM50. *Corrosion Science* 51, no. 8 (2009) 1738–1746.

20. B. Lingampalli, S. Dondapati. Corrosion behaviour of friction stir welded ZM21 magnesium alloy. *Materials Today: Proceedings* 46 (2021) 1464–1469.

21. G. Ma, H. Gao, C. Sun, Y. Gu, J. Zhao, Z. Fang. Probing the anti-corrosion behavior of X65 steel underwater FSW joint in simulated seawater. *Frontiers in Materials* 8 (2022) 723986.

22. H. Zhang, D. Wang, P. Xue, L.H. Wu, D.R. Ni, Z.Y. Ma. Microstructural evolution and pitting corrosion behavior of friction stir welded joint of high nitrogen stainless steel. *Materials & Design* 110 (2016) 802–810.

23. S.D. Meshram, A.G. Paradkar, G.M. Reddy, S. Pandey. Friction stir welding: An alternative to fusion welding for better stress corrosion cracking resistance of maraging steel. *Journal of Manufacturing Processes* 25 (2017) 94–103.

24. D.G. Mohan, S. Gopi. Influence of In-situ induction heated friction stir welding on tensile, microhardness, corrosion resistance and microstructural properties of martensitic steel. *Engineering Research Express* 3, no. 2 (2021) 025023.

25. H.B. Li, Z.H. Jiang, H. Feng, S.C. Zhang, L. Li, P.D. Han, R.D.K. Misra, J.Z. Li. Microstructure, mechanical and corrosion properties of friction stir welded high nitrogen nickel-free austenitic stainless steel. *Materials & Design* 84 (2015) 291–299.

26. R.A. Rajasekaran, A.K. Lakshminarayanan, R. Damodaram, V. Balasubramanian. Stress corrosion cracking failure of friction stir welded nuclear grade austenitic stainless steel. *Engineering Failure Analysis* 120 (2021) 105012.

27. P. West, V.C. Shunmugasamy, C.A. Usman, I. Karaman, B. Mansoor. Part II: Dissimilar friction stir welding of nickel titanium shape memory alloy to stainless steel-microstructure, mechanical and corrosion behavior. *Journal of Advanced Joining Processes* 4 (2021) 100072.

28. D. Trinh, S. Frappart, G. Rückert, F. Cortial, S. Touzain. Effect of friction stir welding process on microstructural characteristics and corrosion properties of steels for naval applications. *Corrosion Engineering, Science and Technology* 54, no. 4 (2019) 353–361.

29. B. Matlan, M. Joharif, H. Mohebbi, S.R. Pedapati, B.A. Mokhtar, M.C. Ismail, S. Kakooei, N.E. Dan. Dissimilar friction stir welding of carbon steel and stainless steel: Some observation on microstructural evolution and stress corrosion cracking performance. *Transactions of the Indian Institute of Metals* 71 (2018) 2553–2564.

30. I. Koley, S. Kundu. Electrochemical behavior of friction stir welded joint of ultra-low carbon steel. *IOP Conference Series: Materials Science and Engineering,* 1248, no. 1 (2022) 012040.

31. A. James, T.H. North, S.J. Thorpe. Localized corrosion of friction stir spot welds in magnesium AZ31 alloy. *ECS Transactions* 41, no. 25 (2012) 193.

32. K. Sanusi, E.T. Akinlabi, E. Muzenda, S.A. Akinlabi. Enhancement of corrosion resistance behaviour of frictional stir spot welding of copper. *Materials Today: Proceedings* 2, 4–5 (2015) 1157–1165.

33. Y. Savguira, T.H. North, and S.J. Thorpe. Microcapillary polarization measurements of friction stir spot welds made in AZ 31 B magnesium alloy. *Materials and Corrosion* 65, no. 11 (2014) 1055–1061.

34. Z.K. Zhang, Y. Yu, J.F. Zhang, X.J. Wang. Corrosion behavior of keyhole-free friction stir spot welded joints of dissimilar 6082 aluminum alloy and DP600 galvanized steel in 3.5% NaCl solution. *Metals* 7, no. 9 (2017) 338.

35. P. Li, Y. Wang, S. Wang, J. Yang, H. Dong, D. Yan. Corrosion behavior of refilled friction stir spot welded joint between aluminum alloy and galvanized steel. *Materials Research Express* 5, no. 9 (2018) 096524.

36. Y. Savguira, S. Busef, T.H. North, S.J. Thorpe. Corrosion protection of friction stir spot welds made in magnesium alloys. *Electrochemical Society Meeting Abstracts* 228, no. 13 (2015) 665–665.

37. F. Mansfeld. Area relationships in galvanic corrosion. *Corrosion* 27, no. 10 (1971) 436–442.

38. F. Mansfeld, J.V. Kenkel. Galvanic corrosion of Al alloys-III. The effect of area ratio. *Corrosion Science* 15, no. 4 (1975) 239–250.

39. Y. Ma, H. Dong, Y. Wang, G. Yang, Y. Xia, P. Li, X. Hao. Effect of Zn coating on microstructure and corrosion behavior of dissimilar joints between aluminum alloy and steel by refilled friction stir spot welding. *Journal of Applied Electrochemistry* 52, no. 1 (2022) 85–102.

40. E.T. Akinlabi, O.M. Ikumapayi, A.S. Osinubi, N. Madushele, O.O. Abegunde, S.O. Fatoba, S.A. Akinlabi. Characterizations of AA5083-H116 produced by friction stir spot welding technique. *Advances in Materials and Processing Technologies* 8, no. sup4 (2022) 2299–2313.

41. B.D. Al-Mutairi, S.A. Abdallah. Optimization of friction stir spot welding process parameters to minimize the corrosion rate in the stirred zones of AA1050-O/AA6061-T4 dissimilar aluminum lap-joints. 1, no. 39 (2019) 74–78.

42. Y. Savguira, T.H. North, S.J. Thorpe. Localized corrosion behaviour in dissimilar AZ31/AZ80 friction stir welds. *ECS Transactions* 64, no. 28 (2015) 1.

43. A. Arabzadeh, B. Korojy, S.M. Mousavizade, S.A. Hosseini. Corrosion investigation of projection friction stir spot welding of Al 2024 sheets in oil and gas industry. *Journal of Oil, Gas and Petrochemical Technology* 9, no. 1 (2022) 39–48.

44. S. Siddharth, T. Senthilkumar. Optimizing process parameters for increasing corrosion resistance of friction stir spot welded dissimilar Al-5086/C10100 joints. *Transactions of the Indian Institute of Metals* 71 (2018) 1011–1024.

45. Y. Gao, Y. Liang, X. Ren, M. Paidar. Pre-hole friction stir spot welding (PFSSW) for dissimilar welding of AA2219 to AA3003 sheets. *Vacuum* 182 (2020) 109688.

46. S. Basak, M. Mondal, S.Y. Anaman, K. Gao, S.T. Hong, H.H. Cho. Gas pocket-assisted underwater friction stir spot welding. *Journal of Materials Processing Technology* 320 (2023) 118100.

47. D.M. Kapinga, K.D. Nyembwe, O.M. Ikumapayi, E.T. Akinlabi. Mechanical, electrochemical and structural characteristics of friction stir spot welds of aluminium alloy 6063. *Manufacturing Review* 7 (2020) 25.

13 Nanoparticle Reinforcements in Friction Stir Spot Welding (n-FSSW)

M. S. Senthil Saravanan, J. Yoganandh, Trijo Tharayil, J. B. Sajin, and Sajivarghese

13.1 INTRODUCTION

Welding and joining techniques play an essential role in both industrial manufacturing and household fabrication works [1–5]. Joining techniques became more advanced as new materials and new applications arose. Welding is classified into fusion and solid-state techniques. Under the solid-state welding technique, friction stir welding (FSW) is a prominent technique for most of the engineering joints. Spot welding of the materials can be performed using friction stir spot welding (FSSW). FSSW is a solid-state joining process used to create high-strength welds in materials, particularly aluminum and its alloys [5–9]. It is a variant of the FSW process, which was developed in the 1990s. High-strength joints with reduced distortions are achieved through FSSW. Thus, the various applications are credited to FSSW. Figure 13.1 shows the various steps in the FSSW process: (a) rotation, (b) plunging, (c) stirring, and (d) drawing out.

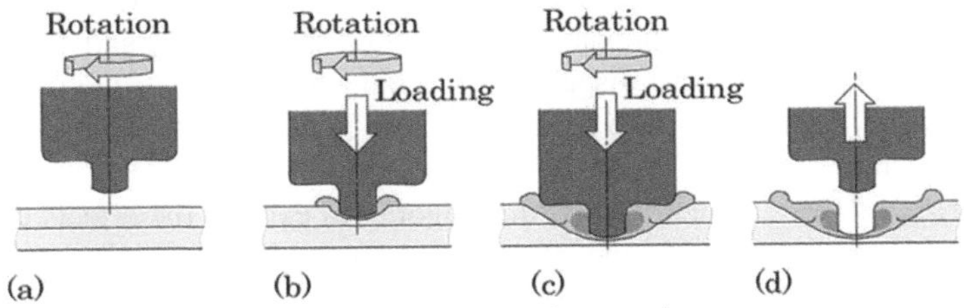

FIGURE 13.1 Steps in FSSW process: (a) rotation, (b) plunging, (c) stirring, and (d) drawing out [10].

DOI: 10.1201/9781003432289-13

13.2 NANOPARTICLE-REINFORCED FSSW

Nanoparticle reinforcements in FSSW involve incorporating nanoparticles into the joint interface to enhance the mechanical properties and performance of the weld [11]. This method is also known as nanoparticle-enhanced FSSW (n-FSSW). The advantages of n-FSSW are as follows:

a. Improved mechanical properties: The nanoparticles can enhance the strength, hardness, and toughness of the weld joint, leading to improved load-bearing capacity and resistance to fatigue and fracture.
b. Enhanced thermal properties: Some nanoparticles possess excellent thermal conductivity, which can improve heat dissipation during welding and reduce the risk of heat-affected zone (HAZ) formation. This can result in better microstructural properties and reduce the chances of defects such as voids or cracks.
c. Grain refinement: The presence of nanoparticles can promote grain refinement within the stir zone. Smaller grain sizes generally contribute to the higher strength and improved mechanical properties of the weld.
d. Reduced defects: Nanoparticles can act as nucleation sites for voids and other defects, reducing the likelihood of defects propagating through the weld. This can improve the overall integrity of the joint.
e. Frictional behavior modification: Nanoparticles can alter the frictional behavior at the tool – work border area, potentially decreasing tool wear and increasing tool life.

The nanoparticles used in n-FSSW are typically chosen based on their size, shape, and chemical composition. Commonly used nanoparticles include carbon nanotubes (CNTs), graphene, metal oxides (such as alumina, silica, or titanium oxide), and other metallic nanoparticles.

13.3 FACTORS AFFECTING N-FSSW

The successful incorporation of nanoparticles in n-FSSW requires careful consideration of factors such as nanoparticle dispersion, distribution, and concentration. Techniques such as ultrasonic dispersion, mechanical mixing, or the use of surfactants can be employed to achieve a homogeneous dispersion of nanoparticles within the joint interface.

Research and development efforts are ongoing to explore the effects of different nanoparticle types, sizes, and concentrations on the quality of welding along with its mechanical properties. The goal is to optimize the n-FSSW process and unlock its full potential for various applications, including lightweight structures in the automotive, aerospace, and marine industries. It's worth noting that while nanoparticle reinforcements in FSSW show promising potential, the technology is still in the developing stage, and further research is required to fully understand the effects and optimize the process parameters for different materials and applications.

13.3.1 Dispersion of Particles in n-FSSW

Nanoparticle dispersion in FSSW results in an equal distribution of nanoparticles within the joint interface. Proper dispersion is essential to ensuring the nanoparticles can effectively enhance the mechanical properties of the weld and provide consistent reinforcement. Achieving nanoparticle dispersion in FSSW can be challenging due to the inherent difficulties in uniformly dispersing nanoparticles within a metal matrix. Here are some common techniques used to improve nanoparticle dispersion.

13.3.1.1 Surfactant-Assisted Dispersion

Surfactants can be used to improve the dispersibility of nanoparticles in the base metal. They act as surface-active agents, reducing interparticle attraction and promoting stable dispersion. Surfactants can be added during the mixing process to enhance the dispersion of nanoparticles.

13.3.1.2 Ultrasonic Dispersion

Ultrasonic energy is often employed to break down agglomerates and promote the dispersion of nanoparticles. Ultrasonic waves create high-frequency vibrations that disrupt nanoparticle clusters, leading to improved dispersion in the metal matrix. This technique is effective for achieving a more uniform distribution of nanoparticles. Figure 13.2 shows a photograph of ultrasonic dispersion equipment.

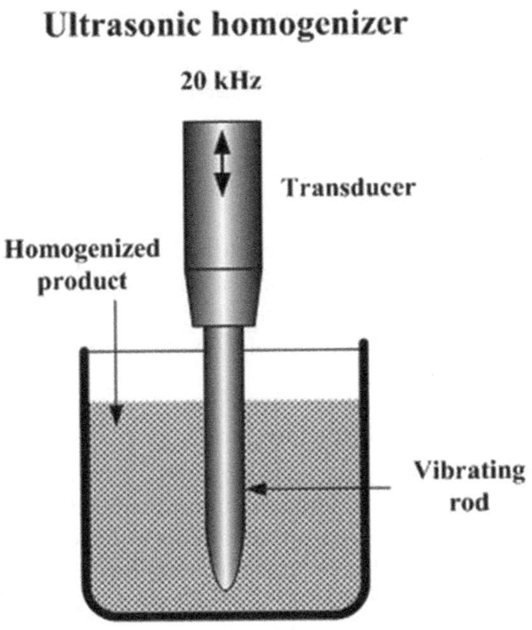

FIGURE 13.2 Photograph of ultrasonic dispersion equipment [12].

13.3.1.3   Mechanical Mixing

Mechanical mixing in FSSW involves the intense stirring action of a specially designed tool to mix and bond the materials being joined. The steps in mechanical mixing are explained as follows:

- Tool design: The FSSW tool contains a shoulder and a pin. The shoulder is a cylindrical component that generates frictional heat and applies downward pressure, while the pin has a threaded or profiled shape that stirs the materials.
- Initial contact: The tool is brought into contact with the workpieces to be joined. The shoulder presses against the top surface of the materials, applying downward force.
- Rotation and translation: The tool starts rotating and simultaneously moves along the desired welding path. The rotation generates heat in between the tool and workpieces due to friction.
- Plastic deformation: The combination of the rotating tool and applied pressure leads to plastic deformation of the materials. The heat softens the material without reaching its melting point, allowing it to become more malleable.
- Stirring action: The pin of the tool, with its threaded or profiled shape, generates vigorous stirring action within the materials. The rotation of the pin creates a mixing effect, causing the materials to be stirred and mixed together.
- Solid-state bonding: As the materials are stirred, the softened material from the top surface and the bottom surface intermix, forming a solid-state bond. This bond occurs without the complete melting of the materials, resulting in a joint with improved mechanical properties.
- Cooling and solidification: After the stirring and mixing process, the tool continues to move along the welding path. As the materials cool down, the joint solidifies and gains strength.

13.3.1.4   In Situ Formation

In some cases, nanoparticles can be formed in situ during the FSSW process. For example, during the joining of aluminum-based alloys, the addition of precursor material can lead to the formation of nanoparticles through reactions with the base metal during welding. This allows for the nanoparticles to be synthesized directly at the joint interface.

13.3.1.5   Alloying During Fabrication

Another approach is to introduce nanoparticles into the base metal during the fabrication process. This can be achieved through powder metallurgy techniques, where nanoparticles are mixed with the base metal powders before consolidation. The subsequent FSSW process then disperses and consolidates the nanoparticles within the joint.

It's important to note that the dispersion of nanoparticles in FSSW is still an area of active research, and the optimal techniques may vary depending on the specific materials and nanoparticles being used. Factors such as nanoparticle size, shape, concentration, and the choice of dispersing agent need to be carefully considered to achieve a uniform dispersion and maximize the benefits of nanoparticle reinforcement in FSSW.

13.3.2 DISPERSION OF NANOPARTICLES IN N-FSSW

In n-FSSW, achieving an equal distribution of nanoparticles within the joint interface is crucial to ensuring consistent reinforcement and improved mechanical properties. The distribution of nanoparticles can significantly affect the resulting weld quality and performance. Several factors influence the distribution of nanoparticles in n-FSSW, including the welding parameters and the method used for introducing nanoparticles. Here are some considerations for achieving a desirable nanoparticle distribution in n-FSSW:

13.3.2.1 Mixing Techniques

Proper mixing techniques are essential to ensuring a uniform dispersion of nanoparticles in the base material. Mechanical mixing methods, such as ball milling or high-energy stirring, can be employed to disperse nanoparticles within the base material before the FSSW process. These methods help break down nanoparticle agglomerates and promote uniform distribution.

13.3.2.2 Tool Design

The design of the FSSW tool can play a role in nanoparticle distribution. Specialized tool designs, such as those with grooves or pockets, can aid in the even distribution of nanoparticles within the joint. These features facilitate the mixing and dispersion of nanoparticles during the stirring action of the tool.

13.3.2.3 Surfactants or Dispersing Agents

The use of surfactants or dispersing agents can assist in achieving a homogeneous distribution of nanoparticles. Surfactants reduce interparticle attraction and help stabilize nanoparticle dispersion. They can be added during the mixing process to enhance dispersion and prevent nanoparticle agglomeration.

13.3.2.4 Ultrasonic Assistance

Ultrasonic energy can be utilized to improve nanoparticle dispersion in n-FSSW. Ultrasonic waves can break down nanoparticle clusters and enhance their distribution within the base material. Ultrasonic-assisted mixing or dispersion processes can help achieve a more uniform distribution of nanoparticles.

It is important to note that achieving a perfectly uniform distribution of nanoparticles throughout the weld joint can be challenging. Factors such as nanoparticle properties, material compatibility, and process parameters must be carefully optimized to maximize the distribution uniformity. Advanced characterization techniques, such as microscopy and elemental mapping, can be used to assess the distribution of nanoparticles within the joint and evaluate the effectiveness of the dispersion methods.

13.3.3 CONCENTRATION OF NANOPARTICLES IN N-FSSW

The concentration of nanoparticles in n-FSSW refers to the amount or percentage of nanoparticles present in the joint interface. The concentration plays a crucial role in determining the level of reinforcement and the resulting properties of the weld. The optimal concentration of nanoparticles in n-FSSW can vary depending on various factors, such as the type of nanoparticles, base material, desired properties, and the specific application. Here are some considerations regarding nanoparticle concentration in n-FSSW.

13.3.3.1 Particle-to-Metal Ratio

The concentration of nanoparticles is often expressed as a ratio of the weight or volume of nanoparticles to the weight or volume of the base material. This ratio can vary depending on the desired level of reinforcement. Higher concentrations typically lead to greater strengthening effects, but there may be diminishing returns or potential negative impacts beyond a certain point.

13.3.3.2 Material Compatibility

The concentration of nanoparticles should be compatible with the base material and the intended application. Excessive concentrations of nanoparticles may adversely affect the weld's mechanical properties, such as ductility or toughness. It is essential to balance the reinforcement benefits with potential trade-offs in other properties.

13.3.3.3 Dispersibility and Agglomeration

The concentration of nanoparticles can affect their dispersibility within the base material. Higher concentrations may increase the likelihood of nanoparticle agglomeration or clustering, which can lead to uneven distribution and reduced effectiveness. It is crucial to ensure proper dispersion and minimize agglomeration through appropriate mixing techniques and the use of dispersing agents, as mentioned earlier.

13.3.3.4 Weldability and Process Considerations

The concentration of nanoparticles should not compromise weldability or process stability. Excessive concentrations can introduce challenges during the FSSW process, such as tool wear, increased heat generation, or difficulties in achieving sound welds. The concentration should be optimized to maintain process feasibility and avoid potential issues.

Determining the optimal nanoparticle concentration in n-FSSW involves a balance between achieving the desired reinforcement and maintaining the weld's overall properties and process feasibility. Experimental studies, numerical simulations, and optimization approaches are typically employed to investigate the effects of different nanoparticle concentrations and find the optimal range for specific applications.

It is important to note that the optimal concentration can vary for different materials, nanoparticles, and applications. Therefore, extensive research and characterization are required to understand the effect of nanoparticle concentration on the resulting weld characteristics and to identify the most suitable concentration for a particular n-FSSW application.

13.3.4 Nanoparticles Used in n-FSSW

In n-FSSW, various nanomaterials can be used to reinforce the weldment and improve the mechanical properties. The choice of nanomaterial depends on factors such as the base material being welded, the desired properties, and the specific application requirements. Here are some commonly used nanomaterials for n-FSSW.

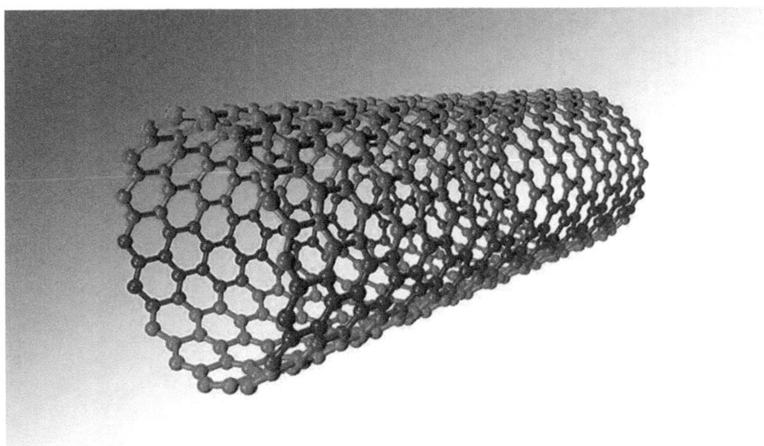

FIGURE 13.3 Photograph of carbon nanotubes.

13.3.4.1 Carbon Nanotubes

CNTs are tubular structures with carbon atoms having exceptional mechanical, thermal, and electrical properties. They have high strength and stiffness, making them effective reinforcements for improving the mechanical properties of the weld. CNTs are incorporated into the weld-joint interface to enhance strength, fatigue resistance, and electrical conductivity. Figure 13.3 shows a photograph of CNTs.

13.3.4.2 Graphene

Graphene is a monolayer of carbon atoms arranged in a 2D honeycomb lattice. It has excellent mechanical, thermal, and electrical properties. Graphene sheets or nanoplatelets can be introduced into the weld interface to enhance strength, toughness, and electrical conductivity. Figure 13.4 shows a photograph of graphene.

13.3.4.3 Metal Oxides

Various nanoparticles of metal oxides can be utilized in n-FSSW, such as aluminum oxide (alumina), titanium oxide (titania), and zinc oxide. Metal oxide nanoparticles offer advantages like high hardness, frictional resistance, and thermal stability. They can improve the mechanical properties, wear resistance, and corrosion resistance of the weld. Figures 13.5 and 13.6 show the TiO_2 powder and SiO_2 powder, respectively.

13.3.4.4 Metallic Nanoparticles

Nanoparticles made of metals, such as copper, silver, or nickel, can be used as reinforcements in n-FSSW. These metallic nanoparticles can enhance the electrical, thermal, and mechanical strengths of weldments. They are particularly useful when joining dissimilar materials. Figures 13.7 and 13.8 show the copper nanoparticles and nickel nanoparticles, respectively.

FIGURE 13.4   Photograph of graphene.

FIGURE 13.5   TiO$_2$ powder.

FIGURE 13.6 SiO$_2$ powder.

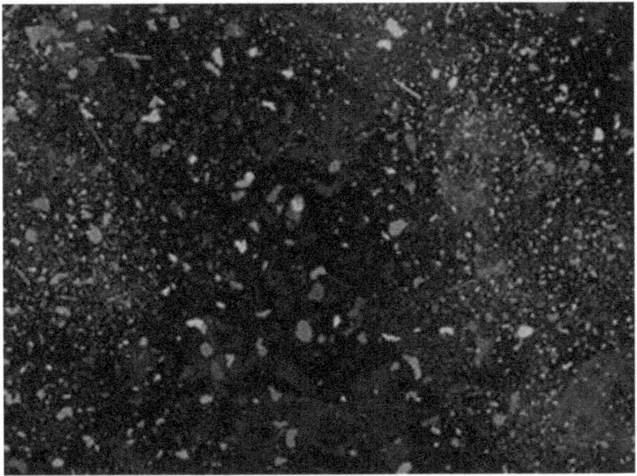

FIGURE 13.7 Copper nanoparticles.

FIGURE 13.8   Nickel nanoparticles.

13.3.4.5   Hybrid Nanomaterials

Hybrid nanomaterials combine multiple types of nanoparticles or incorporate nanoparticles with other reinforcing materials, such as fibers or nanofibers. These combinations can offer synergistic effects, combining the properties of different nanomaterials to achieve desired enhancements in the weld joint. The selection of nanomaterials for n-FSSW should consider factors like compatibility with the base material, dispersibility, and the intended improvements in properties. Additionally, the synthesis, dispersion, and distribution of nanomaterials within the weld joint should be carefully optimized to ensure effective reinforcement and achieve the desired improvements in the weld's mechanical properties. The choice of nanomaterials for n-FSSW is an active area of research, and advancements continue to be made to explore new nanomaterials and optimize their incorporation into the welding process.

13.4   CASE STUDIES

13.4.1   Automotive Industry

n-FSSW has been investigated for joining aluminum alloys in automotive applications, aiming to improve the mechanical properties and performance of lightweight structures. For instance, researchers have explored the incorporation of CNTs in n-FSSW aluminum alloys. It was successfully demonstrated that the CNT addition can considerably enhance the strength, fatigue resistance, and electrical conductivity of the welds. This improvement in properties makes n-FSSW with CNTs a promising technique for lightweight automotive construction.

The incorporation of CNTs in n-FSSW will improve the mechanical strength and workability of welded joints. CNTs are cylindrical carbon structures with exceptional mechanical, thermal, and electrical properties, making them attractive reinforcements in various applications, including welding.

When CNTs are added to the joint interface in n-FSSW, they can provide several benefits:

- Improved mechanical properties: the high strength and stiffness of CNTs make them an ideal material for the enhancement of the mechanical strength of weld joints or weldments. The incorporation of CNTs can increase the strength, hardness, and toughness of the joint, improving its load-bearing capacity and resistance to fatigue and fracture.
- Enhanced electrical conductivity: CNTs possess excellent electrical conductivity. When incorporated into the weld joint, they can improve the electrical conductivity of the joint, which is particularly important in applications where electrical conductivity is desired, such as in electrical connections or components.
- Thermal properties: CNTs also have excellent thermal conductivity. By incorporating CNTs into the joint interface, they can enhance heat dissipation during welding, reducing the risk of HAZ formation. This can lead to improved microstructural properties and a reduction in the occurrence of voids and cracks.
- Grain refinement: The presence of CNTs can promote the refinement of grains or crystallites within the stir zone of the weld. Smaller grain sizes generally contribute to higher strength and improved mechanical properties of the joint. Research efforts have been undertaken to study the effects of CNTs in n-FSSW on various materials, including aluminum alloys and steel. These studies have demonstrated improvements in the tensile strength, hardness, fatigue resistance, and electrical conductivity of the weldments.

Recently, researchers have discovered the use of CNTs to enhance the mechanical properties of the weld joints in the FSSW process. When CNTs are added to the base material, they can improve the mechanical strength, stiffness, and toughness of the joint. The CNTs act as reinforcing agents, providing additional resistance to deformation and enhancing the resistance to the impact load of the weld. The distribution of CNTs in the matrix material plays a crucial role in evaluating the overall mechanical properties of the joint.

During the FSSW process, the CNTs are subjected to the intense stirring action of the tool, leading to their alignment along the weld line. This alignment enhances the load transfer across the joint, resulting in improved mechanical performance. The addition of CNTs in the FSSW process has been investigated in various materials, including aluminum, magnesium, and steel alloys. The reinforcement effect of CNTs depends on factors such as the CNT concentration, dispersion quality, and the interaction between the CNTs and the matrix material. Figure 13.9 shows the hardness variation during carbon fiber reinforcement in Al using FSSW.

Overall, incorporating CNTs as reinforcements in the FSSW process has the potential to enhance the mechanical properties of the weld joint, providing increased strength and improved performance. However, additional research is needed to optimize the CNT dispersion and investigate the long-term reliability of such joints in real-world applications.

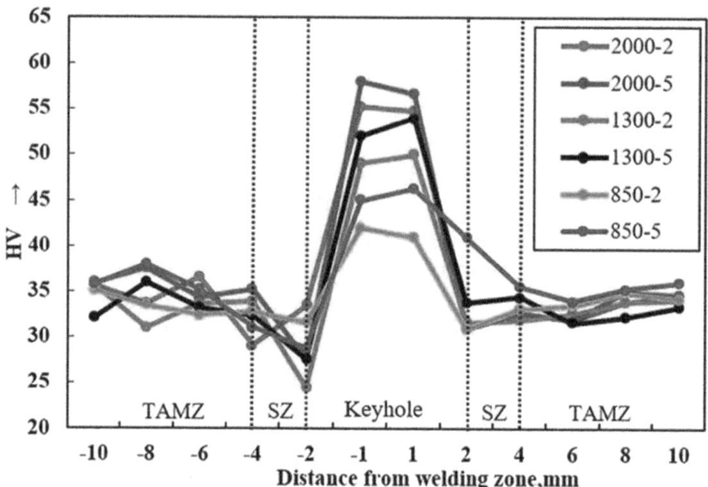

FIGURE 13.9 Hardness variation during carbon fiber reinforcement in Al using FSSW [13].

When carbon fiber was added with Al during FSSW, a maximum tensile shear load of 1779.6 N and a joint efficiency of 14.6% were obtained at a rotational speed of 2,000 rpm and a 2 s dwell time. This is 39.5% greater than the value at a low rotational speed of 850 rpm and a 2 s dwell time. The maximum hardness of 58 HV was attained in the keyhole region at a rotational speed of 2,000 rpm and a dwell time of 5 s, which is 22.4% higher compared to a low rotational speed [13].

Graphene also possesses exceptional mechanical, electrical, and thermal properties, making it an attractive candidate for enhancing the performance of welded joints. Incorporating graphene as reinforcement in the FSSW process can lead to improved mechanical properties, such as increased strength, stiffness, and toughness of the weld joint. The unique structure and properties of graphene allow for efficient load transfer and improved bonding at the graphene–base material interface [14].

Similar to CNTs, graphene can be dispersed in the matrix material prior to the FSSW process. Various techniques, including sonication, ball milling, and chemical functionalization, can be employed to achieve a complete and uniform dispersion of graphene in the base material. During the FSSW process, a strong stirring action of the tool facilitates the alignment of graphene sheets along the weld line. This alignment enhances load transfer and results in improved mechanical performance of the joint. Graphene reinforcement in the FSSW process has been explored in several materials, including aluminum, magnesium, and steel alloys. The concentration of graphene, quality of dispersion, and interaction between graphene and the matrix material play significant roles in determining the effectiveness of reinforcement. The addition of graphene in the FSSW process can lead to improved weld strength, fatigue resistance, and corrosion resistance. The unique properties of graphene, such as its high aspect ratio, excellent mechanical strength, and electrical conductivity, contribute to the enhanced performance of the joint. However, it is worth noting that the dispersion and alignment of graphene in the matrix material can be challenging

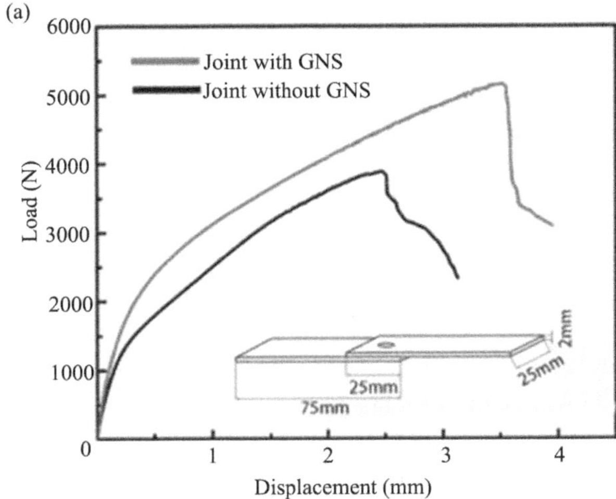

FIGURE 13.10 Tensile strength analysis of graphene nanosheets reinforced AA 2014 alloy [16].

due to its tendency to agglomerate. Proper processing techniques and surface functionalization of graphene are crucial for achieving uniform dispersion and maximizing the reinforcement effect [15].

Further research and optimization are necessary to fully understand the effects of graphene reinforcement in FSSW and its long-term reliability in practical applications. Nonetheless, the incorporation of graphene in FSSW shows great potential for advancing the mechanical properties of welded joints and expanding the range of applications for this solid-state joining technique. Graphene sheets were used to fabricate the joints using AA 2014 aluminum alloy using refilled FSSW. The amount of graphene nanosheets used is 0.6 wt% of the aluminum matrix. In the results, the tensile and shear strength of the joints increased by 31%, and the toughness of the material also improved. The fatigue life of the joint is also extended when graphene nanosheets are added [16]. Figure 13.10 shows the tensile strength analysis of graphene nanosheets reinforced by the AA 2014 alloy.

13.4.2 Aerospace Industry

n-FSSW has also been studied for joining aerospace materials, such as aluminum alloys used in aircraft structures. In one study, aluminum alloy joints were reinforced with alumina (Al_2O_3) nanoparticles using n-FSSW. The addition of alumina nanoparticles resulted in enhanced mechanical properties, including enhanced tensile strength and hardness of the weld joints. This suggests that n-FSSW with nanoparticle reinforcement has the potential to develop lightweight and high-strength structures in the aerospace industry.

These case studies demonstrate the potential benefits of n-FSSW in enhancing the mechanical properties and performance of weld joints. By incorporating nanoparticles, n-FSSW offers opportunities for improved strength, fatigue resistance, electrical conductivity, and other desirable properties in a range of industries. However, it's important to note that further research, optimization of parameters, and characterization efforts are needed to fully understand and exploit the benefits of n-FSSW in different applications.

13.4.2.1 Alumina-Reinforced n-FSSW

Incorporating alumina particles as reinforcement in the FSSW process can enhance the mechanical properties of the weld joint. The addition of alumina particles increases the strength, stiffness, and toughness of the joint, thereby improving its overall performance.

To incorporate alumina reinforcement in the FSSW process, the particles are typically mixed with the base material prior to welding. Techniques such as powder metallurgy or mechanical alloying can be used to attain a uniform dispersion of alumina particles in the matrix material.

During the FSSW process, the alumina particles are subjected to the strong stirring action of the tool. This leads to their redistribution and alignment along the weld line. The aligned alumina particles act as strengthening agents, improving the load transfer and enhancing the mechanical properties of the joint.

The addition of alumina reinforcement in the FSSW process has been studied in various materials, including aluminum, magnesium, and steel alloys. The particle size, distribution, and volume fraction of alumina particles play significant roles in determining the reinforcement effect. The presence of alumina particles in the joint can increase its resistance to deformation, improve fatigue strength, and enhance wear resistance. Furthermore, the high melting point of alumina helps to maintain the thermal stability of the joint during the FSSW process. It is important to note that achieving a uniform dispersion of alumina particles in the matrix material can be challenging, and agglomeration of particles may occur. Proper processing techniques and control of parameters are crucial to ensuring a homogeneous distribution of alumina particles. Further research is needed to optimize the alumina reinforcement in the FSSW process, including investigating the effect of particle size, volume fraction, and distribution on the mechanical properties of the joint. Additionally, long-term durability and reliability studies are required to evaluate the performance of alumina-reinforced FSSW joints in real-world applications.

Nano-Al_2O_3 powders are reinforced with Al 2024 for high hardness and better mechanical properties using keyhole-less FSSW. Two types of welds were analyzed. One with alumina powders reinforced, and one with alumina powder used weldments. The influence of the addition of alumina powders on the weldments was studied using joint strength and microstructure. The results show that the hardness at the stir zone increases with the addition of alumina powders. The joint strength of the weld is increased by 23% by adding the nanoalumina powder [17].

13.4.2.2 Titania-Reinforced n-FSSW

Nano titania, or titanium dioxide nanoparticles (TiO_2 NPs), have gained significant attention as reinforcing agents in various engineering applications, including the FSSW process. The unique properties of nano titania, such as its high surface area, enhanced reactivity, and improved mechanical strength, make it a promising choice for enhancing the performance of weld joints. Incorporating nano titania reinforcement in the FSSW process can lead to several advantages. Firstly, the addition of nano titania can improve the mechanical properties of the weld joint, including increased strength, stiffness, and toughness. The high surface area of nano titania particles allows for effective load transfer and bonding at the interface between the nanoparticles and the base material.

To introduce nano titania reinforcement in the FSSW process, the nanoparticles are typically dispersed in the matrix material prior to welding. Various techniques, such as ultrasonication, ball milling, or chemical functionalization, can be employed to achieve a uniform dispersion of nano titania particles.

During the FSSW process, the intense stirring action of the tool facilitates the alignment and distribution of nano titania particles along the weld line. This alignment enhances the load transfer and mechanical performance of the joint. The incorporation of nano titania reinforcement in the FSSW process has been investigated in various materials, including aluminum, magnesium, and steel alloys. The concentration, size, and distribution of the nano titania particles play a crucial role in determining the reinforcement effect. In addition to mechanical enhancement, nano titania reinforcement can offer other benefits. For instance, titanium dioxide nanoparticles possess photocatalytic properties, which can contribute to self-cleaning and anti-corrosion properties in the weld joint. However, challenges exist in achieving a uniform dispersion of nano titania particles and preventing their agglomeration. Proper processing techniques and control of parameters are necessary to achieve a homogeneous distribution and maximize the reinforcement effect.

Further research is needed to optimize the nano titania reinforcement in the FSSW process, including studying the effects of particle size, concentration, and distribution on the mechanical properties and long-term performance of the joint. Additionally, the influence of nano titania on the microstructure and corrosion behavior of the welded joint should be further investigated to ensure reliable and durable welds in practical applications.

13.4.2.3 ZnO-Reinforced n-FSSW

Nano zinc oxide (ZnO) has shown potential as a reinforcing agent in various engineering applications due to its unique properties, such as its high surface area, excellent mechanical strength, and good thermal stability. In the context of the FSSW process, incorporating nano zinc oxide reinforcement can offer several advantages. The addition of nano zinc oxide reinforcement in the FSSW process can enhance the mechanical properties of the weld joint. It can lead to improved strength, stiffness, and toughness of the joint, resulting in enhanced performance under mechanical loading. To incorporate nano zinc oxide reinforcement in the FSSW process, the nanoparticles are typically dispersed in the matrix material prior to welding. Various techniques, including ultrasonication, ball milling, or chemical functionalization, can be employed to achieve a uniform dispersion of nano zinc oxide particles.

During the FSSW process, the intense stirring action of the tool facilitates the alignment and distribution of nano zinc oxide particles along the weld line. This alignment improves the load transfer and mechanical performance of the joint, enhancing its overall strength and durability. The addition of nano zinc oxide reinforcement in the FSSW process has been explored in various materials, such as aluminum, magnesium, and steel alloys. The concentration, size, and distribution of the nano zinc oxide particles play a crucial role in determining the reinforcement effect. In addition to mechanical enhancements, nano zinc oxide reinforcement can offer other functionalities. For example, zinc oxide nanoparticles possess antimicrobial properties, which can contribute to the prevention of microbial growth and corrosion in the welded joint. However, achieving a uniform dispersion of nano zinc oxide particles and preventing their agglomeration can be challenging. Appropriate processing techniques and optimization of parameters are necessary to achieve a homogeneous distribution and maximize the reinforcement effect.

More research is needed to optimize the nano zinc oxide reinforcement in the FSSW process, including studying the effects of particulate size, concentration, and distribution of the nano zinc oxide on the mechanical properties and long-term performance of the joint. Additionally, the influence of nano zinc oxide on the microstructure, corrosion behavior, and biocompatibility of the welded joint should be further investigated to ensure reliable and safe applications in various industries.

13.4.2.4 Other Metal Oxides Reinforced n-FSSW

The strength and other properties of the joints were improved using iron oxide reinforced with aluminum through FSSW. The addition of Fe_3O_4 nanoparticles to the weld zone through FSSW improves the strength and fracture energy. Figure 13.11 shows the schematic illustrations of the FSSW.

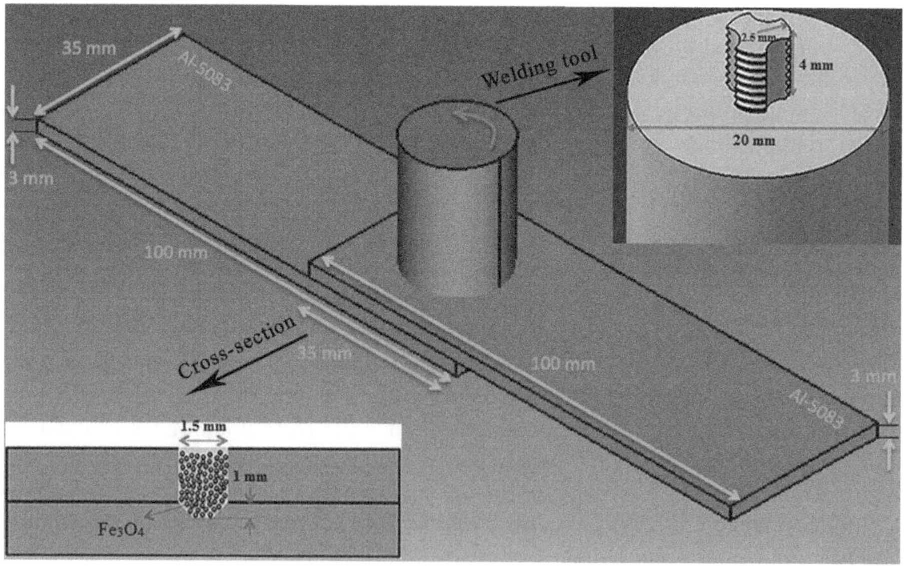

FIGURE 13.11 The schematic illustrations of the FSSW [18].

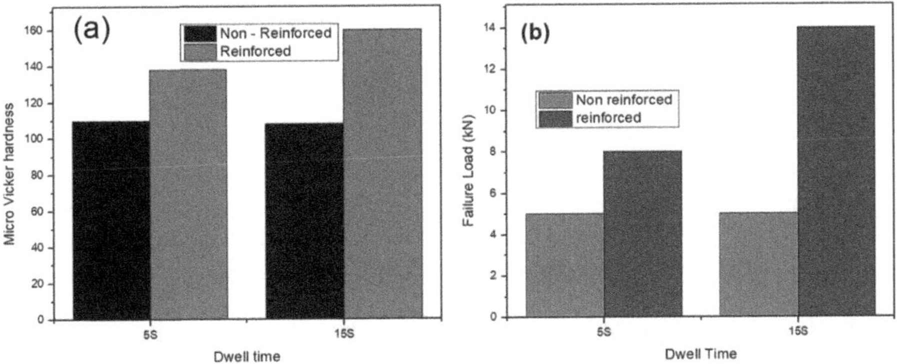

FIGURE 13.12   (a) Hardness variation and (b) fracture energy variation [18].

The strengthening mechanism in this case is well-dispersed Fe_3O_4 nanoparticles along the grain boundaries, and the grain growth is further retarded, which increases the strength. The hardness and tensile strength show considerable improvement. Figure 13.12 shows the (a) hardness variation and (b) fracture energy variation.

13.4.2.5   Metallic Nanoparticles-Reinforced n-FSSW

The reinforcement of FSSW joints with metallic nanoparticles has gained attention due to the enhancement in mechanical and physical properties of the welds. Incorporating metallic nanoparticles as reinforcements in the FSSW process can offer several benefits, depending on the specific material used.

Metallic nanoparticles such as copper (Cu), nickel (Ni), and aluminum (Al) have been studied as potential reinforcements in the FSSW process. These nanoparticles can improve the mechanical strength of the weld joint by increasing its strength, stiffness, and toughness.

To incorporate metallic nanoparticles in the FSSW process, they are typically mixed with the base material before welding. Techniques like ultrasonication, ball milling, or chemical functionalization can be employed to achieve a standardized dispersion of nanoparticles in the matrix material.

During the FSSW process, the nanoparticles experience the strong stirring action of the tool, leading to their distribution and alignment along the weld line. This alignment enhances load transfer and strengthens the joint, improving its overall mechanical performance.

The addition of metallic nanoparticles in the FSSW process has been investigated in various materials, including aluminum, steel, and magnesium alloys. The size, shape, and distribution of the nanoparticles play a substantial role in determining the reinforcement effect.

In addition to mechanical improvements, metallic nanoparticles can also provide other functionalities. For example, copper nanoparticles can enhance electrical conductivity and thermal properties, while nickel nanoparticles can offer improved corrosion resistance.

However, challenges exist in achieving a uniform dispersion of metallic nanoparticles and preventing their agglomeration. Proper processing techniques and control of parameters are necessary to achieve a homogeneous distribution and maximize the reinforcement effect.

Further research is needed to optimize the use of metallic nanoparticles as reinforcements in the FSSW process. This includes studying the effects of nanoparticle concentration, size, and distribution on the mechanical and physical properties of the joint. Long-term durability studies are also necessary to evaluate the reliability and performance of these reinforced welds in practical applications.

13.4.2.6   Hybrid Materials Reinforced n-FSSW

The use of nanofibers as reinforcements in the FSSW process has gained significant interest due to their high aspect ratio, exceptional mechanical properties, and potential for enhancing the performance of welded joints. Nanofibers, such as CNTs and graphene, can offer unique advantages when incorporated into the FSSW process. Nanofiber reinforcement in FSSW can lead to improved mechanical properties of the weld joint, including increased strength, stiffness, and toughness. The high aspect ratio of nanofibers allows for efficient load transfer and enhanced bonding at the interface between the fibers and the base material. To incorporate nanofibers in the FSSW process, they are typically dispersed in the matrix material prior to welding. Various techniques, such as ultrasonication, ball milling, or chemical functionalization, can be employed to achieve a uniform dispersion of nanofibers. During the FSSW process, the strong stirring action of the tool helps to align and distribute the nanofibers along the weld line. This alignment enhances the load transfer, reinforcing the joint and improving its mechanical performance.

Both CNTs and graphene nanofibers have been investigated as reinforcements in FSSW joints. They can provide excellent tensile strength, improve fatigue resistance, and enhance thermal and electrical conductivity. However, achieving a uniform dispersion of nanofibers and preventing their agglomeration can be challenging. Proper processing techniques and control of parameters are crucial to achieving a homogeneous distribution and maximizing the reinforcement effect. Further research is needed to optimize the nanofiber reinforcement in the FSSW process, including studying the effects of fiber concentration, aspect ratio, and distribution on the mechanical properties and long-term performance of the joint. Additionally, the influence of nanofibers on the microstructure, corrosion behavior, and other functional properties of the welded joint should be further investigated to ensure reliable and durable welds in practical applications.

13.5   COMPOSITE JOINTS

Composite joints refer to the joining of dissimilar materials, typically combining a metal matrix with reinforcing materials such as fibers, particles, or laminates. In the context of FSSW, composite joints can be achieved by incorporating reinforcements into the joint region during the welding process. Composite joints enhanced with nanoparticles in the context of FSSW involve incorporating nanoparticles into the joint region during the welding process. The addition of nanoparticles aims to enhance the mechanical, thermal, or functional properties of the welded joint.

Apart from general parameters such as material selection, nanoparticle selection, dispersion, stirring, mixing, and bonding, some specific parameters known as tool rotational speed, traverse speed, and applied pressure need to be optimized for achieving a successful nanoparticle-enhanced composite joint in FSSW. Additionally, further research is needed to investigate the effects of nanoparticle concentration, size, and distribution on the mechanical, thermal, and functional properties of the joint, as well as its long-term durability and reliability. Graphite-reinforced Al spot joints were successfully fabricated by FSSW. Aluminum 5052-H32 alloy sheets were used as the matrix materials during FSSW. Graphite powders are introduced during the stirring process, and graphite/Al MMC is induced in the stir zone. The result of this study suggests that the mechanical properties, especially the toughness, of structural components are enhanced by adopting metal matrix composite (MMC) joints by FSSW [19]. Mostly used aerospace materials, such as carbon fiber-reinforced polyetherimide (PEI) laminate, were joined through FSSW. The joints are strongly bonded in the stirring region and thermo-mechanically affected zone. The strength of the joints is attributed to interdiffusion and fiber interlocking [20].

13.6 CONCLUSION

One of the key advantages of nanoparticle reinforcement in FSSW is the improved strength and hardness of the welded joints. The nanoparticles act as strengthening agents, effectively increasing the load-bearing capacity and resistance to deformation of the weld. This results in joints with superior mechanical properties compared to conventional welding methods.

Here, the different types of nano-reinforcements, process parameters were discussed with a few case studies. In conclusion, nanoparticle-reinforced FSSW holds great promise for advancing welding technologies and enabling the production of high-performance joints. Continued research and development in this field will contribute to further improvements in the mechanical properties, microstructural characteristics, and overall reliability of welded joints, thereby expanding the application potential of this innovative joining technique.

REFERENCES

1. N. E. Thomas, W. M. Needham, J. C. Murch, M. G. Temple-Smith, and P. Dawes. US Patent No. 1995.
2. R. S. Mishra, and Z. Y. Ma, Friction stir welding and processing, *Mater. Sci. Eng. R Rep.* 50(1–2), 1 (2005).
3. R. Nandan, T. Debroy, and H. Bhadeshia, Recent advances in friction-stir welding-process, weldment structure and properties, *Prog. Mater. Sci.* 53(6), 98 (2008).
4. R. S. Mishra, P. S. De, and N. Kumar, Fundamentals of the friction stir process. In: *Friction Stir Welding and Processing*, edited by Rajiv Sharan Mishra, Partha Sarathi De, Nilesh Kumar Springer, New York (2014).
5. M. K. Besharati-Givi, and P. Asadi, *Advances in Friction-Stir Welding and Processing*, Elsevier, Amsterdam (2014).
6. M. Fujimoto, S. Koga, N. Abe, Y. S. Sato, and H. Kokawa, Microstructural analysis of stir zone of Al alloy produced by friction stir spot welding, *Sci. Technol. Weld. Joining.* 13(7), 663 (2008).

7. M. Merzoug, M. Mazari, L. Berrahal, and A. Imad, Parametric studies of the process of friction spot stir welding of aluminium 6060-T5 alloys, *Mater. Des.* 31, 3023 (2010).

8. H. Badarinarayan, F. Hunt, and K. Okamoto, Friction stir spot welding. In: R. S. Mishra and M. W. Mahoney (eds), *Friction Stir Welding and Processing,* ASM International, Materials Park (2007).

9. M. Pouranvari, and S. P. H. Marashi, Critical review of automotive steels spot welding: Process, structure and properties, *Sci. Technol. Welding Joining.* 18(5), 361 (2013).

10. Z. Shena, Y. Dingb, and A. P. Gerlich, Advances in friction stir spot welding, *Critical Rev. Solid State Mater. Sci.* 45, 457–534 (2020)

11. Mubiayi MP, Akinlabi ET. Friction stir spot welding of dissimilar materials: an overview. InProceedings of the world congress on engineering and computer science 2014 (Vol. 2) WCECS 2014, 22-24 October, 2014, San Francisco, USA.

12. https://www.substech.com/dokuwiki/doku.php

13. O. Kalaf, T. Nasir, M. Asmael, B. Safaei, Q. Zeeshan, A. Motallebzadeh, and G. Hussain, Friction stir spot welding of AA5052 with additional carbon fiber-reinforced polymer composite interlayer, *Nanotechnol. Rev.* 10 (1), pp. 201–209, (2021).

14. P. Fanelli, F. Vivio, and V. Vullo, Experimental and numerical characterization of friction stir spot welded joints, *Eng. Fracture Mech.* 81, 17 (2012)

15. A. Sharma, V. Mani Sharma, B. Sahoo, S. Kanta Pal, and J. Paul, Effect of multiple micro channel reinforcement filling strategy on Al6061-graphene nanocomposite fabricated through friction stir processing, *J. Manuf. Processes* 37, 53–70 (2019)

16. S. Wang, X. Wei, J. Xu, J. Hong, X. Song, C. Yu, J. Chen, X. Chen, and H. Lu, Strengthening and toughening mechanisms in refilled friction stir spot welding of AA2014 aluminum alloy reinforced by graphene nanosheets, *Mater. Des.* 186, 108212. (2020).

17. M. Enami, M. Farahani, and M. Farhang, Novel study on keyhole less friction stir spot welding of Al 2024 reinforced with alumina nanopowder, *Int. J. Adv. Manuf. Technol.* 101, 3093–3106 (2019)

18. E. Fereiduni, M. Movahedi, and A. Baghdadchi, Ultrahigh-strength friction stir spot welds of aluminium alloy obtained by Fe_3O_4 nanoparticles, *Sci. Technol. Weld. Join.* 23, 63–70 (2018)

19. C.-S. Jeon, Y.-H. Jeong, S.-T. Hong, Md. Tariqul Hasan, H. N. Tien, S.-H. Hur, and Y.-J. Kwon, Mechanical properties of graphite/aluminum metal matrix composite joints by friction stir spot welding, *J. Mech. Sci. Technol.* 28(2) (2014) 499–504

20. Y. Huang, X. Meng, Y. Xie, Z. Lv, L. Wan, J. Cao, and J. Feng, Friction spot welding of carbon fiber-reinforced polyetherimide laminate, *Compos. Struct.* 189, 627–634 (2018)

<h1>14 Challenges and Application of Friction Stir Spot Welding of Dissimilar Al Joints</h1>

K. Anton Savio Lewise, J. Edwin Raja Dhas, and M. Satheesh

14.1 INTRODUCTION

The Welding Institute (TWI) developed dissimilar friction stir spot welding (DFSSW) in 1991 as a method for joining various base metals, such as aluminum, copper, steel, titanium, magnesium, and other metals [1]. There is no melting since it is based on solid state welding. In order to soften the materials and stir them together utilizing both tool rotational and tool traverse motions, DFSSW is based on the frictional heat produced by a simple tool. Due to the presence of solidification faults when combining them by fusion welding procedures such as porosity coupled with thick intermetallic complexes, it is initially primarily employed for uniting aluminum base metals [2]. In the last ten years, DFSSW has been regarded as an effective way to combine different types of materials. Friction stir welding is widely used for dissimilar connections due to its numerous benefits over other welding techniques, such as its cheap cost and user-friendly operating process [3]. On the other hand, DFSSW has challenges like unfavorable FSSW tool pin performance and a brief life span in the mass production stage, the creation of crack initiation sites such as hook defects, material flow voids, temperature, different materials, and probe holes, as well as limitations on the thickness of materials that can be welded. Due to these difficulties, friction stir spot welding of aluminum alloys has undergone significant revolutionary tool profile improvements.

The necessary supplies and tools for DFSSW include a milling machine, base materials, backing plates (fixtures), and welding tools. Other welding techniques, such as shielded metal arc welding (SMAW), on the other hand, often require a highly skilled operator in addition to expensive equipment. The DFSSW process [4] is illustrated in Figure 14.1.

Researchers have recently experimented with a variety of techniques for combining different aluminum alloys, including friction stir welding, laser welding, electron beam welding, diffusion welding, arc welding, etc. Figure 14.2 displays the total number of research publications on FSSW of different Al alloy joints that have been

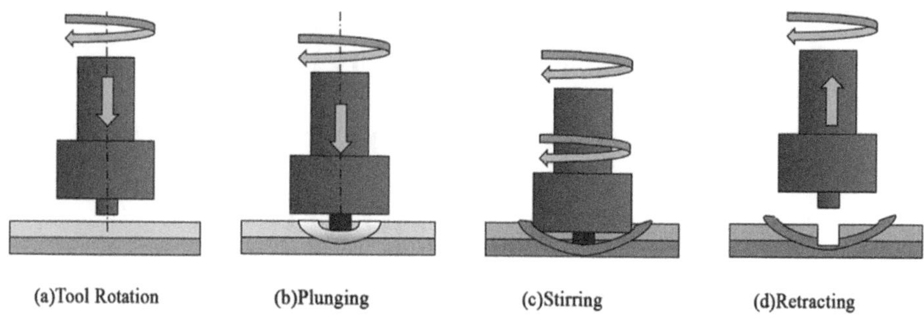

FIGURE 14.1   DFSSW process.

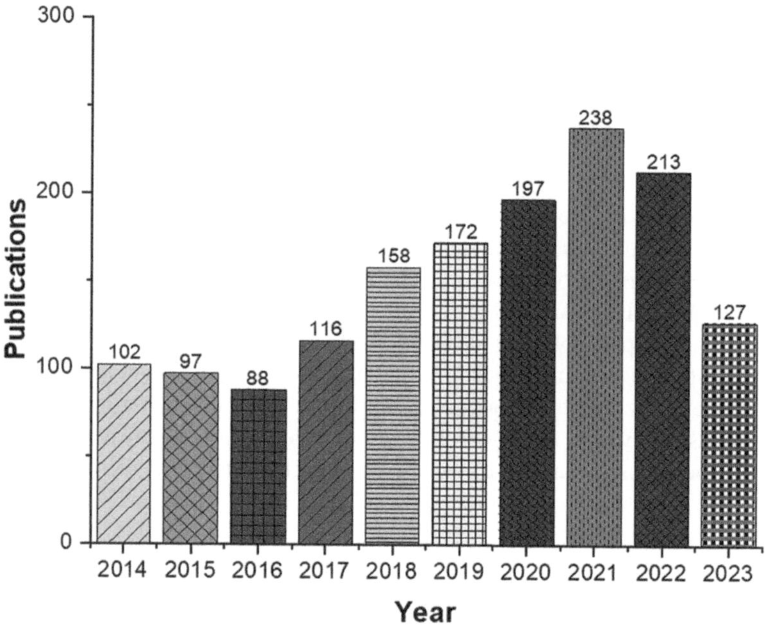

FIGURE 14.2   No. of publications related to dissimilar FSSW process.

published to date (a search from the last 10 years turned up 1295 papers from Science Direct). The great majority of publications were made within the last five years, peaking in 2021.

In addition, Yamamoto et al. [5] examined both the liquation and liquid penetration induced cracking during FSSW of Mg alloy. Local melting and cracking were also observed during the Al7075-T6 FSSW process. Formation of eutectic films was also observed in the high temperature stir zone after the cooling process because of the limited dissolution of melted metals.

Mustafa Kemal Bilici [6] investigated the process parameters of the FSSW of polypropylene using the Taguchi statistical design of the experiment technique.

In order to minimize the problems related to wear, flush, overheating, and melting, the Taguchi approach was used. Finally, 47% accuracy has been obtained when compared to initial and optimal weld parameters.

Miles et al. [7] analyzed the FSSW process in DP 980 steel material with both experimental (welding robot) and numerical methods using Forge software by applying different boundary conditions, such as a rotating tool with a fixed backing plate. It has been noticed that the welding robot can sustain only a 10 kN vertical spindle load, which is 30%–50% less than the required vertical spindle load produced for a perfect weld.

Piccini and Svoboda [8] investigated the dissimilar FSSW of AA6063 with low carbon steel with different welding process parameters. Macro- and micro-welded material characterization is also done along, with tension and hardness tests. And it has been observed that the maximum load-carrying capacity increased when the tool penetration depth increased; there are few challenges related to the melting and solidification conditions of dissimilar materials.

Mubiayi et al. [9] reviewed the dissimilar FSSW of aluminum and copper joints with respect to the various material testing characteristics such as microstructure evolution and mechanical and electrical characteristics. It has been observed that different melting temperatures influence the quality of dissimilar FSSW joints, and many researchers have successfully made dissimilar joints with minimal drawbacks.

Azarniya et al. [10] investigated the effects on microstructural properties of Al-Zn-Mg-Cu alloys and also addressed the current advances and forthcoming visions of these alloys. Most Al alloys have superior properties to other metallic alloys, which is appropriate for the production of physical and engineering components like automotive parts such as wheels, pistons, engine blocks, and steering boxes.

Kang and Kim [11] investigated the mechanical and microstructure properties of Al 7xxx series of alloys and also stated the importance of Al alloys in automotive industries. Mostly the Al alloys are used for reduction of weight in car bodies. Especially 5XXX series of Al-Mg alloys and 6XXX series of Al-Mg-Si alloys are characteristically used for their strength and good formability in the automotive industry. Just in 7XXX series, Al alloys are applied to automobile components.

Zhou et al. [12] reported the recent advances in the usage of aluminum 2xxx and 7xxx series alloys in aerospace applications. They argued that the advent of composite materials has given rise to greater demand for developing alternative materials with better performance and resistance against corrosion, fretting, stress cracking, and fatigue. They highlighted the characteristics of aluminum 2xxx series and 7xxx series as being on par with expectations within the aeronautical industry by offering exceptional mechanical and anti-corrosion properties and better lifecycle costs.

Selamat et al. [13] reported the application of FSWed AA5083-AA5083 Al alloys in automotive industries. The alloy AA6061 consists of Al-Mg-Si, which is low in density and good in mechanical properties, whereas AA5083 consists of Al-Mg, which is high in strength, decent ductility, and superior corrosion-resistant property. It was concluded that FSS-welded AA5083-AA5083 Al alloys were well suitable for external panels in various automobiles.

Osterreicher et al. [14] reported and compared the strength of two 7xxx series Al alloys by using a heat treatment process. The alloys AA7021 and AA7075 were used as the base metal and heated via differential scanning calorimetry. Both alloys have a lesser weight and massive strength properties. Thus, the results obtained from various experiments concluded that AA7075 has the maximum yield strength compared to AA7021 alloy. So, the AA7075 conceivably prompts an expanded utilization of 7xxx alloys in automotive industries.

Rambabu et al. [15] reported the applications of 2xxx and 7xxx Al alloys in the aircraft and spacecraft sectors. These two alloys offer more in the aerospace sector due to their immense weight-saving characteristics, as well as their strength and stiffness. The good corrosion property leads to its application in lower and upper fuselage panels, bulkheads, skins and stringers, lower and upper wings, etc. In spacecraft sectors, the 2xxx and 7xxx Al alloys are the key contenders for liquid propellant launchers in spacecraft.

Boldsaikhan et al. [16] reported the application of Al alloys in the aerospace sector. Previously, Al alloys were used in aerospace using conventional fusion welding techniques. The advantage of FSW is that it does not add any kind of filler material while joining the metals/alloys, and it causes weight reduction. In FSW, there is no material weakening and a lack of fusion produced by solidification and liquefaction. It was concluded that weight reduction is the major key feature for the use of Al alloy in aerospace assembly.

Pujari and Patil [17] investigated the strength, hardness, and physical properties of 7xxx series Al alloys and their uses in the aerospace industry. The 2xxx, 6xxx, and 7xxx alloys are heat-treated alloys. Especially in 7xxx series, Al alloys consist of 4%–8% Zn and 1%–3% Mg. These Zn and Mg have high solubility in the Al element. The Al 7075-T6 alloys are mostly used in aerospace because of their high strength.

Rao et al. [18] investigated the corrosion behavior and application of 7xxx Al alloys. There are eight series of Al alloys, of which 2, 5, and 7 series alloys are vulnerable to stress corrosion and cracking. Comparing with 2 and 5 series Al alloys, 7 series Al alloys have detailed physical and airplane applications because of their superior mechanical characteristics. The 7xxx series Al alloys were utilized in the air force, army, and construction applications, due to their greater rigidity, strength, durability, and superior fatigue characteristics.

Jawalkar et al. [19] stated that the use of Al alloys in aircraft components. In aerospace applications, materials with high strength-to-weight proportions and properties such as brilliant corrosion resistance, light weight, creep resistance, and high thermal strength are required. It is concluded that Al alloys are good for fuselages; wings also support the construction of aircraft and military freight. The manufacture cost, characteristics of performance, and notable fabrication procedures are a couple of explanations behind the utilization of Al alloys in huge amounts for future airplanes.

Wang et al. [20] reported their study on friction stir welding of aerospace-grade aluminum alloys for rocket tank fabrication. The authors reported that aluminum alloys are highly suitable for fabricating high strength structures due to their superior advantages, such as low defects, fewer distortions, and the high mechanical performance capability of the welded aluminum joints. They systematically analyzed the tool design, process parameter optimization, and testing techniques for the tank fabrication prior to the actual tank fabrication process to obtain performance specifications. The authors stated that the use of aluminum 7xxx series alloys with friction stir welding techniques greatly augmented the aerospace equipment production capability of the industry.

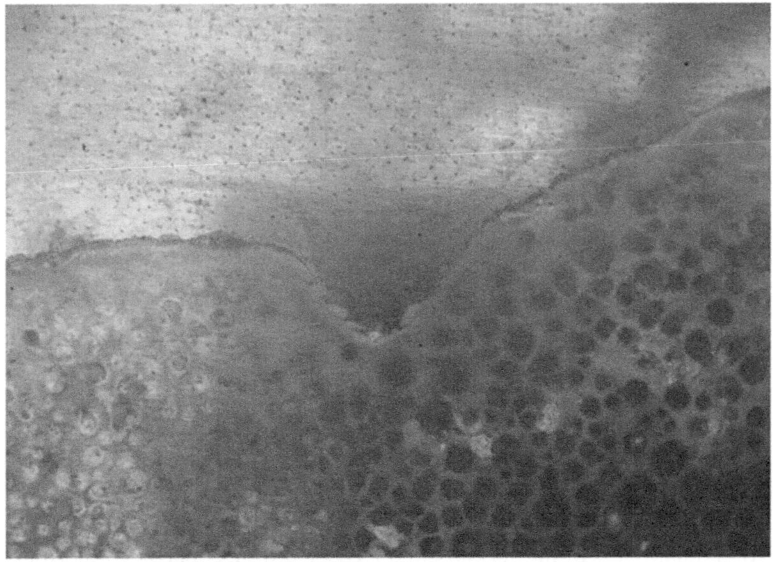

FIGURE 14.3 Corrosion on Al surface.

Corrosion behavior of Al alloys is studied by subjecting them to aqueous solutions with corresponding compositions of electrolytes. They are typically investigated with varying concentrations of acid medium and sodium hydroxide medium at different temperature levels. Al alloys are vulnerable to corrosion when directly exposed to string acidic or alkaline environments with a pH on either side of 4–9. Metal pitting results on Al alloys resulted from violent corrosion, with alkaline environments being capable of breaking down Al alloys much faster compared to acidic surroundings, as clearly shown in Figure 14.3. Al alloys are innately corrosion-resistant but that quality diminishes with increasing alloying elements. Pure Al of the 1xxx series has the best corrosion resistance among the Al alloy series, which gradually diminishes with each passing composition. But the corrosion resistance of these alloys can be greatly enhanced by certain joining techniques and suitable companion alloys (dissimilar joints). Types of Al alloy corrosion are galvanic corrosion (this happens due to the breaking down of metals; for example, Al contacting Cu or Zn) and pitting corrosion (local surface damage that is typically bound to aesthetics; for example, corrosion due to the exposure of salts).

This chapter highlights the challenges and applications of dissimilar FSSW joints with respect to various aspects. Furthermore, significant problems were also stated.

14.2 CHALLENGES IN DISSIMILAR FSSW

The chemistry of the materials, as well as deviations in their melting temperatures, thermal conductivity, coefficient of thermal expansion, and other characteristics, provide a significant challenge to achieving sound welds between different metals. When filler material is utilized during fusion welding operations, the chemistry of

TABLE 14.1

Melting Points of Different Materials

SL. No.	Metal	Melting Temperature
1.	Aluminum alloy	660.3°C
2.	Magnesium	650°C
3.	Carbon steel	1425°C
4.	Stainless steel	1375°C
5.	Copper	1083°C
6.	Ferrous metal	1244°C

the final weld is considered. Table 14.1 shows the melting points of different materials with which it is possible to perform FSSW with aluminum. Few of the basic challenges are listed as follows:

- High disparity within thermal conductivity: Large capacity of thermal conductivity enables the heat flow from the weld arc into the material. This situation leads to a lack of material fusion within the weld zone due to the excess melting of lower thermal conductivity material.
- The co-efficient of thermal conductivity: The broad variation between the co-efficient of thermal conductivity of both weld materials could cause buildup of internal stress along the intermetallic zone at temperature changes during the welding procedure.
- Difference in melting temperature between the two constituent alloys: The formation of porosity defects arises due to the presence of trapped gases forming bubbles that are not dissolved within the molten weld metal as it solidifies post-welding.
- Porosity defects: They are typically larger in magnitude in metal inert gas (MIG) welding when compared to tungsten inert gas (TIG) welding processes. This is the result of the high surface area and volume of filler metal in MIG welding. These defects can be reduced by optimal input factors such as input current, weld arc length, and weld speed. Typically, higher welding speeds lead to greater porosity defects.
- Formation of hot cracks: Hot cracks are surface defects formed by high temperatures and are dependent on solidification of alloys. The probability of hot cracks is influenced by the vulnerability of the base alloy. Hot cracks can be eliminated by suitable selection of filler material and control of input heat. The sensitivity of Al alloys for varying compositions is illustrated in Figure 14.4.
- Cracks due to stress corrosion: These types of cracks are formed due to the concentrated effects of tensile stress buildup in a corrosive operating environment. Vulnerable material microstructure [21], corrosive operating environment, and subjected tensile stresses are causal conditions of stress corrosion cracking. The alloys that possess relatively low Mg content,

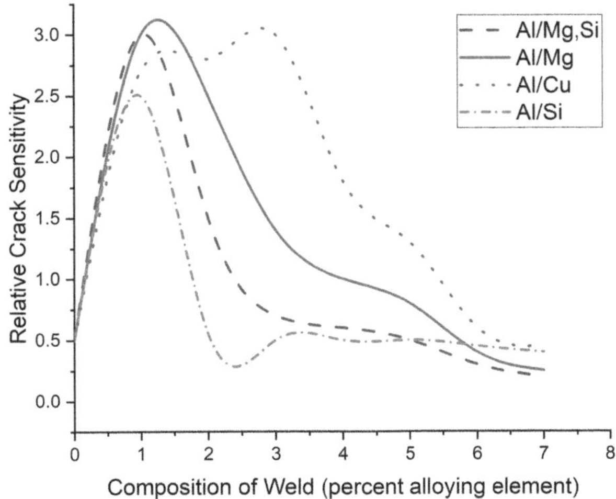

FIGURE 14.4 Al alloy content vs. crack sensitivity.

such as 5052 and 5454, have low chances of exhibiting these defects at high temperatures. Alloys with Mg content greater than 3% are more vulnerable to these types of defects at high temperatures. These defects can be countered by suitable weld joint design and welding parameters.

14.3 FORMATION OF INTERMETALLIC COMPOUNDS

The majority of the FSSW materials with varying base metals, such as Al, steel, Mg, Fe, and Cu, have shown the formation of intermetallic compounds (IMCs). Due to increased diffusion during welding, these IMCs develop and expand quite quickly. IMC production has been considered to be generally unavoidable, while the thickness of IMC may sometimes be decreased by lowering the frictional heat input or by utilizing a third medium to rapidly eliminate the heat. Furthermore, it has been observed that reducing the thickness of the IMC may greatly increase the quality of the weld; selecting optimized process parameters can also be reduced the thickness of the IMC.

14.4 FSSW TOOL WEAR

The FSSW tool/tool pin plays a major role in the whole welding process. The continuous contact between the surface of base metals and the tool profile experiences unavoidable wear. But the wear rate will vary with respect to the hardness of the dissimilar materials that can be used in the FSSW process [9]. When welding between dissimilar materials, tool wear is crucial because at least one of the materials must melt at an elevated temperature. By removing the thick oxide layer that exists on the welded zone, preventing IMC thickening, and forming either a metallurgical bonding or a mechanical interlock, active material mixing between dissimilar materials through pin stirring and scribing is generally accepted as being necessary for considerable mechanical performance.

FIGURE 14.5   Tool shapes: (a) before and (b) after welds with respect to various tool rotational speeds.

This is true even though it has been demonstrated that sound dissimilar welds can be achieved without inserting the welding tool into hard materials; therefore, whenever going for the DFSSW process, it is necessary to study the placement of high melting point material at the top or bottom for spot welding operations. Figure 14.5 displays the FSSW tool shapes before and after the welding process. It has been observed that severe wear has taken place between the pin center and edge.

14.5   INSUFFICIENT MATERIAL MIXING

As previously noted, from the perspective of mechanical performance, correct material mixing between different materials is necessary. Material mixing may be process- and workpiece-dependent due to the differing physical qualities of the materials and the asymmetric heat input between the top and bottom sides of the weld. And it is much more difficult to perform spot weld between softer and harder materials. Additionally, in the FSSW of dissimilar joints, it is sometimes possible to witness a few modes of failure, one via the stir zone and the other through the heat-affected zone [22]. The poor mechanical strength and failure through the stir zone were attributed to insufficient material intermixing caused by slow tool rotational speeds. These kinds of challenges can be minimized by using a few external heat sources and an adequate amount of axial load.

14.6   OPTIMIZED PROCESS PARAMETERS

Factorial designs are methods of experiments that enable the exploration of additional process parameters/factors in an experimental design. These are integral to the results of the DOE, but they are not inclusive of previously assumed parameters.

They are used to manipulate multiple factors in a design system to estimate at varying levels and result in conclusions valid over the entire DOE. By using factorial design, it is possible to simultaneously analyze the effects of input variables. They can also be used to reveal the interaction between the input parameters/variables and these parameters are termed as 'factors' during the course of factorial design.

Response surface methodology (RSM) [23] is a collection of statistical and numerical techniques that can be used to explore the relationship among design and response variables. It is an optimization technique that utilizes second-order polynomial models to obtain optimal responses for a system. The features of RSM enable it to implement optimizations, such as orthogonality, rotatability, and uniformity. RSM can be used to deal with multiple responses to a problem. The challenging nature of these variables is that the optimal value for one response might not be applicable anymore for another response, and vice versa. Thus, multi-objective extensions are used to reduce the variations for a single response with a specific target value with a simultaneous effort to prevent variation in that response from dissonance.

The output of any process is determined by the influence of the said process input parameters. The means of controlling input parameters is the primary concern in any process. Diverse optimization techniques have been developed and have been utilized to obtain optimal performance for a dissimilar FSSW process. The desired output variables are obtained by definitive control of the input variables by specifically developed mathematical models and Design of Experiment (DOE) techniques. In addition, various optimization techniques are widely used in the FSSW process to achieve a sound weld. Few of the familiar techniques are as follows:

- Box-Behnken design
- Central composite design
- Hybrid heuristic methods

So it is necessary to choose the best optimization technique to have a sound DFSSW.

14.7 EFFECT OF CORROSION IN FSSW JOINTS

The effects of different types of corrosion, such as galvanic corrosion and stress corrosion cracking, on structural integrity and service life have previously been observed. Despite the fact that the bulk of structures or metallic components exposed to the air are shielded against corrosion, such as coating or painting, they create additional challenges that bring dissimilar material welding into post-industrial applications. Understanding how the FSSW process influences the galvanic corrosion of the dissimilar weld structure and the corrosion properties of different materials will be crucial because, after the protective coating has worn off, it may speed up the deterioration process. FSSW uses less residual stress than traditional fusion welding processes to create a weld. When two materials have melting points and thermal conductivities that are greatly different from one another, it may be more difficult to reduce residual stress in a dissimilar weld. Additionally, the stress corrosion cracking behavior must be assessed, especially when the dissimilar welds are subjected to thermo-mechanical fatigue.

14.8 APPLICATIONS OF DISSIMILAR AL ALLOY JOINTS

Al alloys offer unique properties and performance advantages owing to their distinctive compositions based on their grades. This feature makes Al alloys versatile and economical in nature for a broad range of applications. The general applications range from high-ductility foil wrap to corrosion-resisting high-strength machine parts. They enjoy high fervor in the engineering sector, second only to steel. Al offers an excellent strength-to-weight ratio, with any service component weighing nearly three times less than a comparable steel component. This overwhelming advantage in weight and offering nearly the same performance makes Al alloys particularly useful in dynamic applications. Many components of aircraft, such as fuselage, rudders, and wing stabilizers, are highly dependent on Al alloys, as are high-performance automotive racing components such as engine blocks, crankcases, and gearbox casings for racing vehicles. Al alloys offer unprecedented performance against grave factors such as corrosion and wear.

Al alloys are typically composed of various elemental metals such as Mg, Zn, zirconium, Cu, etc. Al alloys are graded in various series from 2xxx to 7xxx, based on their composition and major alloying elements. The present study involves the investigation of Al alloys 2024-T3 (Al-Cu) and 7075-T6 (Al-Zn-Mg) and a dissimilar joint of these two alloys made by FSSW and investigated for the mechanical and metallurgical characteristics of the resulting joint.

Al alloy dissimilar joints are mainly valued for their combination of excellent resistance to corrosion, improved mechanical properties, and enhanced toughness. Dissimilar Al alloys offer the best combination of properties of either base alloy, with superior weight savings and reliable performance metrics. Al alloy 2xxx series is generally desired for its high microhardness and strength capacity. Their scope of applications is comparatively narrow compared to other series of Al alloys, but they are exceptionally suitable for mechanical parts, rivets, and accessories. Their capacity can be greatly improved in conjunction with other Al alloys.

2xxx series are heat-treatable alloys, thus being capable of being alloyed with 6xxx and 7xxx series to offer enhanced performance properties. Al alloy 7xxx series possesses better aptitude for precipitation hardening compared to the 2xxx series, which provides better scope for enhanced tensile strength and metallurgical properties. 7xxx series are much stronger than 6xxx series, are inherently harder, possess better corrosion resistance, and are capable of anodizing. The studies regarding dissimilar Al alloy joints and their characterizations are listed below.

Abd El-Hafez et al. [24] investigated the friction stir welded (FSWed) AA2024-T365 and AA5083-H111 alloys and discussed the applications of Al alloys. For the most part, FSWed Al combinations are utilized in marine, car, aviation, and many other applications due to their high-strength, fracture-resistant, and fatigue welds. This paper reported that the FSW of dissimilar AA5083-H111 and AA2024-T351 alloys is suitable for use in the marine and aerospace sectors. Raval and Judal [25] reported the advances in FSWed Al to Mg alloys and their applications. Current aerospace/automobile models mandate reductions in weight and production cost materials. The Al 6xxx alloys contain Mg and silicon, which are heat-treatable alloys, and the Al 5xxx alloys contain manganese and Mg, which are non-heat-treatable alloys. These alloys have high strength, high ductility, and superior corrosion resistance, which are mostly applied for auto, aero, marine, and military applications.

14.9 DISSIMILAR AL ALLOY JOINTS IN AUTOMOBILE APPLICATIONS

Al alloys are capable of being produced into parts of any shape and size, which allows them to be used for automotive body components that are lighter than steel and have comparable strength performance. They are relatively simple to process, with a greater focus on recycling. They are resilient against corrosion and relatively economical to manufacture and process. The automotive industry utilizes Al in variant forms such as alloys, dissimilar alloy joints, and granular composites for a range of purposes, including weight reduction, enhancing efficiency of components, reducing oil consumption, and enhancing the service life of the component/automobile.

Al alloy 2xxx to 7xxx series both possess superior physical and morphological properties and are highly desired in high-performance industries in making suspension setups, gearbox forks, power drive shafts, steering column systems, engine valves, etc. The applications range from the fabrication of engine valves and wheel rims to body parts and chassis construction.

14.10 DISSIMILAR AL ALLOY JOINTS IN AEROSPACE APPLICATIONS

Al alloys are the most commonly used metals in the aviation industry. They are employed anywhere between the fuselage, wing skins, cockpit control, etc. They are processed in myriad ways, such as extrusion, bending, and welding, due to their excellent machinability. The widely used Al alloys in the aerospace and aircraft industries are the 2xxx series (Al-Cu), 7xxx series (Al-Zn), and Al-lithium combination alloys. They primarily find their usage in applications requiring heavy damage tolerance and toughness against fracture. Al 2024-T3 is a mainstay alloy found in aircraft fuselage and main frame structures owing to its outstanding tolerance to damage. Advanced manufacturing methods have led the Al alloy 7xxx series to be widely employed in the aerospace industry, where the most needed property is high strength. They are particularly employed in the fabrication of wing skins, directional stabilizers, and wing stringers. Recent advancements in the 7xxx series have made them achieve higher yield strength and thus are used in thick-size configurations in making aircraft and ship bulkheads. They also possess enhanced fatigue strength and are increasingly used as energy-absorbing structural components.

Al alloys are dominantly used in aircraft, making up to 80% of the total material usage. They innately possess high ductility, resistance to corrosion, a relatively low cost, and are suitable for various manufacturing techniques. These properties cement the ubiquitous nature of Al alloys in the fields of aircraft and aerospace engineering. FSSW offers less initial preparation, ease of assembly, reduced operating steps, and a faster joining speed. Friction stir spot welded aluminum alloy joints are used in repairing rivets in the aircraft wing structure to improve the service life of the specific structure (Figure 14.6a). FSSW is also used in the fabrication of connecting strips that bind rocket booster cylinders in place of traditional rivets (Figure 14.6b). FSSW is typically used by Kawasaki Aerospace Corporation in their helicopters, where the cockpit doors involve joints by this process as a means to save precious weight (Figure 14.6c).

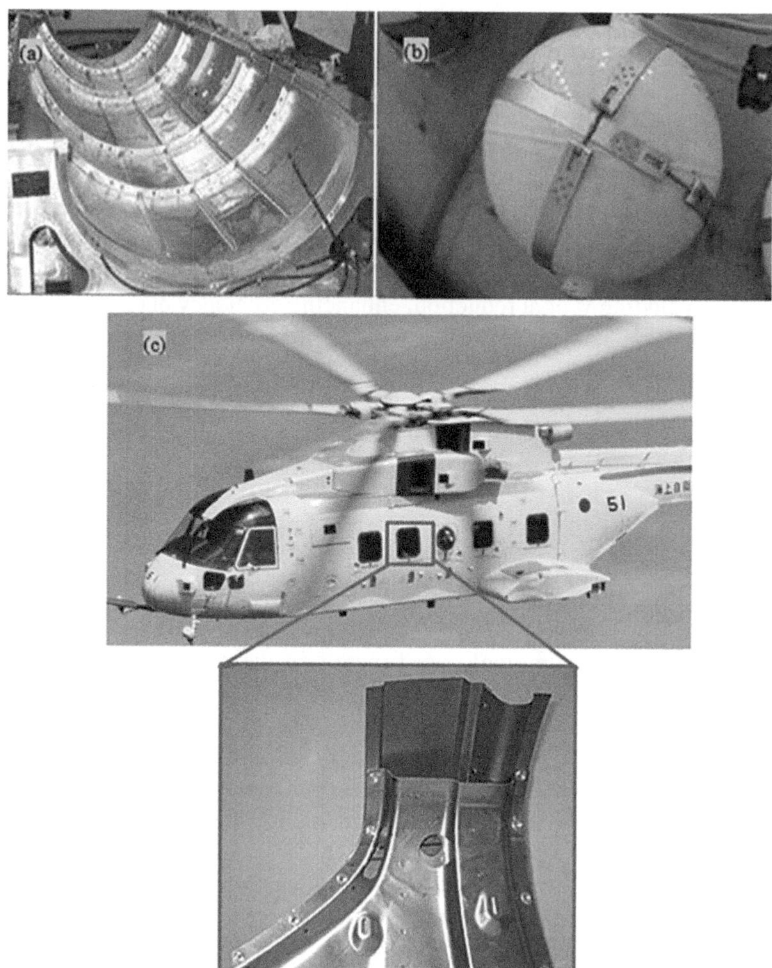

FIGURE 14.6 Applications of dissimilar Al alloys: (a) aircraft wing, (b) rocket booster cylinder, and (c) helicopter cockpit doors.

Commercial jetliners are also increasingly benefitting from FSSW as a means of replacing rivets, as rivets are dangerous in case of failure and their rate of failure is relatively higher than that of FSSW joints. Figure 14.7 shows that structural components such as fuselage, wing structures, and stringers made out of Al typically exhibit high stiffness and strength. Al 7xxx series alloys, such as 7075-T6, are widely used in high-strength aerospace applications in places such as the upper wing skin, which experiences high overdraft, and as directional stabilizers. They have been actively used since World War-II due to their high specific strength and economical availability. However, they are vulnerable to corrosion and typically have a reduced service life. They are now being replaced by advanced 7475 series that possess high yield strength, better resistance to corrosion, and improved fracture toughness.

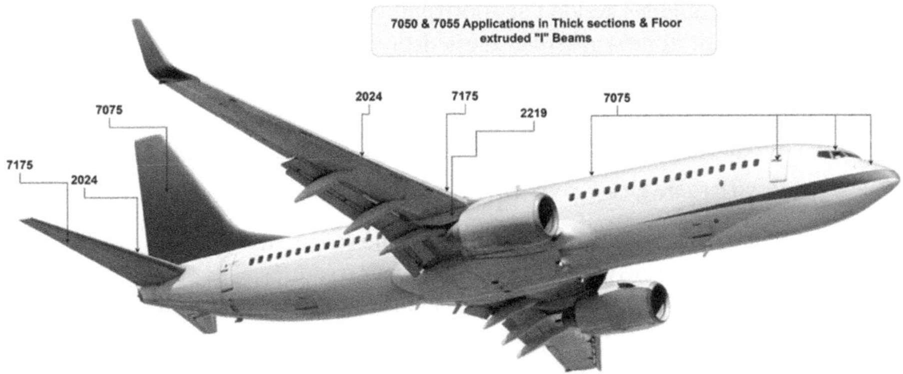

FIGURE 14.7   Applications of Al alloys in various aircraft components.

FSSW presents potential and gives industry optimism that welding of previously non-weldable dissimilar materials is achievable, even if there are still numerous challenges to overcome before it can be completely adopted for welding dissimilar materials in larger industrial areas.

14.11   CASE STUDIES

14.11.1   Case Study (i): Influence of Tool Rotational Speed on the Mechanical and Electrical Properties of the FSSWed AA Sheets [26]

In this study, the effect of rotation speed during the friction stir spot welding of AA 5754-H111 was explored in order to assess the mechanical and electrical features of the welds. Providing high-quality weld connections with less electrical resistance, a crucial need of the electrical sector, is one of this technology's toughest hurdles. The experimental guidelines followed in this investigation were consistent with those in the prior study. At an axial stress of 4 kN, a penetration depth of 0.25 mm, and two typical rotating speeds of 1000 and 4500 rpm, two samples were welded together. Ohm's law was used to determine the precise electric resistance in the electric circuit that was used to measure the electric resistance in the base material and samples welded at 1000 and 4500 rpm.

It was observed that the samples welded at 1000 and 4500 rpm both had greater specific electrical resistance. But the base substance had the lowest specific resistance. Two microstructural findings may be connected with this impact. Strain hardening, which is predominant in the samples welded at 1000 rpm, results in the buildup of dislocations that occur.

It was discovered that the rotating speed has a considerable impact on how electrically resistant the weld joints are. Although it was found that the welded joints' electrical resistance was always higher than the base material's, it does so relative to the base material at low rotation speeds and at higher rotation speeds, rising by roughly 42% and 14%, respectively. Strain hardening caused a significant rise in microhardness in samples welded at 1000 rpm, but due to a typical thermal softening process, the stir zone's microhardness decreased to below the base material in samples welded at 4500 rpm.

14.11.2 Case Study (ii): Evaluation of Refill FSSW as Crack Arrest Features in Co-Consolidated Thermoplastic Laminates [27]

Co-consolidated thermoplastic composite specimens were used for the experimental investigation of Refill FSSW's capacity to prevent fractures. The samples were created using a low-melt polyaryletherketone matrix reinforced by T700 carbon fibers and a crack lap shear (CLS) arrangement. The experimental series with the features was then analyzed after specimens without the crack arrest characteristics had been assessed as a baseline. The application of quasi-static and fatigue loading conditions was done, while C-scanning and optical measurements were used to track the spread of interlaminar damage. Research work was conducted to assess RFSSW's ability to stop cracks under quasi-static and fatigue loading conditions. The selection of CLS specimens allowed for the ultrasonic monitoring of the development of cracks at predetermined loading intervals.

Overall, the approach effectively recorded the behavior of crack propagation along the joint interface, allowing for the contrast between specimens bearing RFSSW features and the reference cracks. The last laminates (500×500 mm) used a 10-ply [0/45/90/45/-45]s quasi-isotropic lay-up. At TORAY, consolidation in a press was used to unite the laminated plates. A 90 mm wide by 25 mm thick Upilex film was sandwiched between the two stacks of ten-ply material to form a pre-crack. The resultant consolidated laminates were then cut into individual specimens using a water-cooled diamond saw set up in a CLS configuration.

This work's major objective was the experimental assessment of the unique RFSSW technology's performance as a crack arrestment feature in thermoplastic laminates. The major objective was to specify how the feature would behave in a mixed-mode CLS specimen that was being loaded under quasi-static and fatigue circumstances. The findings suggested that the joint's capacity to support loads as well as the crack retardation provided by the crack arrest feature (CAF) under quasi-static situations were both constrained. For the CAF-containing specimens, a mean rise of 0.31 kN for the crack entry load was recorded, with no signs of subsequent damage deceleration. The co-consolidated joint was subjected to ever-increasing stress, which is what led to the feature's poor performance. However, taking into account how the joint behaved throughout the fatigue testing, the CAF was able to significantly reduce the interfacial failure.

The CAF considerably reduced the damage time, as seen by the ultrasonic and optical crack monitoring between reference and RFSSW-containing specimens. In the case of the reference specimens, the fracture spread linearly over the 100,000-cycle testing to an average length of 50 mm. Additionally, it was discovered that the feature's influence on the interfacial surface area was greater than its actual diameter, perhaps as a result of a process-induced heat-affected zone. This finding increased the feature's crack-arresting capacity because the deformed co-consolidated interface was transverse to the crack's propagation.

14.12 CONCLUDING REMARKS AND FUTURE RESEARCH DIRECTIONS

This chapter introduces Al alloys and their variants; specific applications; and the need for dissimilar Al alloy joints and their subsequent applications, which are discussed widely with pictures and specifications. The chapter also discussed the Al alloy welding processes, the different challenges involved, FSSW and its mechanism, and the advantages and demerits of the process. The techniques for optimization of the FSSW process and characterization of the FSSWed joints are discussed. This chapter gives a clear understanding of the difficulties present in the dissimilar FSSW joints while also providing ample solutions to minimize the challenges.

This work could be further extended to the FSSW of other dissimilar Al alloys with varying thicknesses of material and multiple process parameters, tool materials, addition of external heat source, etc. Also, this chapter can be expanded with different friction welding techniques such as refill FSSW, friction stir clinching, modified FSSW, pre-hole FSSW, and FSSW processes without tool pins, etc. The FSSW process can be carried out in multiple mediums, like water, or with the addition of other non-flammable fluids. Furthermore, an interlayer can be added to joints during the FSSW process, and the variation in challenges can be observed. Random techniques such as gray relational analysis, ANN analysis, and fuzzy logic approaches could be attempted to optimize process parameters. The upgraded system can be made into an integrated weld-optimized process parameter system by interfacing the FSSW instrument with a PC to enhance automation.

REFERENCES

1. W.M. Thomas, Friction stir butt welding, Int. Pat. Appl. PCT/GB92/02203 (1991).
2. G. Sierra, P. Peyre, F. Deschaux Beaume, D. Stuart, G. Fras, Galvanised steel to aluminium joining by laser and GTAW processes, *Mater. Charact.* 59 (2008) 1705–1715. https://doi.org/10.1016/j.matchar.2008.03.016.
3. S. Kumar, C.S. Wu, G.K. Padhy, W. Ding, Application of ultrasonic vibrations in welding and metal processing: A status review, *J. Manuf. Process.* 26 (2017) 295–322. https://doi.org/10.1016/j.jmapro.2017.02.027.
4. K.A. Savio, Metallurgical and mechanical investigation on FSSWed dissimilar aluminum alloy, *J. Alloy. Metall. Syst.* 2 (2023) 100010. https://doi.org/10.1016/j.jalmes.2023.100010.
5. M. Yamamoto, A. Gerlich, T.H. North, K. Shinozaki, Cracking and local melting in Mg-alloy and Al-alloy during friction stir spot welding, *Weld. World.* 52 (2008) 38–46. https://doi.org/10.1007/BF03266667.
6. M.K. Bilici, Application of Taguchi approach to optimize friction stir spot welding parameters of polypropylene, *Mater. Des.* 35 (2012) 113–119. https://doi.org/10.1016/j.matdes.2011.08.033.
7. M. Miles, U. Karki, Y. Hovanski, Temperature and material flow prediction in friction-stir spot welding of advanced high-strength steel, *JOM* 66 (2014) 2130–2136. https://doi.org/10.1007/s11837-014-1125-6.
8. J.M. Piccini, H.G. Svoboda, Effect of pin length on friction stir spot welding (FSSW) of dissimilar aluminum-steel joints, *Proc. Mater. Sci.* 9 (2015) 504–513. https://doi.org/10.1016/j.mspro.2015.05.023.

9. M.P. Mubiayi, E.T. Akinlabi, M.E. Makhatha, Current state of friction stir spot welding between aluminium and copper, *Mater. Today Proc.* 5 (2018) 18633–18640. https://doi.org/10.1016/j.matpr.2018.06.208.

10. A. Azarniya, A.K. Taheri, K.K. Taheri, Recent advances in ageing of 7xxx series aluminum alloys: A physical metallurgy perspective, *J. Alloys Compd.* 781 (2019) 945–983. https://doi.org/10.1016/j.jallcom.2018.11.286.

11. M. Kang, C. Kim, A review of joining processes for high strength 7xxx series aluminum alloys, *J. Weld. Join.* 35 (2017) 79–88. https://doi.org/10.5781/jwj.2017.35.6.12.

12. L. Zhou, R.X. Zhang, G.H. Li, W.L. Zhou, Y.X. Huang, X.G. Song, Effect of pin profile on microstructure and mechanical properties of friction stir spot welded Al-Cu dissimilar metals, *J. Manuf. Process.* 36 (2018) 1–9. https://doi.org/10.1016/j.jmapro.2018.09.017.

13. N.F.M. Selamat, A.H. Baghdadi, Z. Sajuri, A.H. Kokabi, Friction stir welding of similar and dissimilar aluminium alloys for automotive applications, *Int. J. Automot. Mech. Eng.* 13 (2016) 3401–3412. https://doi.org/10.15282/ijame.13.2.2016.9.0281.

14. J.A. Österreicher, G. Kirov, S.S.A. Gerstl, E. Mukeli, F. Grabner, M. Kumar, Stabilization of 7xxx aluminium alloys, *J. Alloys Compd.* 740 (2018) 167–173. https://doi.org/10.1016/j.jallcom.2018.01.003.

15. V.V.K. Prasad Rambabu, N. Eswara Prasad, R.J.H. Wanhill, *Aerosp. Mater. Mater. Technol.* 1 (2017) 586. https://doi.org/10.1007/978-981-10-2134-3.

16. E. Boldsaikhan, S. Fukada, M. Fujimoto, K. Kamimuki, H. Okada, B. Duncan, P. Bui, M. Yeshiambel, B. Brown, A. Handyside, Refill friction stir spot joining for aerospace aluminum alloys, *Miner. Meter. Mater. Ser.* (2017) 237–246. https://doi.org/10.1007/978-3-319-52383-5_23.

17. V. Haliyal, M. Engg, A review on GTAW technique for high strength aluminium alloys (AA 7xxx series), *Int. J. Eng.* 2 (2013) 2477–2490.

18. A.C.U. Rao, V. Vasu, M. Govindaraju, K.V.S. Srinadh, Stress corrosion cracking behaviour of 7xxx aluminum alloys: A literature review, *Trans. Nonferrous Met. Soc. China (English Ed.)* 26 (2016) 1447–1471. https://doi.org/10.1016/S1003-6326(16)64220-6.

19. Yashpal, C.S. Jawalkar, S. Kant, Uso Dell'Al nei componenti aerospace, *J. Mater. Sci.* 3 (2015) 2015.

20. P. Li, Y. Wang, S. Wang, J. Yang, H. Dong, D. Yan, Corrosion behavior of refilled friction stir spot welded joint between aluminum alloy and galvanized steel, *Mater. Res. Express Accept.* 29 (2018) 1–27.

21. N. Jeyaprakash, C.H. Yang, M. Duraiselvam, G. Prabu, Microstructure and tribological evolution during laser alloying WC-12%Co and Cr3C2-25%NiCr powders on nodular iron surface, *Res. Phys.* 12 (2019) 1610–1620. https://doi.org/10.1016/j.rinp.2019.01.069.

22. A. Bist, J.S. Saini, V. Sharma, Comparison of tool wear during friction stir welding of Al alloy and Al-SiC metal matrix composite, *Proc. Inst. Mech. Eng. Part E J. Process Mech. Eng.* 235 (2021) 1522–1533. https://doi.org/10.1177/09544089211005994.

23. N. Jeyaprakash, M. Duraiselvam, S.V. Aditya. Numerical modeling of Wc-12% Co laser alloyed cast iron in high temperature sliding wear condition using response surface methodology. *Surf. Review Lett.* 26(7) (2018) 1–19. https://doi.org/10.1142/S0218625X19500094.

24. H.A. El-Hafez, A. El-Megharbel, Friction stir welding of dissimilar aluminum alloys, *World J. Eng. Technol.* 6 (2018) 408–419. https://doi.org/10.4236/wjet.2018.62025.

25. S.K. Raval, K.B. Judal, Recent advances in dissimilar friction stir welding of aluminum to magnesium alloys, *Mater. Today Proc.* 22 (2019) 2665–2675. https://doi.org/10.1016/j.matpr.2020.03.398.

26. Zlatanovic, D. Labus, S. Balos, J. Pierre Bergmann, S. Rasche, M. Pecanac, S. Goel, Influence of tool geometry and process parameters on the properties of friction stir spot welded multiple (Aa 5754 H111) aluminium sheets, *Materials* 14(5) (2021) 1–26. https://doi.org/10.3390/ma14051157.

27. I. Sioutis, K. Tserpes, E. Tsiangou, H. Boutin, F. Allègre, and L. Blaga, Experimental evaluation of refill friction stir spot welds (RFSSW) as crack arrest features in co-consolidated thermoplastic laminates, *Compos. Struct.* 309 (2023) 1–10. https://doi.org/10.1016/j.compstruct.2023.116754

15 Advances in Friction Stir Spot Welding

K. Ananthakumar, N. Jayanth,
M. Satthiyaraju, and T. Arunnellaiappan

15.1 INTRODUCTION

Friction stir welding (FSW) is a solid-state material fusion method and was invented by Thomas et al. at The Welding Institute (TWI), Abbington, UK (1991). It was primarily used on the 2000 and 7000 series aluminum and Al-Li alloys [1], which were notoriously difficult to weld. The solid-state welding method can prevent these problems because the temperature generated in FSW is lower than the base material's bulk solidus temperature [2–4]. These alloys are subjected to liquation cracking, solidification, or porosity during conventional welding. This reduces thermal distortions and cracking problems brought on by heat input. The automobile, aerospace, marine, and rail industries have all adopted FSW, which is one of the vital advancements in material joining in recent years [5–8].

Based on the fundamentals of linear FSW, friction stir spot welding (FSSW) was created without the tool moving laterally. By inserting a revolving tool at a particular pace until the shoulder touches the top sheet, the goal was to spot weld 2 or more similar or different sheets, whereas a static anvil gives support for the loads acting axially. A brief dwell stage is frequently helpful when achieving the correct temperature needed for plastic flow, mainly for materials with large melting points. The volume of the deformed material also grows during this dwell phase, which further encourages material flow and fusing of the top and bottom sheet materials. During the plunge and dwell stages, the material surrounding the tool gets softened and deformed plastically due to large strain rates. The tool's vertical movement encourages the flow of plasticized material in both axial and circumferential directions, which also aids in disrupting the layer of oxide at the weld interface [9].

The tool is quickly retracted after the weld has been created, either when the intended depth of plunge is attained or after a dwell time. A ring-shaped, solid-state metallurgical bond is formed by combining the pressure of forging, heating, and plasticized material flow. This bond has a refined microstructure of grains and grain boundaries with high angles that are close to the revolving tool. As with the FSW, connecting with the FSSW involves forging and extruding the material under large strains. The intricate interconnections in the concurrent thermo-mechanical processes influence the heating and cooling speeds, plastic flow, and deformation, which result in the recrystallization phenomenon. Although the real welding duration is very short in FSSW, the process dynamics are substantially more

DOI: 10.1201/9781003432289-15

complicated because they all center on the momentary phases of tool plunging, material mixing, and tool retraction. During the brief cycle period (between 2 and 5 s) [9], there are rapid heating and cooling rates that can also result in the creation of non-equilibrium phases. Due to the lack of a traverse movement and the absence of a tool moving front or back, the FSSW technique can be regarded as axisymmetric. The three microstructures, namely stir zone (SZ), thermo-mechanically affected zone (TMAZ), and heat-affected zone (HAZ) are formed by FSSW, which is identical to FSW. Based on the heat cycle and level of plastic deformation, each of these zones exhibits unique microstructural characteristics. FSSW is gaining popularity as a possible alternative to the traditional single-spot welding techniques like self-piercing rivets (SPR), RSW, and clinching in the aerospace and automotive sectors because it has advantages from an economic and environmental perspective (reduced energy needs). Weld uniformity and electrode life are the key process restrictions during the RSW of aluminum [12,13].

The main drawback of SPR is the high cost of consumable rivets, the small number of joint configurations that can be made with each gun, and the recycling of aluminum when steel rivets are used [14,15]. As FSSW is not susceptible to surface conditions or tool wear, it can consistently create welds with equivalent or higher strength than RSW. Similar to linear FSW, FSSW avoids flaws connected with porosity, solidification, hot cracking, thermal deformation, and production of bulk intermetallic by keeping temperatures less than the bulk melting point. Additionally, FSSW has advantages for welding polymers, copper, Al alloys, Mg alloys, steels, metal matrix composites, and dissimilar metal/composites combinations.

15.2 CLASSIFICATION OF FSSW

Over the last 20 years, a number of FSSW process variations were investigated, including traditional FSSW, refill FSSW (RFSSW), pinless FSSW, and additional short traverse FSW variations like stitch FSW, swept FSW, and swing FSW. The three most frequently and thoroughly studied FSSW types at the moment are conventional, refill, and pinless FSSW procedures, as shown in Figure 15.1. These processes are actually spot welds compared to short traverse seam welds. As a result, the emphasis of this chapter will be on those welding methods in terms of material flow, optimization of parameters, heat generation, microstructural advancement, and mechanical characteristics [15].

15.2.1 TRADITIONAL FSSW

The traditional FSSW process has been used in the automotive industry ever since Sakano et al. and Iwashita et al. originally described it in 2001 and 2003, respectively. The FSSW process can be broken down into three major steps, as given in Figure 15.2. The rotation of the pin is done at a preset speed and then lowered into the overlapped sheets at a preset rate and depth to encourage material to run away from the shoulder of the tool. The components of the two metal sheets are then allowed to mix together by applying a dwell period of a specific time period (often a few seconds) [14]. The keyhole geometry is dependent on the swept volume

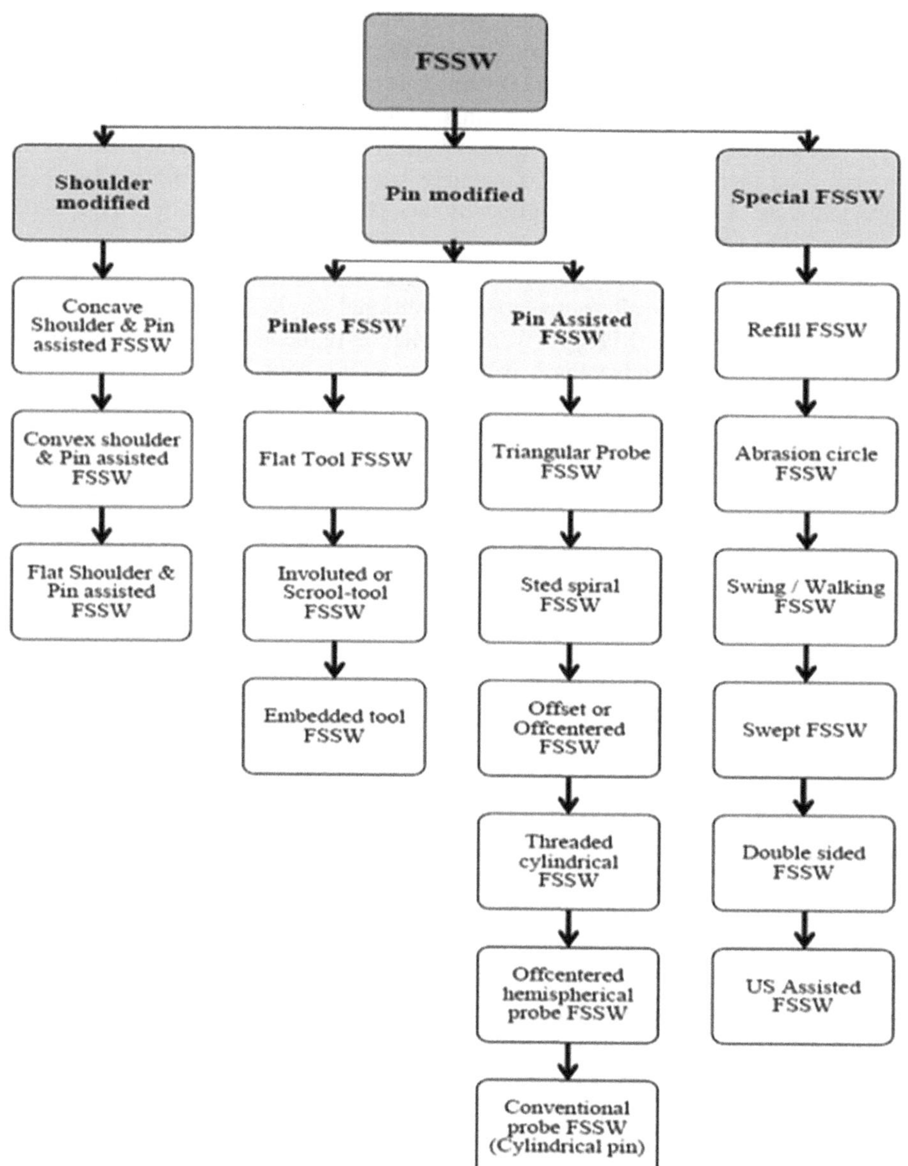

FIGURE 15.1 Recent innovations in FSSW.

of the pin and is left behind by the withdrawal of the pin tool at the center of the fusion zone. For the 2005 Mazda MX-5 sports vehicle, the aluminum trunk lid was attached to the steel bolt retainers using conventional FSSW, which was initially utilized in the Mazda RX-8 back door panels in 2003. Additionally, Toyota has applied it to the hood and deck lid of Prius hybrid automobiles. It was predicted that using ordinary FSSW saved 90% of the energy and reduced capital expenditure by 40%

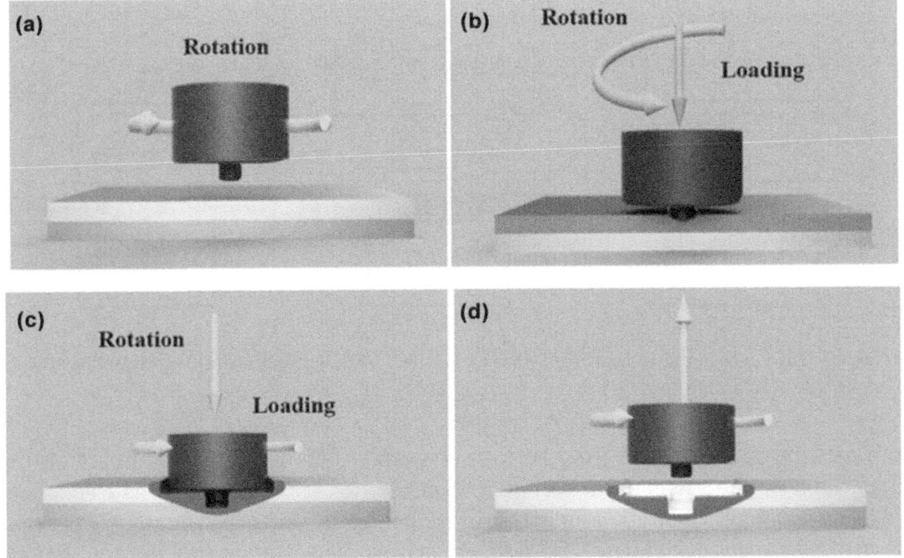

FIGURE 15.2 Schematic representation of traditional FSSW process: (a) rotation, (b) plunge, (c) stir (dwell), and (d) drawn out.

while compared to the RSW of Al alloys. This is because it eliminates the requirement for numerous parts of equipment, such as a sizable power source, cooling unit, and electrode dresser.

It must be noticed that the weld center still has an esthetically unfavorable keyhole depression. In addition to altering the appearance of the surface, the keyhole reduces the size of the weld, decreases the fused region, and makes a possible corrosion concentration point where electrolyte could gather. During FSSW, the top sheet may thin owing to extreme tool penetration and upward flow of material around the pin (also called "hooking"), which has been observed to weaken the joint.

15.2.2 Pinless FSSW

The keyhole made due to the tool pin at the weld center is one of the main shortcomings of conventional FSSW, as was previously mentioned. This may reduce the surface that is metallurgically linked, which may result in reduced failure energies from fracture across the weld line. Additionally, tools without pins have been taken into consideration to prevent this and speed up the formation of a bigger bonded area. Pinless FSSW was created in 2009, and in recent years, interest in it has grown significantly.

The pinless FSSW process is comparable to the traditional FSSW method which involves joining by briefly agitating the top sheet material with a pinless tool before retreating (Figure 15.3). However, pinless FSSW has a very different weld production mechanism than traditional FSSW. Here, the plastic flow is very intricate, and the top sheet saw the majority of the grain refining. The lower sheet material is displaced upward at the perimeter of the weld due to the upper sheet material being forced

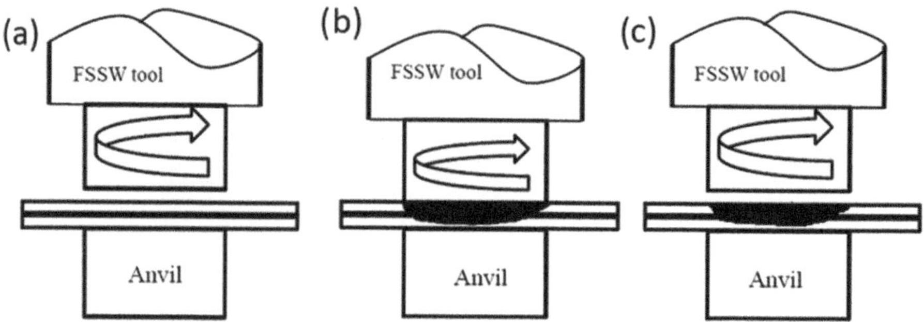

FIGURE 15.3 Schematic representation of pinless FSSW process: (a) rotation, (b) plunge and stir (dwell), and (c) drawn out.

towards the weld's center. Tool surface characteristics had a major impact on the material flow pattern, and Bavakos et al. evaluated a variety of shapes. The necessity of a key shoulder penetration for achieving dependable weld performance for the pinless FSSW technique should also be emphasized. Because the welding tool has no pin, it is fully possible to skip the keyhole when joining thin material (1 mm), which increases the effective bonded area and increases the likelihood of failure by nugget pullout around the weld's perimeter [15].

Short weld cycle periods (less than 1 s) can be used to make the weld, and this quick thermal cycle can prevent cracking and reduce heat input at the HAZ. Because of this, the mechanical properties of pinless spot welds can be on par with or even better than those of FSSWs in the traditional way. Given that the rising flow of material in the bottom sheet is less prominent in this situation, connecting thick sheets (1 mm) requires substantially deeper tool penetration. As a result, the upper layer beneath the tool thins due to the increased penetration of the tool. To make up for this, it was suggested to use double-sided FSSW, which would enable the connecting of thicker sheets by utilizing a spinning anvil and pinless tool. Improved mechanical strength is possible with the rotating anvil FSSW, and the axial or reaction force on the spot welding frame is also minimized.

15.2.3 Refill FSSW

Scientists at the Helmholtz-Zentrum Geesthacht in Germany created and filed a patent for RFSSW in 1999 to attach two or more lightweight material sheets made up of Al and Mg alloys, collectively in the lap arrangement form. The material is plasticized and displaced to form the joint by a back extrusion-like process. RFSSW's primary benefit is that there is no keyhole produced and that the weld technically matches the actual material surface. The tool used in the RFSSW technique is shown in Figure 15.3, which consists of a sleeve, an inner pin, and an outside fixed clamp ring. The job of the clamping ring is to hold the work parts tightly together while welding and to stop them from separating or coming out as the plasticized material is moved about by the sleeve and pin. Using the same motor, the sleeve and pin can revolve in the same direction at uniform speed, and they can be separately raised and lowered.

According to whatever tool component is used as the plunging part, the RFSSW method is classified into two types: the sleeve-plunge and pin-plunge methods. A graphic representation of the sleeve-plunge version is given in Figure 15.3. In order to plasticize the materials, the sleeve is first inserted into the sheet to a preset depth. Meanwhile, the pin is raised to a suitable height to produce a cylindrical cavity that can hold the plasticized material that was ejected by the sleeve. When the revolving sleeve and the pin return to their original places after a predetermined dwell period, the softened material is forced back into the weld, resulting in a spot weld with little or no surface indentation.

The key benefit of the sleeve-plunge method over the pin-plunge method is the stronger joint thanks to the bigger nugget size. This variation, meanwhile, requires more dive force. While joining times are equivalent to the latter, this replenishment approach can produce overlap shear strengths that are favorably compared with standard FSSW, SPR, and RSW. Additionally, RFSSW provides benefits and an innovative method for combining lightweight materials like polymer, Mg, and Al alloys, as well as dissimilar pairings of non-ferrous metals, ferrous to non-ferrous metals, and polymer to metals. There are a number of uses for the RFSSW process, including filling the keyhole left by FSW, producing skin-strengthened panels, mending fatigue cracks, replacing RSW, replacing fasteners or rivets, and as a tacking welding preparatory procedure for FSW.

15.2.4 Short Traverse FSW Method

It is important to quickly discuss other techniques for improving the welded region and, consequently, the mechanical characteristics of the weld. Different short traverse FSW versions, such as stitch (or walking), swept, and swing FSW, have been investigated. As seen in Figure 15.4, the tool is inserted inside the sheets to a preset depth and then moved along a predetermined path (the joining path might have

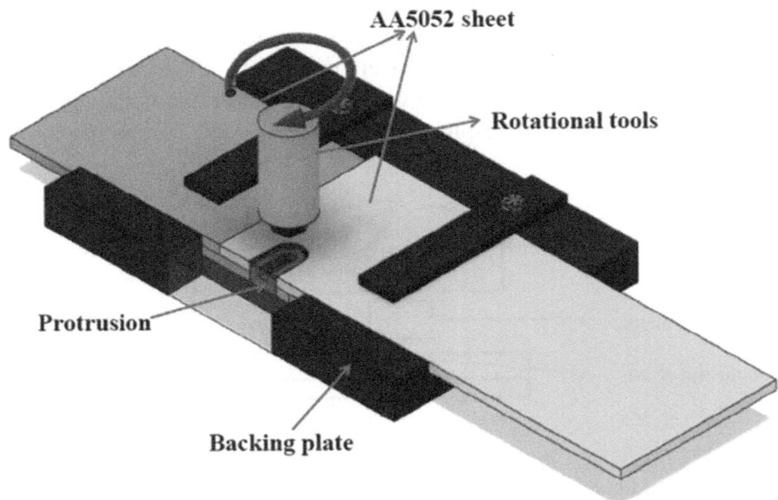

FIGURE 15.4 Schematic representation of PFSSW.

different geometries) for a brief distance. As a result, a bigger junction was made, which might increase the weld strength. GKSS proposed the stitch FSW process and swing FSW was derived from stitch FSW established by Hitachi. Researchers, namely, Wichita State University (Octaspot TM pattern) and TWI (Squircle TM pattern), developed the swept FSW method during this time.

15.3.1 Fundamentals of FSSW

In order to compete with RSW methods, FSSW, a novel spot welding technology, is utilized to unite overlapping work parts. The three stages of conventional FSSW are plunge, dwell, and retraction. Figure 15.1 depicts a schematic representation of the traditional FSSW technique. As the revolving pin touches the surface of the upper sheet during the tool plunge phase, a mixture of plastic deformation, frictional heating, and viscous dissipation begins. The material flow brought on by the tool pin features causes the temperature to rise steadily during the dwell stage, and the SZ region volume also rises. Figure 15.5 displays the macro-scale cross-sections for traditional FSSW with dwell periods varying between 0.4 and 2.0 s. Weld size and mixed zones can be seen to drastically increase with dwell times.

15.3.2 Heat Generation in FSSW Process

Since the FSSW method cannot be represented as a steady-state process like the linear FSW, modeling the FSSW process is extremely difficult. With extremely momentary material flows and temperature distributions, FSSW entails a non-steady state. Convective phenomena are frequently overlooked while designing the FSW technique and solving the mass transfer governing equations, unless the FSW plunge stage is taken into account. The boundary conditions and a few physical mechanisms in the FSSW process are not completely understood or quantified, which makes it difficult to model the process. Especially the contact conditions, heat generation at the interface of the tool or workpiece, and heat transfer at the interface of workpiece/backing plate are not precisely defined by changing forces. There is also disagreement over the workpiece material's constitutive behavior at elevated strain rates and temperatures, despite the existence of a few competing hypotheses.

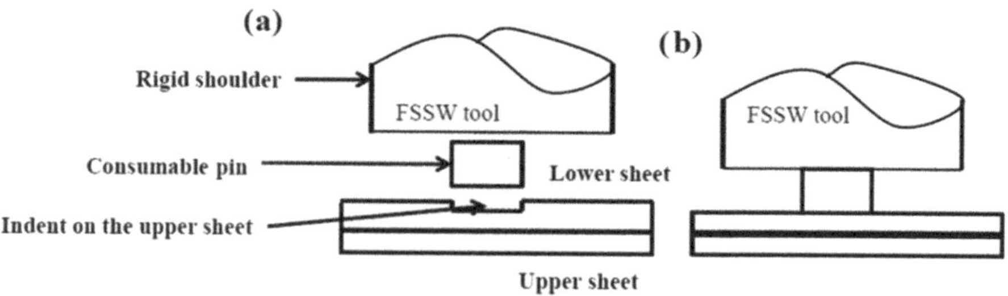

FIGURE 15.5 consumable pin FSSW: schematic representation of (a) the parts for welding and (b) consumable pin on the upper sheet.

The momentum and energy conservation equation in particular requires careful choice of the proper constitutive conversation equations. As a result, early models frequently ignored the material flow brought on by stirring and merely calculated the equation of heat transfer to determine the distribution of temperature. There are currently more complex models available that use simulations based on solid mechanics, fluid dynamics, or a combination of both types of models. These three models match the Eulerian and Langragian simulations and their mixture from the perspective of numerical analysis. While the execution of these methodologies varies greatly, they all serve as a foundation for examining the impact of joining factors and the shape of the tool on the welding process by determining a number of features [17].

15.3.3 Process Parameters and Tool Design

The achievement of the FSSW method is based on the tool design characteristics, namely the shape or material of the tool, and the factors, namely plunge rate and depth, rotational speed, and dwelling time. These factors have a significant impact on heat produced, flow of material, weld reliability, torque, axial force, and mechanisms of tool degradation. Defects like voids, improper bonding, an incomplete refill, or insufficiently bonded sections may be the result of less heat generation and inadequate material flow. However, with more heat input, precipitate dissolution causes grain expansion or additional softening of the HAZ around it. Understanding the mechanisms affecting these aspects is essential for exploring the best tool design and process parameters. The impact of pre- and post-heat treatment on joint strength must also be taken into account.

15.3.3.1 Tool Design

The tool's main purposes are to accelerate heating, promote material flow, and prevent plasticized material from overflowing into the weld zone. FSSW was categorized according to the sorts of tools and movements mentioned in the first part. In FSW, a fixed pin tool is a one piece that includes both the pin and shoulder. This is how the tool utilized in traditional FSSW was derived. Due to the unchanging pin length, the tool pin can only fuse a workpiece with an even thickness. The primary objectives of tool design in FSSW and butt FSW differ considerably. When it comes to FSW, the horizontal mixing process—which involves moving material from one workpiece to another in the welding plane so that it flows across the vertical interface line—is prioritized more. Additionally, because they may result in excessive top sheet thinning and the oxide trapped in the surface between the overlapping surfaces, several tool designs of FSW, namely cylindrical and triflute pin tools and the patented whorl, are not suited for lap joints. This suggests that FSW tools need to be optimized before being used in FSSW. In reality, tools for FSSW, such as pinless and RFSSW tools, have steadily developed into wholly new technologies [16].

The pinless FSSW tool, which is made up of just a shoulder, is an intense case of the fixed pin tool. In RFSSW, the sleeve length or adjustable pin enables attaching workpieces of various densities and keyhole refilling. To combine materials like polymers, Al and Mg alloys, steel tools with low melting points are now a very popular and dependable tool material. Due to a shortage of accessible and long-lasting tool materials, FSSW is yet to be widely used for metals with large melting points, such as steel alloys.

15.3.3.2 Tool Geometry

The shoulder restricts the plasticized material surrounding the pin, creates heat from deformation and friction, and establishes the flow of material in a comparatively slim layer beneath the lower shoulder surface. The axial force is applied to the workpiece's upper surface via the shoulder to make forging action necessary for welding consolidation. Along with the SZ, a broader region under the tool shoulder also experiences plastic flow. It is proposed that the primary characteristic of FSW is the generation of heat beneath the big tool shoulder region. However, because FSSW is a very transient welding process and a significant role is played by the shoulder only for a small period of time when the tool pin is fully inserted inside the workpiece, the heat production mechanism for FSSW differs somewhat from that during linear FSW. The size, form, and surface characteristics of the tool, as well as other geometrical elements, are thought to be crucial welding parameters. The dimensions and shape of the tool shoulder have a considerable impact on hook geometry, plastic flow, and heat generation rate. The majority of shoulder dimension optimization of the FSSW tool to date has been done experimentally using trial-and-error techniques. Larger shoulder diameters may enhance heat generation, the fused area, and joint strength, according to certain reports [38]. The diameters of tools often employed in traditional FSSW of lightweight materials such as Mg and Al alloys are listed below.

15.3.4 FSSW OF SIMILAR ALUMINUM ALLOYS

The summary in Table 15.1 provides information about the kind of sheet material utilized, thickness of sheet, tool material, shape, and inferences made for the FSSW of equivalent aluminum.

15.3.4.1 FSSW of 2xxx Series Similar Aluminum Alloys

Aluminum 2xxx alloys exhibit high strength and can be heat treated, which are mostly used in heavy-duty vehicles, aerospace, and structural applications. The welding of 2xxx series aluminum is effectively done via FSSW by several researchers. The chosen studies on FSSW of 2xxx series similar aluminum alloys are presented in this section.

Vinay et al. [17] investigated the effect of post-weld heat treatment (PWHT) on the mechanical behavior, microstructure, and crack propagation rate of FSSWed AA2024 sheets. The welded AA2024 sheets were PWHT at varying temperatures (190°C and 200°C for 10h). The PWHT at 200°C for 10h retrieved the strength lost after FSSW. It was found that an increase in the aging temperature improved the value of microhardness and ductility. But, the tensile strength of the welds decreased with an increase in the aging temperature.

TABLE 15.1

A Summary of the Tool Materials, the Welding Parameters and Inferences for FSSW Similar Aluminum Alloys

Material Type and Thickness	Tool Material and Dimensions	Welding Parameters	Inference	References
AA2024-T3	H13 steel shoulder and pin diameter: 18 and 6 mm Pin length: 5.8 mm	Rotating speed (931 rpm), traverse speed (70 mm/min) PWHT at aged temperatures 190°C and 200°C	Increase in the ageing temperature improved hardness value and ductility. But, the tensile strength decreased.	[17]
AA 2024-T3 (0.6 mm)	AISI M4 high-speed steel, clamping ring, shoulder and probe diameter: 12, 6, and 4 mm	Rotational speed (1250 rpm), plunge depth and speed (0.8 mm and 0.7 mm/s), welding time (2.8 s), welding force (10 kN)	Refill-FSSW with sealant improved the joint strength and fatigue properties.	[18]
AA5052-H32 (2 mm)	H13 steel Flat shoulder diameter: 20 mm Pin length and diameter: 5 and 3.2 mm	Dwell time (1, 2 and 3 s), rotational speed (500 rpm), depth of plunge (3.2 mm), tilt angle (0°) and plunge rate 0.1 mm/s	Maximum SZ hardness value and tensile shear load were attained at 2 s dwell time and 500 rpm tool rotation speed.	[19]
AA5052-H32 (2 mm)	H13 steel flat shoulder diameter: 20 mm	Dwell time (2 s) rotational speed (500, 1000 and 1500 rpm)	Hardness value decreases due to the rise in rotational speed.	[20]
AA5754 (2 mm) pure copper (0.1 mm)	Shoulder and pin diameter: 15 and 5 mm Length of the pin: 3.5 mm	Welding time (15 s) rotational speed (1000, 1500 and 2000 rpm)	Better mechanical properties and micro hardness value are attained for 1000 rpm rotation speed and the addition of copper interlayer improved joint strength at higher speeds.	[21]
AA5052 (1 mm)	H13 pinless tool shoulder diameter: 16 mm	Plunging depth (0.2 mm) and dwell time (6 s), rotation speed (500, 800, 1250 and 1600 rpm)	Protrusion FSSW method produced smooth weld surfaces that reduced the stress concentration and improved the corrosion resistance.	[22]

(Continued)

TABLE 15.1 (*Continued*)

A Summary of the Tool Materials, the Welding Parameters and Inferences for FSSW Similar Aluminum Alloys

Material Type and Thickness	Tool Material and Dimensions	Welding Parameters	Inference	References
AA6061-T6 (1.5 mm)	H13 grade steelflat shoulder tool with 12 mm diameter, consumable pin height and diameter: 2.5 and 6 mm	Rotational speed (557 rpm),plunge rate (20 mm/min), dwelling time (10 s)	The consumable pin improved the joint strength when compared to traditional FSSW and the best weld quality is obtained for AA6061-T6 pins.	[23]
AA6061 (2 mm)	H13 steel shoulder diameter: 16.4 mm, tapered pin having diameter reducing from 3.7 to 2.4 mm	Rotational speeds (700, 900 and 1100 rpm)	Underwater welded joints are free from defects and PWHT enhanced the strength of those joints.	[24]
AA7075-T6 (0.6 mm)	Hotvar material, clamping ring, shoulder and probe diameters of 9, 6, and 4 mm	rotating speed (3000 rpm), depth of plunge (0.7 mm), welding time (2.8 s)	The ideal temperature for joining is 80% of the melting point and excess temperature reduced the joint strength.	[25]
AA5052-H112 (3 mm)	H13 steel shoulder of 12 mm diameter with cylindrical and step pins	Dwell time (5 s), rotating speed (900, 1400, and 1800 rpm) plunge rate and depth (4 mm/min and 0.1 mm)	The weld strength was affected by geometry of pin and rotating speeds. Higher weld strength is obtained for cylindrical pins	[26]
Semi-solid metal cast aluminum alloy - SSM6061(4 mm)	High-strength steel (SKH 57 tool) with 20 mm shoulder diameter, pin diameter and height: 5.2 and 3.2 mm	Rotation speeds (380, 760, 1240, 2700, and 3500 rpm), dwell time (7, 14, 21, and 28 s), depth and rate of plunge (2.8 and 18 mm/min)	The maximum hardness value is attained for rotating 3500 rpm speed and 28 s dwell time.	[27]
AA7050-T7451, Alclad 2024-T3	Diameter of clamping ring: 16 mm, shoulder:7 mm, and probe: 4 mm	Rotational speed (1200 rpm), depth of plunge (2.2 mm), force (13.5 kN), dwell time (3 s)	Higher fatigue limit and ultimate static strength were obtained for primed configurations.	[28]

(Continued)

TABLE 15.1 (*Continued*)

A Summary of the Tool Materials, the Welding Parameters and Inferences for FSSW Similar Aluminum Alloys

Material Type and Thickness	Tool Material and Dimensions	Welding Parameters	Inference	References
AA7075-T6 (2 mm)	H13 steel tool, diameters of the clamping ring: 14.5 mm, outer sleeve: 9 mm and inner pin: 6.4 mm	Rotational speed (2200 rpm), dwell time (3 s), sleeve penetration depth (1.4–2.4 mm), plunging rate (2.8–4.8 mm/s)	The tool penetration depth is the significant parameter that affected hardness, grain size, shear strength and weld deformation volume.	[29]
AA6082-T6 (1 and 2 mm)	H13 steel tool of shoulder diameter: 20 mm, pin diameter and length: 5 mm and 2.6 mm	Dwelling time (3 s), rotational speed (400, 600, 800, and 1000 rpm)	Highest hardness and tensile shear load were attained at rotating speed of 600 rpm.	[30]
2219 aluminum alloy (2 and 8 mm)	clamping ring, sleeve and pin diameter: 15, 9 and 6 mm	Rotational speed (1800 rpm), welding time (3.6 s), and plunge depth (2.2 mm)	The SZ exhibited exceptional corrosion resistance properties, followed by top and bottom plate.	[31]
7050-T74511 aluminum alloy (2 mm)	Diameters of the stir: 6 mm, sleeve: 9 mm and clamping ring: 13 mm	Rotational speed (1800 rpm), plunge depth (2.9 mm) dwelling time (7 s) and shear strength (7.2 kN)	The Al alloy's corrosion potential is reduced by Refill-FSSW.	[32]
AA6061-T6 (1.6 mm)	Diameter of the shoulder: 10 mm, pin: 3 mm	Rotational speed (800–1200 rpm), depth of plunge (2.2 and 2.6 mm), rate of plunge (0.15 mm/s)	The predicted temperature values by upper bound method were slightly larger than the experimental and FEM results.	[33]
AA 5754-H111 (0.3 mm)	H13 hot-work steel diameter of the shoulder: 12 mm	Rotational speeds (1000 rpm and 4500 rpm), dwell time (5 s), axial load (4 kN), depth of penetration (0.25 mm)	Better average shear failure load and micro hardness values are obtained at rotational speed of 1000 rpm.	[34]
6082 Al alloy (2 mm)	Diameters of clamping ring: 13.5 mm, sleeve: 9 mm and pin: 5.2 mm	Rotational speed (1950 rpm), refilling time (3, 4 and 5 s), dwelling time (3 s), plunge speed and depth (60 mm/min, 1.6 mm)	The maximum tensile strength is obtained at 5 s refilling time. Also, it can remove the crack due to enough diffusion time.	[35]
6005 Al alloy (3 mm)	Diameter of the pin: 5.2 mm, sleeve: 9 mm and clamping ring: 13.5 mm	Rotational speed (1600, 1800 and 2000 rpm), tool plunging and retracting speed (60 mm/min), depth (2.6 mm)	Maximum hardness and finer grains were obtained at SZ for 1600 rpm rotational speed.	[36]

TABLE 15.2

Chemical Compositions for AA2024-T3 (wt.%)

Cu	Cr	Fe	Mg	Mn	Si	Zn	Al
3.8–4.9	0.1	0.5	1.2–1.8	0.3–0.9	0.50	0.25	Bal.

TABLE 15.3

Mechanical Properties of AA2024-T3

Elasticity Modulus (GPa)	Yield Strength (MPa)	Tensile Strength (Mpa)	Elongation (%)	Poisson Ratio
73.1	345	485	18	0.33

Matteo et al. [18] studied the mechanical behavior of multi spot AA 2024-T3 joints prepared via RFSSW by adding sealant, mostly used in the aviation industry. The elemental compositions and mechanical properties of AA2024-T3 are specified in Tables 15.2 and 15.3, respectively. It was observed that lap joints produced by RFSSW with sealant improved the mechanical performance of the welds when compared to welds made without sealants. The sealant also has beneficial effects on fatigue properties.

Zou et al. [31] fabricated 2219-O (top plate) of thickness 2 mm and 2219-C10S (bottom plate) of thickness 8 mm using RFSSW and examined the microstructure, mechanical, and corrosion performance of the welds. It was found that the SZ contains refined and equiaxed grains, which are caused by dynamic recrystallization. The SZ exhibited good corrosion resistance; it was followed by the top plate (2219-O) and the bottom plate (2219-C10S). The mechanical properties such as tensile force and tear force decreased after dipping in the NaCl solution for 48 hrs.

15.3.4.2 FSSW of 5xxx Series Similar Aluminum Alloys

AA 5052 H32 (cold-worked aluminum alloy) exhibits superior corrosion resistance, formability, and weldability in marine and industrial environments. Hence, it is generally used in the shipbuilding, automotive, and aerospace industries. The chemical compositions and mechanical properties of the AA 5052 H32 alloy are specified in Tables 15.4 and 15.5. Ahmed et al. [19] studied the effect of dwell time (1, 2, and 3 s) on the mechanical and microstructural properties of FSSWed AA5052-H32 strips of 2 mm thickness and found that the maximum SZ hardness value and tensile shear load were attained at 2 s dwell time and 500 rpm tool rotation speed. The higher hardness value is due to the high grain refinement in the SZ when compared to the base metal (BM).

The study on the effect of variation in the rotation speeds (1500, 1000, and 500 rpm) on FSSW AA 5052 H32 joints was carried out. It was found that the SZ hardness value is larger than the BM for all the speeds of tool rotation. The hardness value decreased with a rise in the speed of rotation (500–1000 rpm). The grain size at the SZ decreased with a decrease in the heat input. The weld made at 500 rpm rotational speed and 2 s dwell time attained the highest tensile shear load (4330 N). This value decreased to 2569 N at the rotation speed of 1500 rpm [20].

TABLE 15.4

Chemical Compositions for AA 5052 H32 (wt.%)

Al	Mg	Fe	Mn	Si	Ti	Cu	Zn	Cr	V
Bal.	2.490	0.258	0.091	0.127	0.017	0.001	0.200	0.194	0.001

TABLE 15.5

Mechanical Properties of AA5052-H32 Alloy

Hardness (Hv)	Yield Tensile Strength (MPa)	Ultimate Tensile Strength (Mpa)
69 ± 2	193 ± 3	229 ± 4

TABLE 15.6

Chemical Composition of 5754 Aluminum Alloy Sheets

Al	Cr	Mn	Mg
95.8	0.23	0.21	3.76

TABLE 15.7

Mechanical Properties of 5754 Aluminum Alloy Sheets

Tensile strength (MPa)	Yield Strength (MPa)	Elongation (%)
207	97	20

Gassaa et al. [21] evaluated the mechanical behavior of FSSWed 5754 aluminum sheets. The chemical compositions and mechanical characteristics of the 5754 aluminum alloy are given in Tables 15.6 and 15.7. The impact of rotation speed on weld strength is compared with riveted joints. Also investigated was the influence of the addition of pure copper interlayers on the joint strength of FSSWed 5754 aluminum sheets. Tensile shear stress revealed that better mechanical properties were obtained at lower and intermediate rotational speeds for the joints without interlayers when compared to riveted joints. The strength and microhardness of FSSWed joints made at larger speeds were improved with the addition of a copper interlayer. The higher microhardness values are obtained at a 1000 rpm rotation speed. The failure mode obtained for the lowest rotation speed is complete nugget pull-out, which shows good bonding of sheets. The working principle of FSSW involves three steps, as given in Figure 15.1.

Farmanbar et al. [22] studied the protrusion FSSW (PFSSW) of 5052 aluminum alloy sheets of 1 mm thickness. Factors such as plunge depth (0.2 mm) and dwell time (6 s) were kept as constant. The tool rotation speed was varied (500, 800, 1250, and 1600 rpm) to obtain the most favorable condition for better mechanical and micro-structural characteristics. It was observed that this method produced smooth weld surfaces, which would decrease the stress concentration and increase the corrosion resistance. The highest load and failure energy were obtained for sample protrusion FSSW at 6 s dwell time and 500 rpm tool rotation speed. The schematic representation of PFSSW is shown in Figure 15.4.

Tiwan et al. [26] evaluated the effect of the pin shape (cylindrical and stepped) and speed of rotation on the mechanical behavior and microstructure of FSSWed AA 5052-H112 joints. It was observed that the higher joint strength was obtained for the cylindrical pin at 1400 rpm.

15.3.4.3   FSSW of 6xxx Series Similar Aluminum Alloys

Bhardwaj et al. demonstrated the exit-hole-free FSSW of AA6061-T6 sheets using a consumable pin and compared the joint strength of FSSWed AA6061-T6 using pins made up of three different materials, namely mild steel (MS), AA6061-T6, and oil hardened non-shrinking die steel (OHNS). The chemical composition of sheet and pin materials is given in Table 3.8. The tensile characteristics of AA6061-T6 at various temperatures (RT, room temperature, 100°C, 200°C, and 400°C) are given in Table 15.9. The schematic representation of FSSW having a consumable pin is given in Figure 15.5. The lap shear test revealed that the use of consumable pin improved the weld strength when compared to conventional FSSW. The best weld quality is obtained for AA6061-T6 pins when compared to the other two pins.

TABLE 15.8:

Chemical Composition of Sheet and Pin Materials

%	Al	C	Mn	Si	S	P	Cr	Cu	Mg	Zn	Fe
AA6061-T6	Bal.	Nil	0.09	0.46	Nil	Nil	Nil	0.10	0.40	0.06	0.30
MS	Nil	0.25	0.65	0.26	0.018	0.022	Nil	Nil	Nil	Nil	Bal.
OHNS	Nil	1.02	0.58	0.24	0.018	0.016	1.10	Nil	Nil	Nil	Bal.

TABLE 15.9

Tensile Properties of AA6061-T6

Property	RT	100°C	200°C	400°C
Ultimate tensile stress (MPa)	330	285	251	190
0.2% offset yield stress (MPa)	276	242	220	156
Total elongation (%)	12	13	13.5	16
Strain hardening exponent	0.1	0.1	0.1	0.1
Strength coefficient (MPa)	415	378	300	205

Shekhawat et al. [24] produced FSSW of 2 mm thick AA6061 sheets using two mediums: water and air, at varying tool rotation speeds (700, 900, and 1100 rpm). The load-bearing capacity and hardness value increased with a rise in the tool rotation speeds from 700 to 1100 rpm for both mediums. The underwater welded joints have a higher hardness value and load-bearing capacity when compared to joints made in air. This is caused by the refinement of grains in the weld nugget zone.

Meengam et al. [27] studied the influence of the depth-to-width (D/W) ratio of the welding zone on the weld excellence of the SSM6061 aluminum alloy using the FSSW process. The better mechanical properties were obtained with a higher D/W ratio. The maximum hardness value was attained for a tool rotating at 3500 rpm with a 28 s dwell time.

Ahmed et al. [30] performed FSSW of thin AA6082-T6 sheets of different thicknesses (1 and 2 mm) at different rotation speeds (400, 600, 800, and 1000 rpm) and dwell times of 3 s. The grain refinement at the SZ improved the mechanical properties of the welded joints. The maximum SZ hardness (95 ± 2 HV0.5) and tensile shear load (4300 ± 30 N) is attained at a 600 rpm tool rotation speed.

15.3.4.4 FSSW of 7xxx Series Similar Aluminum Alloys

Janga et al. [25] reported the effect of tool rotation speed and plunge rate on refill-FSSW of 0.6 mm thick AA7075-T6 sheets via a validated 3D thermomechanical model and found that the ideal temperature for joining is 80% of the melting point, and excess temperature reduced the joint strength. The elemental composition of AA7075-T6 is given in Table 15.10. The steps involved in RFSSW are shown in Figure 15.6.

Shen et al. [29] evaluated the static behavior, fatigue behavior, and microstructure of RFSSWed Al 7075-T6 alloy joints made with a customized sleeve and found that defect-free welded joints are made by this method. The grain sizes, hardness, weld deformation volume, and shear strength were influenced by tool penetration depth. It was concluded that the RFSSW is suitable for joining non-weldable aerospace aluminum alloys.

Zhang et al. [32] studied the microstructural and corrosion properties of RFSSWed 7050 aluminum alloy. The chemical composition of AA 7050 is given in Table 15.11. The welding process parameters and mechanical properties of the RFSSW joint are given in Table 15.12.

Based on the differences in the microstructure, the welded joint can be divided into five different zones, including the SZ, BM, TMAZ, HAZ-I, and HAZ-II. The SZ has fine grains, and the other zones have coarse grains. The intergranular and exfoliation corrosion tests revealed that high corrosion sensitivity was exhibited by TMAZ and

TABLE 15.10

Chemical Composition of AA7075-T6

Al	Mn	Cr	Mg	Cu	Zn	Fe
Bal.	0.03	0.20	2.32	1.39	5.39	0.16

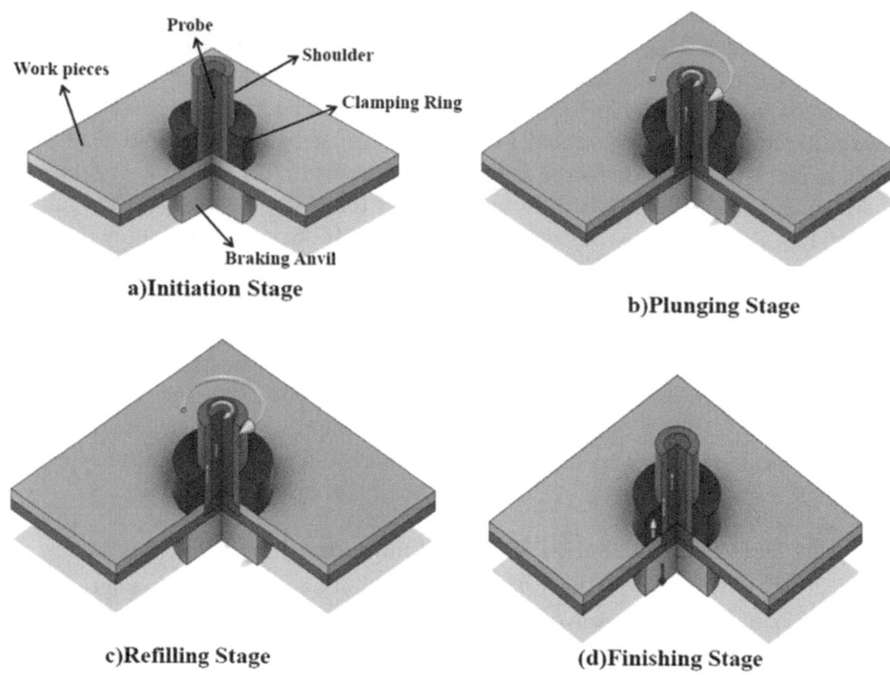

FIGURE 15.6 The steps involved in RFSSW process.

TABLE 15.11

Chemical Composition of AA7050

Al	Cu	Mg	Zn	Fe	Ti	Si	Mn
Bal.	2.45	1.91	5.93	0.14	0.03	0.05	0.64

TABLE 15.12

Welding Parameters and Mechanical Property of the RFSSW Joint

Rotation Speed (rpm)	Depth of Plunge (mm)	Dwell Time (s)	Shear Strength (kN)
1800	2.9	7	7.2

HAZ-II. But the HAZ-I and SZ have low corrosion sensitivity. Hence, the RFSSW improved the corrosion resistance of the 7050 aluminum alloy in the join area [32].

Similar aluminum alloys were welded using different methods, such as FSSW, RFSSW, FSSW using consumable pins, protrusion FSSW, etc. for automobile, marine, and aerospace applications. A defect-free weld is possible using these methods when compared to conventional methods. The FSSW can be used as a replacement for riveted joints. The better mechanical properties and corrosion resistance behavior are obtained for lower rotational speeds and shorter dwell times in most cases. This will lead to a reduction in the cost and time of the welding.

15.4 CONCLUSION

The light-weight alloys were welded using different methods, such as FSSW, RFSSW, FSSW using consumable pins, protrusion FSSW, etc. for automobile, marine, and aerospace applications. A defect-free weld is possible using these methods when compared to conventional methods. The FSSW can be used as a replacement for riveting and resistance spot welding. The welding parameter optimization, mechanical characteristics, and microstructural evolution by various researchers have shown that better mechanical properties and corrosion resistance behavior are obtained for lower rotational speeds and shorter dwell times in most of the cases. This will lead to a reduction in the weight, cost, and time of the welding.

REFERENCES

1. N. E. Thomas, W. M. Needham, J. C. Murch, M. G. Temple-Smith, and P. Dawes, US Patent No. 1995.
2. Z. Shen, Y. Ding, and P. A. Gerlich, Advances in friction stir spot welding, *Critical Reviews in Solid State and Materials Sciences*, https://doi.org/10.1080/10408436.2019.1 671799,(2019)
3. R. S. Mishra, and Z. Y. Ma, Friction stir welding and processing, *Material Science Engineering: R: Report*, 50(1–2), 1 (2005).
4. R. S. Mishra, P. S. De, and N. Kumar, Fundamentals of the friction stir process. In: *Friction Stir Welding and Processing*, edited by Rajiv Sharan Mishra, Partha Sarathi De, Nilesh Kumar, Springer, New York (2014).
5. H. Zhao, Z. Shen, M. Booth, J. Wen, L. Fu, and A. Gerlich, Calculation of welding tool pin width for friction stir welding of thin overlapping sheets, *International Journal Advanced. Manufacturing Technology*, 2018 Sep;98:1721-31.
6. M. K. Besharati-Givi, and P. Asadi, *Advances in Friction-Stir Welding and Processing*, Elsevier, Amsterdam (2014).
7. Z. Shen, Y. Chen, M. Haghshenas, and A. Gerlich, Role of welding parameters on interfacial bonding in dissimilar steel/aluminum friction stir welds, *Engineering. Science. and Technology*, 18, 270 (2015).
8. Z. Shen, Y. Chen, M. Haghshenas, T. Nguyen, J. Galloway, and A. Gerlich, Interfacial microstructure and properties of copper clad steel produced using friction stir welding versus gas metal arc welding, *Material Characterization*, 104, 1 (2015).
9. R. S. Mishra, M. W. Mahoney, S. X. McFadden, N. A. Mara, and A. K. Mukherjee, High strain rate superplasticity in a friction stir processed 7075 Al alloy, *Scripta Materilia.* 42(2), 163 (1999).
10. M. Fujimoto, M. Inuzuka, S. Koga, and Y. Seta, Development of friction spot joining, *Welding in the World*, 49(3–4), 18 (2005).
11. Q. Yang, S. Mironov, Y. S. Sato, and K. Okamoto, Material flow during friction stir spot welding, *Material Science and Engineering: A*, 527(16–17), 4389 (2010).
12. S. Lathabai, M. J. Painter, G. M. D. Cantin, and V. K. Tyagi, Friction spot joining of an extruded Al-Mg-Si alloy, *Scripta Material*, 55(10), 899 (2006).

13. H. Liu, Y. Zhao, X. Su, L. Yu, and J. Hou, Microstructural characteristics and mechanical properties of friction stir spot welded 2A12-T4 aluminum alloy, *Advance Material Science Engineering*, 2013, 1 (2013).

14. A. Gerlich, G. Avramovic-Cingara, and T. H. North, Stir zone microstructure and strain rate during Al 7075-T6 friction stir spot welding, *Metallurgical and Materials Transactions A,* 37(9), 2773 (2006).

15. A. Gerlich, P. Su, and T. H. North, Tool penetration during friction stir spot welding of Al and Mg alloys, *Journal of Material Science*, 40(24), 6473 (2005).

16. J. Shen, U. F. H. Suhuddin, M. E. B. Cardillo, and J. F. dos Santos, Eutectic structures in friction spot welding joint of aluminum alloy to copper, *Applied Physics Letter*, 104(19), 191901 (2014).

17. M. Bernardi, U. F. Suhuddin, B. Fu, J. P. Gerber, M. Bianchi, I. Ostrovsky, and B. Klusemann, (2023). Fatigue behaviour of multi-spot joints of 2024-T3 aluminium sheets obtained by refill Friction Stir Spot Welding with polysulfide sealant, *International Journal of Fatigue*, 172, 107539. https://doi.org/10.1016/j.ijfatigue.2023.107539

18. M. M. Ahmed, M. M. El-Sayed Seleman, A. M. E. S. Sobih, A. Bakkar, I. Albaijan, K. Touileb, and A. Abd El-Aty, Friction stir-spot welding of AA5052-H32 alloy sheets: Effects of dwell time on mechanical properties and microstructural evolution, *Materials*, 16(7), 2818 (2023). https://doi.org/10.3390/ma16072818

19. M. M. Ahmed, M. M. El-Sayed Seleman, I. Albaijan, and A. Abd El-Aty, Microstructure, texture, and mechanical properties of friction stir spot-welded AA5052-H32: Influence of tool rotation rate, *Materials*, 16(9), 3423 (2023). https://doi.org/10.3390/ma16093423

20. R. Gassaa, L. Hemmouche, R. Badji, L. Gilso, L. Rabet, and O. Mimouni, Effect of rotational speed and copper interlayer on the mechanical and fracture behaviour of friction stir spot welds of 5754 aluminium alloy, *Metallurgical Research & Technology*, 120(1), 118 (2023). https://doi.org/10.1051/metal/2023014

21. N. Farmanbar, S. M. Mousavizade, M. Elsa, and H. R. Ezatpour (2019). AA5052 sheets welded by protrusion friction stir spot welding: High mechanical performance with considering sheets thickness at low dwelling time and tool rotation speed. *Proceedings of the Institution of Mechanical Engineers, Part C: Journal of Mechanical Engineering Science*, 233(16), 5836–5847. https://doi.org/10.1177/0954406219850202

22. Bhardwaj, N., Narayanan, R. G., and Dixit, U. S. (2023). Exit-hole-free friction stir spot welding of aluminum alloy sheets using a consumable pin, *Journal of Materials Engineering and Performance*, 32(5), 2119–2138. https://doi.org/10.1007/s11665-022-07253-x

23. R. S. Shekhawat, V. N. Nadakuduru, and K. B. Nagumothu, Microstructures and mechanical properties of friction stir spot welded Al 6061 alloy lap joint welded in air and water, *Materials Today: Proceedings*, 41, 995–1000 (2021). https://doi.org/10.1016/j.matpr.2020.06.065

24. V. S. R. Janga, M. Awang, M. F. Yamin, U. F. Suhuddin, B. Klusemann, and J. F. D. Santos, Experimental and numerical analysis of refill friction stir spot welding of thin AA7075-T6 sheets, *Materials*, 14(23), 7485 (2021). https://doi.org/10.3390/ma1423748510.

25. M. N. Ilman, Microstructure and mechanical properties of friction stir spot welded AA5052-H112 aluminum alloy, *Heliyon*, 7(2), e06009 (2021). https://doi.org/10.1016/j.heliyon.2021.e06009

26. C. Meengam, Y. Dunyakul, and S. Kuntongkum, A study of the essential parameters of friction-stir spot welding that affect the D/W ratio of SSM6061 aluminum alloy, *Materials*, 16(1), 85 (2023). https://doi.org/10.3390/ma16010085

27. P. Homola, R. Růžek, A. R. McAndrew, and J. De Backer, Effect of primer and sealant in refill friction stir spot welded joints on strength and fatigue behaviour of aluminium alloys, *International Journal of Fatigue*, 168, 107455 (2023). https://doi.org/10.1016/j.ijfatigue.2022.107455

28. Z. Shen et al. (2019). Microstructure, static and fatigue properties of refill friction stir spot welded 7075-T6 aluminium alloy using a modified tool, *Science and Technology of Welding and Joining*. https://doi.org/10.1080/13621718.2019.1572300

29. M. M. Ahmed, M. M. El-Sayed Seleman, E. Ahmed, H. A. Reyad, K. Touileb, and I. Albaijan, Friction stir spot welding of different thickness sheets of aluminum alloy AA6082-T6, *Materials*, 15(9), 2971 (2022). https://doi.org/10.3390/ma15092971

30. Y. Zou, W. Li, Y. Xu, X. Yang, Q. Chu, and Z. Shen, Detailed characterizations of microstructure evolution, corrosion behavior and mechanical properties of refill friction stir spot welded 2219 aluminum alloy, *Materials Characterization*, 183, 111594 (2022). https://doi.org/10.1016/j.matchar.2021.111594

31. D. Zhang, J. Dong, J. Xiong, N. Jiang, J. Li, and W. Guo, Microstructure characteristics and corrosion behavior of refill friction stir spot welded 7050 aluminum alloy, *Journal of Materials Research and Technology*, 20, 1302–1314 (2022), https://doi.org/10.1016/j.jmrt.2022.07.152

32. D. S. Jo, Y. H. Moon, and J. H. Kim, Upper bound analysis of friction stir spot welding of 6061-T6 aluminum alloys, *The International Journal of Advanced Manufacturing Technology*, 120(11–12), 8311–8320 (2022). https://doi.org/10.1007/s00170-022-09294-x

33. D. Labus Zlatanovic, J. P. Bergmann, S. Balos, D. Pejic, P. Sovilj, and S. Goel, Influence of rotational speed on the electrical and mechanical properties of the friction stir spot welded aluminium alloy sheets, *Welding in the World*, 66(6), 1179–1190 (2022). https://doi.org/10.1007/s40194-022-01267-8

34. L. Yinghui, and Y. Deyun, Effect of refilling time on the microstructure and mechanical properties refill friction stir spot welded 6082 Al alloy butt joint, *Transactions of the Indian Institute of Metals*, 75(10), 2673–2682 (2022). https://doi.org/10.1007/s12666-022-02637-y

35. Y. Liu, J. Bi, and D. Yang, Microstructure and mechanical properties of refill friction stir spot welded 6005 Al alloy butt joints, *JOM*, 74(7), 2838–2845 (2022). https://doi.org/10.1007/s11837-022-05341-w

36. J. R. Davis, *Aluminum and Aluminum Alloys*, ASM international, 1993.

37. Z. Shen, Y. Chen, J. S. C. Hou, X. Yang, and A. P. Gerlich, Influence of processing parameters on microstructure and mechanical performance of refill friction stir spot welded 7075-T6 aluminium alloy, *Science and Technology of Welding and Joining*, 20(1), 48–57 (2015). https://doi.org/10.1179/1362171814Y.0000000253

38. V. K. Yadav, V. Gaur, and I. V. Singh, (2020). Effect of post-weld heat treatment on mechanical properties and fatigue crack growth rate in welded AA-2024, *Materials Science and Engineering: A*, 779, 139116. https://doi.org/10.1016/j.msea.2020.139116

16 Nano Particles Reinforcements Via Friction Stir Processing (FSP)

*Harikishor Kumar, Rabindra Prasad,
Parshant Kumar, and Lokasani Bhanuprakash*

16.1 FRICTION STIR PROCESSING

Friction stir processing (FSP) is a groundbreaking material processing technique that has revolutionized the way we engineer and enhance the properties of metals and alloys. Developed in the early 1990s by Thomas W. Nelson and Wayne Thomas at the Welding Institute (TWI) in the United Kingdom [1], FSP has since evolved into a widely adopted method with a broad range of applications in industries such as aerospace, automotive, and marine engineering. This innovative process relies on the concepts of frictional heating and plastic deformation to modify the microstructure and properties of materials, offering numerous advantages over traditional methods such as casting, forging, and welding.

At its core, FSP involves a specially designed rotating tool that is plunged into the surface of a workpiece material. The tool's rapid rotation generates frictional heat, softening the materials without melting [2]. As the tool traverses along the surface, it stirs and mixes the material, causing a dynamic recrystallization (DRX) of the grain structure. The result is a refined microstructure with improved mechanical properties and enhanced homogeneity. The schematic of the process is shown in Figure 16.1.

One of the most significant advantages of FSP is its ability to modify materials with minimum heat input, reducing the risk of thermal distortion and residual stress common in traditional welding processes. This makes it ideal for joining dissimilar materials that are otherwise challenging to weld, such as aluminum to copper or different grades of the same material. Furthermore, FSP can be applied to different materials, including but not limited to aluminum, copper, magnesium, titanium, and composites.

The aerospace industry has been one of the most prominent beneficiaries of FSP. Aerospace components subjected to extreme conditions during flight require materials with exceptional strength, durability, and fatigue strength. FSP allows manufacturers to tailor the properties of critical components such as wing panels, engine components, and landing gear, increasing their performance and lifespan. By eliminating traditional defects such as porosity and cracks, FSP contributes to the safety and reliability of

DOI: 10.1201/9781003432289-16

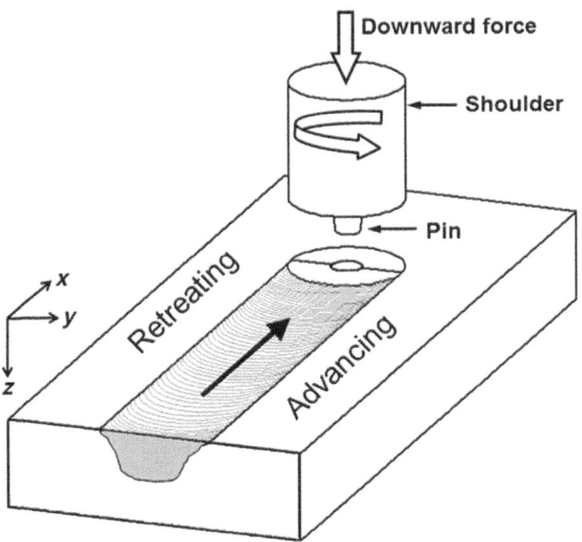

FIGURE 16.1 Schematic of friction stir processing.

aircraft. In addition to aerospace, the automotive industry has also embraced FSP as a means to enhance vehicle performance and light weight. By processing light-weight materials such as aluminum and magnesium, manufacturers can reduce fuel consumption and emissions, aligning with stringent environmental regulations [2]. FSP has been used to improve the structural integrity of car frames, transmission components, and heat exchangers, ensuring enhanced efficiency and safety. The marine sector is yet another field where FSP has demonstrated its potential. The corrosive environment of seawater poses a significant challenge to materials used in shipbuilding and offshore structures. FSP can modify the surface properties of materials, increasing their resistance to corrosion and wear. This extends the life span of marine components and reduces maintenance costs, making it an attractive choice for shipbuilders and offshore engineers.

Beyond these industries, FSP has found applications in the development of novel materials with superior properties. By tailoring the microstructure and introducing additives during the process, researchers can create materials with enhanced strength, ductility, and thermal conductivity. Tailoring the microstructure and incorporating particles externally in the materials for optimum properties have also been done by the researchers, but through different techniques [3–7]. These advanced materials hold promise in fields such as nuclear engineering, where materials must withstand extreme conditions, and the renewable energy system, where efficient heat exchangers and energy storage solutions are crucial.

16.2 METHODS TO REINFORCE PARTICLES

The synthesis of Al-SiC composites via FSP was first documented in the first publication by Mishra et al. [8]. As a result, several research projects have been sparked that are centered on employing FSP to fabricate composite materials at both the surface and bulk levels. In the following sections, the research used by various researchers to introduce particles into the matrix through FSP is discussed. Also, the technique followed to introduce the particles will be discussed.

16.2.1 Inserting Particles by Drilling Groove on the Plate

Mishra et al. used a procedure in their original study where they coated the 5083 Al plate surface with a coating made up of SiC powder and some additives. They used FSP to integrate SiC particles into the stirred zone (SZ) after drying the layer. Morisada et al. made an AZ31 composite reinforced with multiwall carbon nanotubes (MWCNT) via FSP in a different work. In this instance, they first machined a groove on the AZ31 plate that was 1 mm wide and 2 mm deep (1 mm × 2 mm), and then they filled it with MWCNTs. Following that, FSP was performed along the groove at varied traverse rates of 25–100 mm/min and a constant rotational speed of 1500 rpm [9]. The groove approach was also used by Mishra et al. to produce composites made of AZ31-SiC and 5083 Al-Fullerene [10,11]. Several studies have now embraced this strategy, which is shown in Figure 16.2.

Figure 16.3 shown below is the optical micrograph for the composite Cu-SiC fabricated by FSP following this method.

It can be seen from the images that the Si particles are uniformly distributed in the Cu matrix throughout the SZ. Also, the interface between the matrix and the reinforcement is clean. The groove method has been more popular for producing different metal matrix composites (MMCs) using FSP [12–20]. The particle SiO_2 was reinforced in AZ61 Mg plates by Lee et al. to fabricate bulk composites (in a 6 mm deep groove) [12,13]. The SiO_2 particles were 5% by volume and were processed through single pass FSP. Results showed the particles were distributed uniformly throughout the matrix without any noticeable defects, such as agglomeration

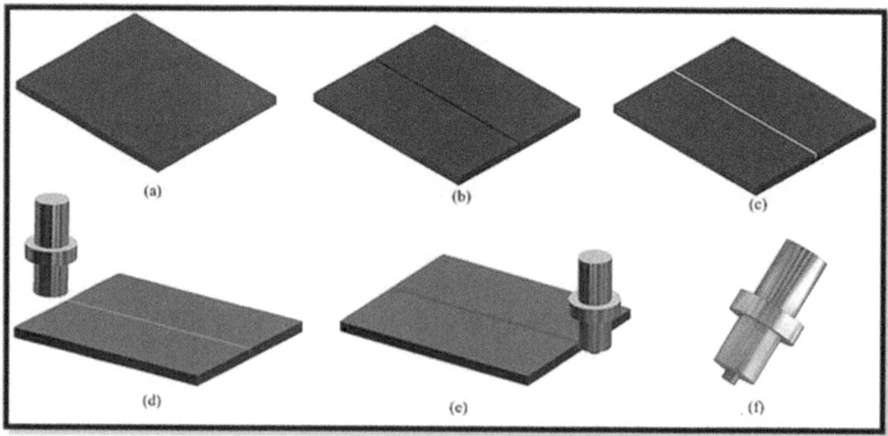

FIGURE 16.2 Schematic diagrams for composite fabrication via FSP: (a) plate of known dimension; (b) groove making on the surface of plate; (c) filling of particles in the groove; (d) closing of groove with pin-less tool; (e) processing of plate; and (f) tool with pin used for processing.

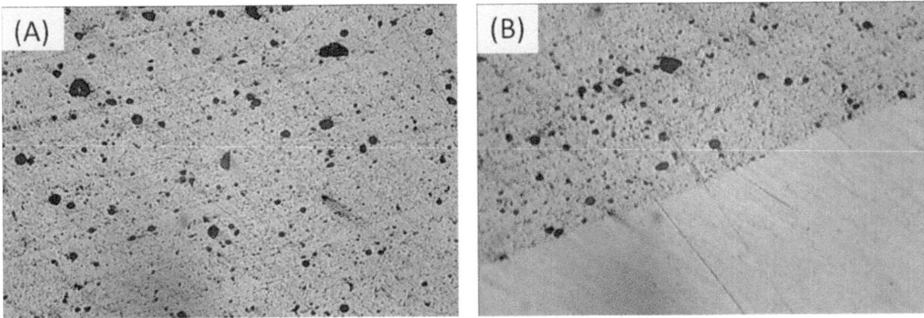

FIGURE 16.3 Images (optical microscopy) of (a) stir zone and (b) interface between matrix and reinforcement.

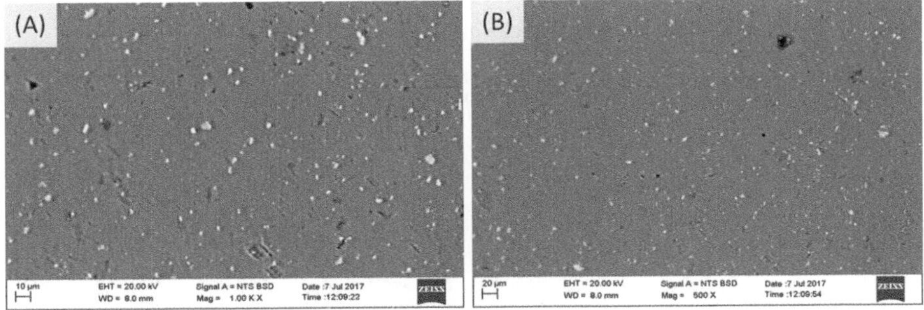

FIGURE 16.4 Images (SEM) of composite (Al/Mica) manufactured by FSP: (a) single pass and (b) multi-pass.

of particles. A similar study was conducted where mica was reinforced in the Al matrix following the groove technique through FSP. The SEM image of the composite (Al/Mica) is shown in Figure 16.4.

The groove approach was also used by Sun et al. to create bulk Mg-nano-SiC composites [14]. They used a blunt tool (without a pin) to first close the groove before processing, as seen in Figure 16.2d. Additionally, it has been proposed that the original plate can also be covered with a thin sheet (Figure 16.5a) to seal the groove [16], so the wastage of particles during processing can be avoided. It is to be noted that the cover plate should be of the same nature as the base plate. The schematic for the same is shown in Figure 16.5.

Hollow tools have also been used to reinforce the particles in the materials (Figure 16.6). A hole at the center of the tool is created, and the particles to be reinforced are packed inside. As the tool traverses in the desired direction, particles get mixed into the surface of the base materials. This hollow tool may be consumable as well as non-consumable, depending on whether the purpose is to just reinforce the particle or to coat the tool materials as well.

Reports are available where the groove method has been used to introduce particles into the matrix by single- and multi-pass FSP. This technique of groove making on the surface of a plate to reinforce particles in the matrix is quite simple and

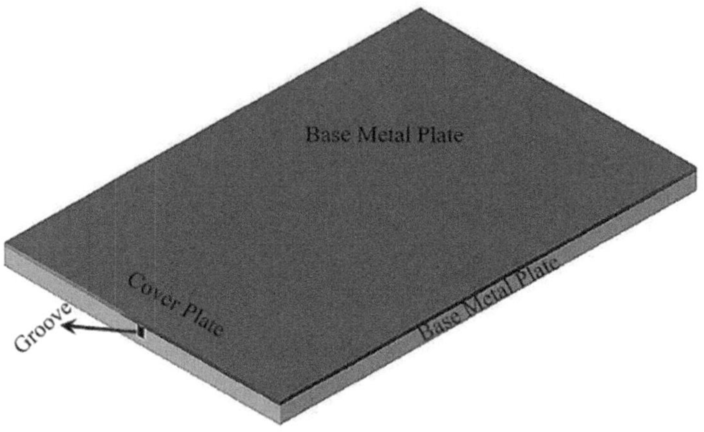

FIGURE 16.5 Schematic diagram for closing of groove with thin plate.

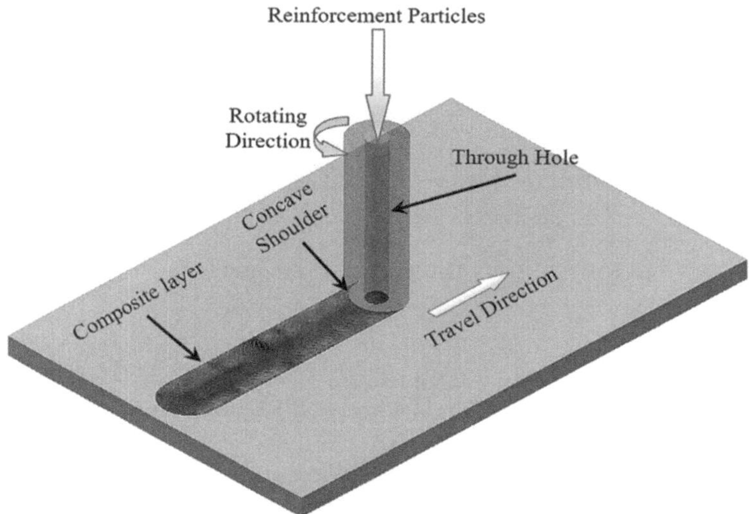

FIGURE 16.6 Schematic for reinforcement of particles through hollow tool.

has been found to produce composites with uniform particle distribution. Also, as evident from Figure 16.4b (Al/Mica composite), with an increased number of FSP passes, particles have dispersed more uniformly in the matrix.

16.2.2 Inserting Particles by Drilling Hole in the Plate

Similar methods include using several drilled holes on the plate to embed reinforcing particles, as schematically shown in Figure 16.7.

Al 1100/NiTi composite was fabricated by FSP following the hole-making technique on the surface of the plate [21]. Four tiny holes, each measuring 1.6 mm in

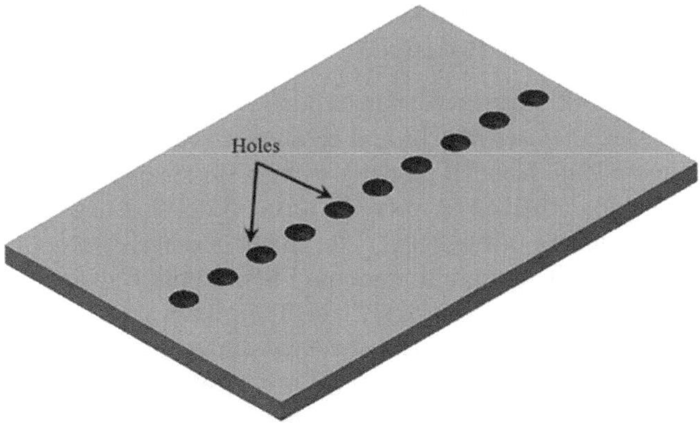

FIGURE 16.7   Schematic of the reinforcing by hole-making technique.

diameter and 76 mm in depth and located around 0.9 mm below the surface of the Al 1100 plate, were machined. All of the holes could be processed by FSP in a single pass since they were all equally distributed across a width that was smaller than the shoulder width. The holes were carefully filled with NiTi particles and processed by FSP at a tool rotating speed of 1000 rpm and bed travel speed of 25 mm/min. The final composite has NiTi particles distributed evenly throughout the Al matrix.

This approach of making holes on the surface of the plate has been used to produce composites of 6061 Al with NiTi particle additions through many passes of FSP [22]. This technique has worked well to include reinforcing particles into MMCs, which entail drilling a series of holes in the processing plate. In Figure 16.7, the idea of this method is demonstrated by the creation of holes within the shoulder diameter, their subsequent filling with reinforcing powder, and their subsequent FSP exposure. In a study conducted by Dixit et al., a hole on the surface of the plate was machined to reinforce NiTi in Al 1100 [21]. The number of holes was four at equal spacing, with a width of 1.6 mm each and 76 mm deep. The holes were then filled with NiTi particles, closed with a pinless tool, and processed with the tool rotating at 1000 rpm and traversing at 25 mm/min. The particles were uniformly dispersed throughout the matrix. Similarly, Liu et al., used the drill-hole method to mix up 6 vol% MWCNT with aluminum and saw robust adhesion between the nanotubes and the matrix of aluminum [23]. Al-Al$_2$O$_3$-CNT hybrid composites were made by Du et al. using MWCNT and nanoscale Al$_2$O$_3$ particles [24]. TiO$_2$ nanofibers have also been incorporated into Al using the drill-hole approach [25], and functionally graded Al-TiC composites have been created using the friction stir additive manufacturing process [26]. Although the approach has been followed to fabricate MMCs by various researchers, it is important to note that different studies may use varying numbers, shapes, and sizes of holes.

16.2.3 FABRICATION THROUGH POWDER METALLURGY AND FRICTION STIR PROCESSING

In some instances, FSP and powder metallurgy (PM) have been coupled in the production of MMCs. Here, matrix and reinforcements will be in powder form, which will be mixed and blended properly using a suitable ball milling setup. They are then compressed and sintered to create a tiny billet that has the required reinforcing evenly distributed throughout. After the first sintering process, the billet is intended to have enough strength to undergo FSP without fracture. The billet goes through final consolidation during FSP, which eliminates any remaining porosity and ensures that the reinforcement is distributed uniformly owing to material flow. The predicted result of mixing PM and FSP is a completely dense composite with a uniform reinforcement distribution. Lee et al. used this method to manufacture an Al composite reinforced with Si particles and reported the eradication of porosity in the SZ and the uniform dispersion of Si particles. Liu et al. also created an Al-CNT composite using the same technique; they then hot-rolled the composite while aligning the CNTs with the direction of rolling to create high-performance composites [27].

16.2.4 INTRODUCTION OF PARTICLES VIA IN SITU METHOD

Instead of adding reinforcement particles separately, creating them in situ during the manufacturing process has several benefits, including an interruption-free interface, better bonding between reinforcements and the matrix, and the production of smaller, more thermodynamically stable reinforcement particles. The evolution of heat during the processing of materials via FSP and the beneficial effects of frictional heating have been used to treat in situ composites. The in situ response has been initiated using two main methodologies. The base plate is first loaded with precursor powder via grooves or drilled holes, and then FSP is added. The component powders interact with the Al matrix and/or one another during FSP to create reinforcement particles. Once the frictional heat has started the process, it maintains its exothermic character. By putting Ni powder in a groove and applying FSP, S. Zhang et al. used this technique to create an $Al-Al_3Ni$ composite [28]. Al_3Ni particles were created by the in situ interaction between Ni and Al during FSP. Ke et al. created a comparable $Al-Al_3Ni$ composite by filling drilled holes with Ni powder using a similar technique [29]. Additionally, polymer-derived ceramics (PDC) in a Cu matrix have been created using this technique [30,31]. The PDC composite was created by first employing the groove approach to insert a polymer precursor through FSP, then heat-treating the polymer to turn it into ceramic, then doing another cycle of FSP. The exothermic reactions that occur during FSP have been exploited by researchers by using these in situ techniques to harness the reactive qualities of the materials and produce the required composite architectures and attributes. PM is a different choice for the in situ technique. In this method, the matrix powder and precursor powders are mixed together, and the combination is then solidified using PM. An in situ composite is produced when FSP is used to convert the resultant PM billet. $Al-Al_3Ti-Al_2O_3$ composites were successfully created with this technique using reactive FSP on an

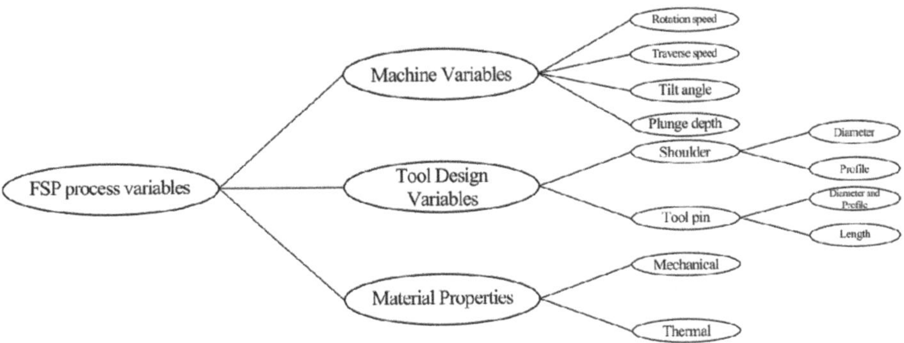

FIGURE 16.8   Schematic of the parameters involved during FSP.

Al-TiO$_2$ billet created via PM. Al$_3$Ti and Al$_2$O$_3$ particles were created inside the Al matrix during FSP as a result of the interaction between the components (Al and TiO$_2$) [32,33]. Similar techniques were also used by Hsu et al. to process Al-Al$_3$Ti nanocomposites [34]. Numerous in situ Al-based composites have been created using the reactive FSP approach [35–38].

16.3   INFLUENCE OF PROCESSING CONDITIONS

Complex material flow and plastic deformation are seen during FSP, and these phenomena are controlled by both process factors and tool geometry. Notably, factors such as temperature, strain, and strain rate may be significantly influenced by the shape of the tool shoulder and pin, as well as their varied characteristics. These elements, in turn, may have an impact on the final microstructure and particle dispersion in FSP-made composites. To assure the creation of defect-free composites using FSP, it is essential to have a thorough grasp of these factors in order to make wise decisions during the manufacturing process. The schematic of the factors involved during processing is shown in Figure 16.8.

16.3.1   PROCESSING CONDITIONS RELATED TO MACHINE

Heat is produced as a result of friction between the tool and the workpiece, which causes the material to soften and flow in the region of the pin periphery as the tool along with the bed moves in a forward direction. The amount of heat received by the materials during processing is determined by the ratio of tool rotation and the traverse of the bed. Heat input will be higher if the tool rotational speed is higher, which will cause more deformation and adiabatic heating. However, in the case of traverse speed, when its value is low, it produces more heat because it remains at a particular place for a longer period of time. Therefore, a larger ratio of tool rotation to tool traverse speed causes more heat generation and absorption in the material. The equation published by Arbegast and Hartley [39] can be used to predict the maximum temperature achieved in the SZ. The maximum temperature during friction stir welding (FSW) or FSP of different aluminum alloys has been shown to commonly fall between 0.6 and 0.9 times the melting point (T_m).

FSP, as previously indicated, causes considerable plastic deformation, with reported stresses reaching as high as 40 [40]. According to Buffa et al.'s hypothesis, effective strain distribution during FSW is asymmetric throughout the weld line, with the advancing side experiencing the largest strain [41]. A significant amount of plastic deformation occurred during FSW/FSP, as shown by the highest computed strain rate for FSW of 160/s. Lim et al. [21] looked into the impact of rotational speeds between 1500 and 2500 rpm on the dispersion of carbon nanotubes (CNTs) in the aluminum matrix during FSP. According to their study, increasing the rotating speed led to better nanotube dispersion inside the matrix. Researchers researching various MMCs reinforced with various particles employing FSP have obtained similar results [17,42–45]. In light of this, it can be said that an increase in rotating speed often results in improved particle dispersion. While samples processed at 800 rpm did not display onion ring patterns, those processed at 1000 and 1200 rpm did. As the ratio of forward velocity to spinning speed grew, so did the distance between the onion rings. It was thought that the passage of material from the warmer top zone to the cooler bottom zone caused these bands or rings to develop. But slower rotating speeds limited material flow, which meant there wasn't enough heat input to produce onion rings. Another important aspect to take into account while analyzing the particle distribution in FSP composites is traversal speed. Farazi and Asadi showed that while lowering the traverse speed, the dispersion of Al_2O_3 particles in AZ31 Mg was improved. The reason for the enhanced particle dispersion with slower traverse speeds was determined to be the greater heat input brought on by the higher tool rotation and traverse speed ratio. Greater heat input caused the material to soften more, which allowed for free flow around the pin and better stirring action. For tiny particles in particular, a single pass of FSP has been shown to be inadequate to ensure uniform reinforcing distribution. In order to produce greater particle dispersion across the matrix, multi-pass FSP has been frequently used in research. With two or more FSP passes, several experiments have shown a considerable improvement in particle dispersion. Additionally, it has been suggested that when the advancing side turns to the retreating side in the following pass, the tool's rotation direction can be adjusted to balance material flow and provide more uniform particle dispersion [17].

16.3.2 Processing Condition Related to Tool Geometry

In the FSP process, the tool's shape has a big impact on heat production, applied stress and torque, material flow, and deformation surrounding the tool. Therefore, it is vital to take into account the geometry of the tool, including the pin's size, shape, and profile, as well as the shoulder's diameter, profile, and surface angle. Because most of the heat produced during FSW/FSP is caused by friction between the shoulder and the workpiece, the shoulder's diameter is a crucial consideration. The shoulder also aids in holding the plasticized material in position during processing. To produce a solid weld, a concave shoulder surface is frequently chosen. To suit differing materials and processing circumstances, several shoulder profiles have evolved over time. Different profiles, such as fillet, hollow, and spiral, have been used to improve weld structure quality [46–49]. These profiles improve the bonding between the tool shoulder and the materials to be processed. Also, prevent the ejection of deformed materials from behind the shoulder; these profiles serve their intended purpose. The tool pin configuration, which includes its profile, shape, and size, has a big impact on how much material flows during FSP. Many other pin shapes have been used, including conical, triangular, three-flat, and cylindrical [50]. Additionally, recent reports on hexagonal, pentagonal, and square cross-section pins have also been reported [51]. To improve material flow and mixing, intricate pin designs comprising threads, whorls, and triflutes have been created [2]. It has been reported that a trifluted tool pin is better than a cylindrical pin in terms of material flow [8,46,51]. How material flow is affected during processing is determined by the dynamic-to-static volume ratio. It shows the volume of material affected by the tool shoulder and the tool pin itself. Pins with unusual properties have greater ratios than conventional cylindrical pins, which generally have a ratio of 1.1:1 [2].

16.4 MICROSTRUCTURE EVOLVEMENT

DRX is triggered by thermomechanical processing, which can be induced by FSP to improve grain architectures [2,52–54]. The refining of grains results from this recrystallization process. Continuous dynamic recrystallization (CDRX), discontinuous dynamic recrystallization (DDRX), and geometric dynamic recrystallization (GDRX) are examples of DRX processes in aluminum (Al) alloys [55,56]. Furthermore, dynamic recovery (DRV) is essential for the recovery and amplification of Al grains. These processes take place during the FSW/FSP of Al alloys and are controlled by variables including temperature, strain, and processing conditions. Al alloys experience DRV because of their high stacking fault energy. It has been found that the dislocation rearrangement mechanisms involved in CDRX and GDRX are strongly connected to the grain refining process during FSP [53,57]. CDRX develops over time when dislocations from high-temperature deformation move to sub-grain boundaries, causing them to become more misoriented and finally morph into grain barriers.

On the other hand, DDRX requires a concentration of free dislocations, which can only build up in extremely stressed areas and happen through the traditional recrystallization method. Although the presence of particles can affect nucleation and grain development, the basic mechanism of grain refining during FSP is unaffected in composites. The effect of nano-Al_2O_3 particle reinforcement on matrix (AA 6061) grain size was made via FSP, and it was found that the composite had smaller grains (2.5 m) than the FSPed alloy (5.9 m) without reinforcement [58]. Khodabakhshi et al. [59] and Du et al. [24] presented findings that were comparable to each other regarding the grain-refining impact of nanoparticles on MMCs. The pinning action of particles on the grain boundaries, which prevents grain development, was thought to be the cause of the composites' reduced grain size. Recrystallization in composites can also be affected by particle-stimulated nucleation (PSN), which actually increases the nucleation rate and hence lowers the grain size of the metal matrix. PSN, on the other hand, is restricted to particles with a critical size (0.1 µm) and certain processing parameters, such as low temperature and high strain rates, which encourage dislocation buildup close to the particles.

When the temperature of the processed composites is increased, like in solution heat treatment, the fine-grained microstructure created by FSP is frequently vulnerable to aberrant grain development because the pinning effect drastically declines at high temperatures [60]. According to certain reports, the evenly dispersed, strong pinning action of small particles can even stop aberrant grain formation in MMCs that have undergone high-temperature annealing following FSP [54,58]. Dislocation climb, cross slip, and annihilation processes are quite difficult in materials with low stacking fault energy, like copper. According to Kumar et al. for copper-PDC composites [31], this can lead to an increase in the dislocation density, which will favor DDRX. The extra dislocations produced as a result of the differing flow dynamics between the softer copper matrix and the stronger ceramic particles support DDRX by increasing the dislocation density. It should be emphasized; however, that depending upon the extent of deformation and temperature rise experienced during processing, DDRX can also happen during the FSP of aluminum and its composites [52].

16.5 MECHANICAL PROPERTIES

In comparison to the base materials and FSP-treated alloys, MMCs made from FSP often have better mechanical characteristics. A number of elements, such as grain refinement and reinforcement strengthening, contribute to this improvement in characteristics such as hardness and strength. Researchers like Liu et al. have studied how the hardness of the composite Al/MWCNT varied as a function of the volume of reinforcement, and they have shown that the inclusion of CNTs significantly improves hardness, especially at higher CNT levels. Various particle-reinforced composites made using FSP have consistently shown this trend of increased mechanical characteristics. A better balance of strength and ductility is also generated by FSP composites as compared to standard MMCs by way of improved ductility.

A clear matrix-reinforcement interface without interruption in the form of detrimental reaction products, strong interface interaction, and even sputtering of particles are a few of the variables that contribute to the increased ductility. Re-crystallization and the pinning effect are used to improve the grain structure while creating surface composites using FSP. The pinning effect is a phenomenon wherein smaller-sized reinforced particles prevent grain boundaries from moving, which strengthens the material. These reinforcements may be affixed to the borders of the grains or may be buried within them. Under rising tensile stresses during tensile testing, dislocations inside the grains have a tendency to migrate towards the grain interiors. Dislocation lines create loops around the reinforcements, acting as barriers to slip, increasing the dislocation density and necessitating greater stresses for material deformation. In general, the strengthening processes utilized in composites made using FSP are identical to those used in traditional MMCs. The following are important strengthening mechanisms observed in MMCs fabricated by reinforcing particles: work hardening because of the varied nature of deformation of matrix and reinforcements, quench hardening brought on by dislocation generation caused by differences in the coefficient of thermal expansion (CTE) between the particle and matrix, and Orowan strengthening [61]. Researchers may disagree on the relative significance of these four strengthening methods, though. For metal matrix nanocomposites, certain researchers, including Zhang and Chen, offered estimates, whereas Clyne's technique [62] was used for microcomposites, and the findings nearly matched experimental studies. Sanaty-Zadeh and Rohatgi [63] found that Hall-Petch strengthening was the main mechanism at work even in microcomposites utilizing Al_2O_3 and Y_2O_3 as reinforcements for a Mg matrix. All four strengthening processes were taken into account by Zarghani et al. [64] while creating Ti/Al_2O_3 nanocomposites via FSP. They created surface composites with three distinct groove widths: 0.8, 1.2, and 1.6 mm. They used reinforcements with particle sizes of 80 and 20 nm. The findings showed that the Hall-Petch effect had the most impact on the increase in strength, whereas the load-bearing effect had the least.

16.6 CONCLUDING REMARKS AND FUTURE DIRECTION

The effectiveness of FSP as a method for improving grain structures and changing microstructures has been thoroughly investigated and shown. By combining several types of reinforcements, such as ceramics, metallic particles, and carbon nanostructures, into metallic matrices using FSP, it has proved its adaptability in processing composites. With various techniques, including the groove technique, hole-making, PM, FSP combining, and in situ procedures, this methodology offers significant potential for generating both bulk and surface composites, including nanocomposites and hybrid composites. The uneven distribution of particles, which can negatively impact the microstructure and characteristics of the final composites, has been a significant obstacle in the fabrication of MMCs. Several tool designs have been created to provide more uniform particle dispersion during FSP as a solution to this problem. Furthermore, it has been found that the processing variables have a significant impact on the microstructure and characteristics of the treated composites, and using multipass FSP often results in better particle dispersion.

In general, as compared to the base matrix without reinforcement, treated composites show improved mechanical characteristics. The main processes for the increased strength in these composites are grain refining and Orowan strengthening. The characteristics of the composites are subsequently impacted by the distribution of reinforcing particles, which has a significant impact on how the microstructure evolves throughout processing. Numerous studies have looked at how reinforcing particles affect the development of microstructures. It is frequently seen that the presence of particles causes a pinning effect, which produces smaller grain sizes. In composites created by FSP, researchers have tried to establish relationships between microstructure and characteristics.

Higher hardness is frequently seen in surface composites created with FSP, improving wear resistance when compared to the base material without reinforcement. These surface composites go through a wear process, and the sort of wear they suffer depends on things such as the applied force and sliding distance. The reinforcements' ability to support weight as well as their lubricating impact greatly lower the pace at which the surface composite layers wear out. Studies integrating various particles into metal matrices show that FSP has interesting potential for the creation of non-equilibrium composites. Positive reports on the effective strengthening of fibers have surfaced [65,66]. Additionally, FSP has shown that it is capable of creating functionally graded composites [67], employing a variety of reinforcements both on the surface and inside the bulk of the material.

Despite the fact that FSP has mostly been used with softer metals with lower melting points, such as aluminum and magnesium, there is rising interest in exploring its possibilities with harder substances, such as titanium and steel. Despite the fact that there are few studies on the FSP of these stronger materials and their composites, there is a considerable demand to investigate their processing options. In such circumstances, it is crucial to take tool wear into account, demanding careful material selection and tool design. Wrought-based, tougher alloys, tungsten carbide, and polycrystalline cubic boron nitride (PCBN) are suggested alternatives. The utilization of polymer-based materials for the creation of non-metallic composites via FSP is another thing that may be looked into. FSP of polymers and processing polymer-based composites have shown potential in preliminary investigations. FSP offers the opportunity for targeted property enhancements in certain spots on engineering components because it predominantly impacts the material's surface. For the purpose of surface alloying, other highly soluble metals can also be added via FSP. In conclusion, FSP has the potential to open up a brand-new area for surface engineering.

REFERENCES

1. W. M. Thomas, J. C. Nicholas, J. C. Needham, M. G. Murch, P. Temple-Smith, and C. J. Dawes, *Stir Friction Welding,* British Patent 9125978 (1991).
2. R. S. Mishra and Z. Y. Ma, "Friction stir welding and processing," *Materials Science and Engineering: R: Reports,* vol. 50, no. 1–2, pp. 1–78, 2005, doi:10.1016/j.mser.2005.07.001.
3. P. Murugesan, V. Satheeshkumar, N. Jeyaprakash, G. Prabu, and C.-H. Yang, "Nano-level mechanical and tribological behavior of additively manufactured AlSi10Mg plates," *Archives of Civil and Mechanical Engineering,* vol. 23, no. 1, pp. 62–62, 2023, doi:10.1007/s43452-023-00605-x.
4. N. Jeyaprakash, G. Prabu, and C.-H. Yang, "The Influence of Different Phases on the Microstructure and Wear of Inconel-718 Surface Alloyed with AlCuNiFeCr Hard Particles Using Plasma Transferred Arc," *Journal of Materials Engineering and Performance,* vol. 31, no. 12, pp. 9921–9934, 2022, doi:10.1007/s11665-022-06982-3.
5. N. Jeyaprakash, G. Prabu, C.-H. Yang, M. P. Banda, and E. Mohan, "Structural and Nanohardness of Laser Melted CM247LC Nickel-Based Superalloy," *Transactions of the Indian Institute of Metals,* vol. 76, no. 2, pp. 287–295, 2023, doi:10.1007/s12666-022-02734-y.
6. N. Jeyaprakash, C.-H. Yang, and S. S. Karuppasamy, "Laser cladding of colmonoy-6 particles on Inconel 625 substrate: Microstructure and corrosion resistance," *Surface Review and Letters,* vol. 29, no. 08, 2022, doi:10.1142/S0218625X22501025.
7. N. Jeyaprakash, C.-H. Yang, P. Susila, and S. S. Karuppasamy, "Laser cladding of NiCrMoFeNbTa particles on inconel 625 alloy: Microstructure and corrosion resistance," *Transactions of the Indian Institute of Metals,* vol. 76, no. 2, pp. 599–612, 2023, doi:10.1007/s12666-022-02701-7.
8. R. S. Mishra, Z. Y. Ma, and I. Charit, "Friction stir processing: a novel technique for fabrication of surface composite," *Materials Science and Engineering: A,* vol. 341, no. 1-2, pp. 307–310, 2003, doi:10.1016/S0921-5093(02)00199-5.
9. Y. Morisada, H. Fujii, T. Nagaoka, and M. Fukusumi, "MWCNTs/AZ31 surface composites fabricated by friction stir processing," *Materials Science and Engineering: A,* vol. 419, no. 1-2, pp. 344–348, 2006, doi:10.1016/j.msea.2006.01.016.

10. Y. Morisada, H. Fujii, T. Nagaoka, and M. Fukusumi, "Effect of friction stir processing with SiC particles on microstructure and hardness of AZ31," *Materials Science and Engineering: A*, vol. 433, no. 1–2, pp. 50–54, 2006, doi:10.1016/j.msea.2006.06.089.

11. Y. Morisada, H. Fujii, T. Nagaoka, K. Nogi, and M. Fukusumi, "Fullerene/A5083 composites fabricated by material flow during friction stir processing," *Composites Part A: Applied Science and Manufacturing*, vol. 38, no. 10, pp. 2097–2101, 2007, doi:10.1016/j.compositesa.2007.07.004.

12. C. Lee, J. Huang, and P. Hsieh, "Mg based nano-composites fabricated by friction stir processing," *Scripta Materialia*, vol. 54, no. 7, pp. 1415–1420, 2006, doi:10.1016/j.scriptamat.2005.11.056.

13. C. J. Lee and J. C. Huang, "High strain rate superplasticity of mg based composites fabricated by friction stir processing," *Materials Transactions*, vol. 47, no. 11, pp. 2773–2778, 2006, doi:10.2320/matertrans.47.2773.

14. K. Sun, Q. Y. Shi, Y. J. Sun, and G. Q. Chen, "Microstructure and mechanical property of nano-SiCp reinforced high strength Mg bulk composites produced by friction stir processing," *Materials Science and Engineering: A*, vol. 547, pp. 32–37, 2012, doi:10.1016/j.msea.2012.03.071.

15. W. Wang, Q.-y. Shi, P. Liu, H.-k. Li, and T. Li, "A novel way to produce bulk SiCp reinforced aluminum metal matrix composites by friction stir processing," *Journal of Materials Processing Technology*, vol. 209, no. 4, pp. 2099–2103, 2009, doi:10.1016/j.jmatprotec.2008.05.001.

16. D. K. Lim, T. Shibayanagi, and A. P. Gerlich, "Synthesis of multi-walled CNT reinforced aluminium alloy composite via friction stir processing," *Materials Science and Engineering: A*, vol. 507, no. 1-2, pp. 194–199, 2009, doi:10.1016/j.msea.2008.11.067.

17. S. Sahraeinejad, H. Izadi, M. Haghshenas, and A. P. Gerlich, "Fabrication of metal matrix composites by friction stir processing with different Particles and processing parameters," *Materials Science and Engineering: A*, vol. 626, pp. 505–513, 2015, doi:10.1016/j.msea.2014.12.077.

18. S. A. Alidokht, A. Abdollah-zadeh, S. Soleymani, and H. Assadi, "Microstructure and tribological performance of an aluminium alloy based hybrid composite produced by friction stir processing," *Materials & Design*, vol. 32, no. 5, pp. 2727–2733, 2011, doi:10.1016/j.matdes.2011.01.021.

19. D. Lu, Y. Jiang, and R. Zhou, "Wear performance of nano-Al2O3 particles and CNTs reinforced magnesium matrix composites by friction stir processing," *Wear*, vol. 305, no. 1-2, pp. 286–290, 2013, doi:10.1016/j.wear.2012.11.079.

20. A. Dolatkhah, P. Golbabaei, M. K. Besharati Givi, and F. Molaiekiya, "Investigating effects of process parameters on microstructural and mechanical properties of Al5052/SiC metal matrix composite fabricated via friction stir processing," *Materials & Design*, vol. 37, pp. 458–464, 2012, doi:10.1016/j.matdes.2011.09.035.

21. M. Dixit, J. W. Newkirk, and R. S. Mishra, "Properties of friction stir-processed Al 1100-NiTi composite," *Scripta Materialia*, vol. 56, no. 6, pp. 541–544, 2007, doi:10.1016/j.scriptamat.2006.11.006.

22. D. R. Ni, J. J. Wang, Z. N. Zhou, and Z. Y. Ma, "Fabrication and mechanical properties of bulk NiTip/Al composites prepared by friction stir processing," *Journal of Alloys and Compounds*, vol. 586, pp. 368–374, 2014, doi:10.1016/j.jallcom.2013.10.013.

23. Q. Liu, L. Ke, F. Liu, C. Huang, and L. Xing, "Microstructure and mechanical property of multi-walled carbon nanotubes reinforced aluminum matrix composites fabricated by friction stir processing," *Materials & Design*, vol. 45, pp. 343–348, 2013, doi:https://doi.org/10.1016/j.matdes.2012.08.036.

24. Z. Du, M. J. Tan, J. F. Guo, G. Bi, and J. Wei, "Fabrication of a new Al-Al$_2$O$_3$-CNTs composite using friction stir processing (FSP)," *Materials Science and Engineering: A*, vol. 667, pp. 125–131, 2016, doi:10.1016/j.msea.2016.04.094.

25. L. Zhang et al., "A metal matrix composite prepared from electrospun TiO_2 nanofibers and an Al 1100 alloy via friction stir processing," *ACS Applied Materials & Interfaces*, vol. 1, no. 5, pp. 987–991, 2009, doi:10.1021/am900095x.

26. A. Sharma, V. Bandari, K. Ito, K. Kohama, R. M., and H. S. B.V, "A new process for design and manufacture of tailor-made functionally graded composites through friction stir additive manufacturing," *Journal of Manufacturing Processes*, vol. 26, pp. 122–130, 2017, doi:10.1016/j.jmapro.2017.02.007.

27. Z. Y. Liu, B. L. Xiao, W. G. Wang, and Z. Y. Ma, "Analysis of carbon nanotube shortening and composite strengthening in carbon nanotube/aluminum composites fabricated by multi-pass friction stir processing," *Carbon*, vol. 69, pp. 264–274, 2014, doi:10.1016/j.carbon.2013.12.025.

28. S. Zhang et al., "In-situ formation of Al3Ni nano particles in synthesis of Al 7075 alloy by friction stir processing with Ni powder addition," *Journal of Materials Processing Technology*, vol. 311, p. 117803, 2023, doi:10.1016/j.jmatprotec.2022.117803.

29. L. Ke, C. Huang, L. Xing, and K. Huang, "Al-Ni intermetallic composites produced in situ by friction stir processing," *Journal of Alloys and Compounds*, vol. 503, no. 2, pp. 494–499, 2010, doi:10.1016/j.jallcom.2010.05.040.

30. P. Ajay Kumar, R. Raj, and S. V. Kailas, "A novel in-situ polymer derived nano ceramic MMC by friction stir processing," *Materials & Design*, vol. 85, pp. 626–634, 2015, doi:10.1016/j.matdes.2015.07.054.

31. P. Ajay Kumar, D. Yadav, C. S. Perugu, and S. V. Kailas, "Influence of particulate reinforcement on microstructure evolution and tensile properties of in-situ polymer derived MMC by friction stir processing," *Materials & Design*, vol. 113, pp. 99–108, 2017, doi:10.1016/j.matdes.2016.09.101.

32. Q. Zhang, B. L. Xiao, Q. Z. Wang, and Z. Y. Ma, "In situ Al_3Ti and Al_2O_3 nanoparticles reinforced Al composites produced by friction stir processing in an Al-TiO_2 system," *Materials Letters*, vol. 65, no. 13, pp. 2070–2072, 2011, doi:10.1016/j.matlet.2011.04.030.

33. Q. Zhang, B. L. Xiao, W. G. Wang, and Z. Y. Ma, "Reactive mechanism and mechanical properties of in situ composites fabricated from an Al-TiO_2 system by friction stir processing," *Acta Materialia*, vol. 60, no. 20, pp. 7090–7103, 2012, doi:10.1016/j.actamat.2012.09.016.

34. C. J. Hsu, C. Y. Chang, P. W. Kao, N. J. Ho, and C. P. Chang, "Al-Al_3Ti nanocomposites produced in situ by friction stir processing," *Acta Materialia*, vol. 54, no. 19, pp. 5241–5249, 2006, doi:10.1016/j.actamat.2006.06.054.

35. G. L. You, N. J. Ho, and P. W. Kao, "In-situ formation of Al_2O_3 nanoparticles during friction stir processing of $AlSiO_2$ composite," *Materials Characterization*, vol. 80, pp. 1–8, 2013, doi:10.1016/j.matchar.2013.03.004.

36. G. L. You, N. J. Ho, and P. W. Kao, "The microstructure and mechanical properties of an Al-CuO in-situ composite produced using friction stir processing," *Materials Letters*, vol. 90, pp. 26–29, 2013, doi:10.1016/j.matlet.2012.09.028.

37. C. J. Hsu, P. W. Kao, and N. J. Ho, "Intermetallic-reinforced aluminum matrix composites produced in situ by friction stir processing," *Materials Letters*, vol. 61, no. 6, pp. 1315–1318, 2007, doi:10.1016/j.matlet.2006.07.021.

38. C. J. Hsu, P. W. Kao, and N. J. Ho, "Intermetallic-reinforced aluminum matrix composites produced in situ by friction stir processing," *Materials Letters*, vol. 61, no. 6, pp. 1315–1318, 2007, doi:10.1016/j.matlet.2006.07.021.

39. W. J. Arbegast and P. J. Hartley, "Friction stir welding," In: *Trends in Welding Research : Proceedings of the 5th International Conference*, Pine Mountain, Georgia, USA, 1998, ASM International.

40. P. Heurtier, C. Desrayaud, and F. Montheillet, "A Thermomechanical analysis of the friction stir welding process," *Materials Science Forum*, vol. 396–402, pp. 1537–1542, 2002, doi:10.4028/www.scientific.net/MSF.396-402.1537.

41. G. Buffa, J. Hua, R. Shivpuri, and L. Fratini, "A continuum based fem model for friction stir welding-model development," *Materials Science and Engineering: A*, vol. 419, no. 1–2, pp. 389–396, 2006, doi:10.1016/j.msea.2005.09.040.

42. M. Bahrami, N. Helmi, K. Dehghani, and M. K. B. Givi, "Exploring the effects of SiC reinforcement incorporation on mechanical properties of friction stir welded 7075 aluminum alloy: Fatigue life, impact energy, tensile strength," *Materials Science and Engineering: A*, vol. 595, pp. 173–178, 2014, doi:10.1016/j.msea.2013.11.068.

43. M. Azizieh, A. H. Kokabi, and P. Abachi, "Effect of rotational speed and probe profile on microstructure and hardness of AZ31/Al_2O_3 nanocomposites fabricated by friction stir processing," *Materials & Design*, vol. 32, no. 4, pp. 2034–2041, 2011, doi:10.1016/j.matdes.2010.11.055.

44. A. Devaraju, A. Kumar, and B. Kotiveerachari, "Influence of rotational speed and reinforcements on wear and mechanical properties of aluminum hybrid composites via friction stir processing," *Materials & Design*, vol. 45, pp. 576–585, 2013, doi:10.1016/j.matdes.2012.09.036.

45. G. Faraji and P. Asadi, "Characterization of AZ91/alumina nanocomposite produced by FSP," *Materials Science and Engineering: A*, vol. 528, no. 6, pp. 2431–2440, 2011, doi:10.1016/j.msea.2010.11.065.

46. S. Hirasawa, H. Badarinarayan, K. Okamoto, T. Tomimura, and T. Kawanami, "Analysis of effect of tool geometry on plastic flow during friction stir spot welding using particle method," *Journal of Materials Processing Technology*, vol. 210, no. 11, pp. 1455–1463, 2010, doi:10.1016/j.jmatprotec.2010.04.003.

47. A. Scialpi, L. A. C. De Filippis, and P. Cavaliere, "Influence of shoulder geometry on microstructure and mechanical properties of friction stir welded 6082 aluminium alloy," *Materials & Design*, vol. 28, no. 4, pp. 1124–1129, 2007, doi:10.1016/j.matdes.2006.01.031.

48. W. M. Thomas, E. D. Nicholas, and S. D. Smith, "Friction stir welding-tool developments. In: *Aluminum Joining Symposium,* New Orleans, Louisiana, USA, 2001, TWI Ltd.

49. W. M. Thomas, K. I. Johnson, and C. S. Wiesner, "Friction stir welding-recent developments in tool and process technologies," *Advanced Engineering Materials*, vol. 5, no. 7, pp. 485–490, 2003, doi:10.1002/adem.200300355.

50. R. Rai, A. De, H. K. D. H. Bhadeshia, and T. DebRoy, "Review: friction stir welding tools," *Science and Technology of Welding and Joining*, vol. 16, no. 4, pp. 325–342, 2011, doi:10.1179/1362171811Y.0000000023.

51. C. Venkata Rao, G. Madhusudhan Reddy, and K. Srinivasa Rao, "Influence of tool pin profile on microstructure and corrosion behaviour of AA2219 Al-Cu alloy friction stir weld nuggets," *Defence Technology*, vol. 11, no. 3, pp. 197–208, 2015, doi:10.1016/j.dt.2015.04.004.

52. J.-Q. Su, T. W. Nelson, and C. J. Sterling, "Microstructure evolution during FSW/FSP of high strength aluminum alloys," *Materials Science and Engineering: A*, vol. 405, no. 1–2, pp. 277–286, 2005, doi:10.1016/j.msea.2005.06.009.

53. K. V. Jata and S. L. Semiatin, "Continuous dynamic recrystallization during friction stir welding of high strength aluminum alloys," *Scripta Materialia*, vol. 43, no. 8, pp. 743–749, 2000, doi:10.1016/S1359-6462(00)00480-2.

54. D. Yadav and R. Bauri, "Effect of friction stir processing on microstructure and mechanical properties of aluminium," *Materials Science and Engineering: A*, vol. 539, pp. 85–92, 2012, doi:10.1016/j.msea.2012.01.055.

55. S. Gourdet and F. Montheillet, "A model of continuous dynamic recrystallization," *Acta Materialia*, vol. 51, no. 9, pp. 2685–2699, 2003, doi:10.1016/S1359-6454(03)00078-8.

56. F. J. Humphreys and M. Hatherly, *Recrystallization and Related Annealing Phenomena*, 1st edn. Pergamon, 1995.

57. P. B. Prangnell and C. P. Heason, "Grain structure formation during friction stir welding observed by the 'stop action technique'," *Acta Materialia*, vol. 53, no. 11, pp. 3179–3192, 2005, doi:10.1016/j.actamat.2005.03.044.

58. J. F. Guo et al., "Effects of nano-Al2O3 particle addition on grain structure evolution and mechanical behaviour of friction-stir-processed Al," *Materials Science and Engineering: A*, vol. 602, pp. 143–149, 2014, doi:10.1016/j.msea.2014.02.022.

59. F. Khodabakhshi, A. Simchi, A. H. Kokabi, M. Nosko, F. Simanĉik, and P. Švec, "Microstructure and texture development during friction stir processing of Al-Mg alloy sheets with TiO2 nanoparticles," *Materials Science and Engineering: A*, vol. 605, pp. 108–118, 2014, doi:10.1016/j.msea.2014.03.008.

60. I. Charit and R. S. Mishra, "Abnormal grain growth in friction stir processed alloys," *Scripta Materialia*, vol. 58, no. 5, pp. 367–371, 2008, doi:10.1016/j. scriptamat.2007.09.052.

61. D. J. Lloyd, "Particle reinforced aluminium and magnesium matrix composites," *International Materials Reviews*, vol. 39, no. 1, pp. 1–23, 1994, doi:10.1179/ imr.1994.39.1.1.

62. D. Hull and T. W. Clyne, *An Introduction to Composite Materials*, 3rd ed. Cambridge University Press, 2019.

63. A. Sanaty-Zadeh, "Comparison between current models for the strength of particulate-reinforced metal matrix nanocomposites with emphasis on consideration of Hall-Petch effect," *Materials Science and Engineering: A*, vol. 531, pp. 112–118, 2012, doi:10.1016/j. msea.2011.10.043.

64. A. Shafiei-Zarghani, S. F. Kashani-Bozorg, and A. P. Gerlich, "Strengthening analyses and mechanical assessment of Ti/Al2O3 nano-composites produced by friction stir processing," *Materials Science and Engineering: A*, vol. 631, pp. 75–85, 2015, doi:10.1016/j. msea.2015.02.038.

65. S. M. Arab, S. Karimi, S. A. J. Jahromi, S. Javadpour, and S. M. Zebarjad, "Fabrication of novel fiber reinforced aluminum composites by friction stir processing," *Materials Science and Engineering: A*, vol. 632, pp. 50–57, 2015, doi:10.1016/j.msea.2015.02.032.

66. A. Mertens et al., "Influence of fibre distribution and grain size on the mechanical behaviour of friction stir processed Mg-C composites," *Materials Characterization*, vol. 107, pp. 125–133, 2015, doi:10.1016/j.matchar.2015.07.010.

67. M. Salehi, H. Farnoush, and J. A. Mohandesi, "Fabrication and characterization of functionally graded Al-SiC nanocomposite by using a novel multistep friction stir processing," *Materials & Design*, vol. 63, pp. 419–426, 2014, doi:10.1016/j.matdes.2014.06.013.

17 Effect of Tool Geometry on Friction Stir Spot Welded Joints

N. Jeyaprakash, Sundara Subramanian Karuppasamy, and Che-Hua Yang

17.1 INTRODUCTION

One of the manufacturing processes that has attained technical advancements along with human evolution is welding technology. The welding technique is a fabrication process that is used to manufacture parts/components by either depositing materials in a layer-by-layer fashion or joining two or more metallic/non-metallic components. In most cases, welding is preferred as a joining tool where two or more parts are joined together, which will pave the way for the fabrication of new components [1–3]. Usually, in welding, a weld bead is formed at the faying site, which provides the necessary support (joint) between the parent materials. This weld bead is deposited/formed in numerous ways based on the type of energy source used. Moreover, depending on the energy source's type, the welding technique is classified into two major categories: fusion and solid-state welding [4–6].

As the name implies, fusion welding involves the fusion of the parent materials, which are fused together in order to form the weld bead. In this process, the test specimens are heated until they reach their melting point and joined together at the faying site. After solidification, a permanent joint can be obtained between the test specimens [7,8]. The fusion welding can be carried out in the presence or absence of the filler material. If the filler material is used, both the test specimen and the filler material are heated to form the permanent joint. Since this process demands heating the test specimen, a wide range of energy (heat) sources are preferred to perform fusion at the faying site. They are (i) electron beams, (ii) electric arcs, (iii) inert gases, (iv) high energy lasers, and so on. Moreover, this process encounters several drawbacks, such as (i) greater distortion effect, (ii) maximal heat-affected zone, (iii) can't be used to join dissimilar materials (Al-Cu, Al-Mg, polymers, and ceramics), (iv) high residual stress, (v) generate harmful fumes during the melting of parent materials, and so on [9,10].

On the other hand, the solid-state welding process is reported to produce high-quality weld joints without intensive heating of the test specimen. This process mostly uses external pressure to produce weld joints [11]. In some cases, the amount of heat required is relatively less when compared to fusion welding. Moreover, in such cases, the test specimens are not melted, as in fusion welding. Some of the

DOI: 10.1201/9781003432289-17

"

major advantages of solid-state welding are as follows: (i) the parent (base) materials are less affected during this process and their mechanical characteristics do not change after the welding, (ii) minimal heat-affected zone since the base material does not undergo melting, (iii) lower levels of distortion and residual stress, (iv) could be adapted for joining dissimilar materials, and so on [12,13].

In recent years, lightweight materials have been drastically implemented in the industrial sector due to their outstanding characteristics, and it is believed that these materials could replace the heavy metallic components made from iron and steel. Since the percentage contribution of these materials is rising day by day, there is a demand for new welding techniques that could provide permanent joints in these materials.

Friction stir spot welding (FSSW) belongs to the category of solid-state welding and is good for creating permanent joints in similar (Al-Al), dissimilar (Al-Cu, Al-Mg, and Al-Fe) alloys, polymers, and ceramics. Here, a rotating tool is plunged against the test specimen, and the generated heat during plunging will pave the way for the plasticized flow of materials between the test specimens. After attaining an adequate dwell time, the FSSW tool is unloaded, leaving the spot weld bead at the faying site. Thus, this process is reported to produce high-quality spot welds in lightweight materials [14–16]. Moreover, researchers work on different types of tool geometries that could be used for producing highly efficient spot welds via the FSSW process. This chapter is aimed at discussing the various types of tool geometries and their influence in relevant literature.

17.2   FSSW: AN OVERVIEW

In the FSSW process, the spot welds are made in the test specimens by arranging them in any one of the two configurations, namely lap and butt configurations. Mostly, the lap configuration is preferred over the butt configuration. In lap configurations, the test specimens are arranged one on one by overlapping them, and this configuration is good for joining specimens with different thicknesses [17,18]. On the other hand, in butt configuration, the test specimens are placed in the same plane where the spot weld is formed at the faying site of the specimens. Moreover, this type of configuration is preferred when using specimens with the same thickness [19,20]. Thus, the spot welds are made by arranging the test specimens in the above-mentioned configurations.

Generally, a spot weld bead is accomplished by a series of step-by-step procedures. These steps are often referred to as the FSSW process step, which combines a series of four processes [21,22]. They are (i) loading, (ii) plunging, (iii) stirring, and (iv) unloading (Figure 17.1), which are explained as follows:

   i. Loading: The FSSW tool is rotated at the required rpm and directed toward the test specimen.
   ii. Plunging: The rotating tool is pressed (plunged) against the test specimen, and during plunging, heat is generated at the tool-base metal interface due to friction.

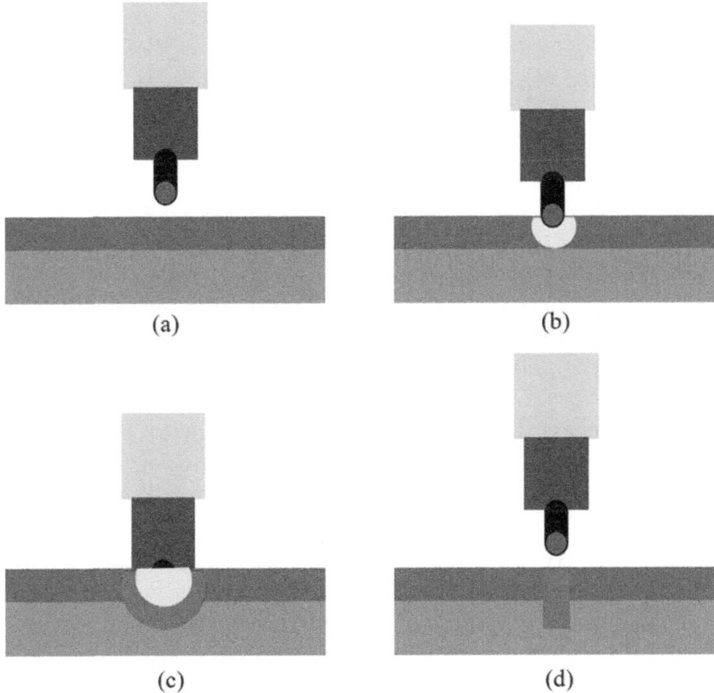

FIGURE 17.1 FSSW process: (a) loading/rotation, (b) plunging, (c) stirring, and (d) unloading.

iii. Stirring: After plunging to a certain depth, the frictional heat and continuous rotation (stirring) of the FSSW tool will result in the formation of plasticized material flow at the interface.

iv. Unloading: After stirring the tool to the desired dwell time, the FSSW tool is unloaded, and the plasticized material flow is solidified to form the spot weld between the test specimens.

Thus, using these steps, spot welds are created with a fine microstructure, a low distortion effect, a defect- or pore-free weld bead, and so on.

In the FSSW process, spot welds can be created in two operating modes. They are (i) force control mode and (ii) displacement control mode. In both modes, several factors are kept under control in order to produce the spot weld.

17.2.1 Force Control Mode

Under this mode, the FSSW tool's rpm, dwell time, and force exerted on the specimen's surface are effectively controlled in order to produce the spot welds. Here, the FSSW tool's rpm depends on the calculated value of the normal force, and after attaining the predefined depth, the force exerted is said to be constant, and the continuous stirring at the predefined dwell time will create the required plasticized material flow, thereby forming the weld bead at the faying site [23]. Afterward, the FSSW tool is retracted, leaving the weld bead at the faying site.

17.2.2 DISPLACEMENT CONTROL MODE

As the name implies, in this mode, the spot welds are created by controlling the tool's displacement and rpm. After displacing the tool to a certain depth, the normal force acting during the tool displacement process is recorded. Also, the researchers found that the recorded force value is at its minimum when the tool starts to plunge the test specimen and hits its maximum value when the tool's shoulder reaches the specimen's surface [24]. Thus, in these two modes, the spot welds are created via the FSSW process.

In the FSSW process, there are some process parameters that need to be optimized for producing high-quality spot welds. These parameters are explained briefly as follows:

i. Rpm of the FSSW tool: The rotating tool plays a major role in deciding the quality of the weld bead because this rotation creates the necessary heat during plunging. At the faying site, the generated heat melts the parent materials, thereby creating the plasticized flow of material at this site. Moreover, this frictional heat determines the grain growth and recrystallization of the weld bead. Hence, there is a need to optimize the rpm of the FSSW tool in order to obtain a fine microstructure with hardening precipitates. In addition, the frictional heat generation differs while using various materials to produce the spot weld joints. In general, the finer microstructure was reported in the weld bead formed at higher rpm [25,26].

ii. Dwell time: The dwell time is the staying time of the FSSW tool in the stir zone. After plunging to the required depth, the tool starts to stir, thereby producing the frictional heat that will pave the way for the plasticized material flow. This dwell time determines the amount of plasticized materials, which in turn determines the bonding strength of the weld bead. Higher dwell time results in better bonding (joint) strength, followed by improving the service life of the weld bead [27]. But in some materials, a higher dwell time results in a greater amount of intermetallic precipitates, which will degrade the mechanical characteristics offered by the weld bead [28].

iii. Tool displacement: The tool displacement is also referred to as the plunge rate or the tool's pin displacement. This is defined as the tool's plunging depth on the test surface. In the refill FSSW, the plunging depth creates the cavity, which is formed during the plunging process. After unloading the tool, the shoulder moves downward, thereby filling this cavity with the plasticized material. Thus, higher plunging depth results in a wider cavity, and along with a higher dwell time, a greater amount of plasticized material is formed and filled in this cavity, thereby improving the joint strength of the spot welds [29,30]. Thus, this section describes the FSSW process, modes of operation, joint configurations, and process parameters that are needed to be considered while creating spot welds via the FSSW process.

17.3   FSSW TOOL MATERIALS

As said earlier, the FSSW process is a variant of the FSW process. The major difference between the FSSW and the FSW is that in the FSSW, the tool does not move in the transverse direction. In addition, to obtain a highly efficient spot weld, the tool materials should have characteristics such as (i) abrasion resistance, (ii) wear and tear resistance, (iii) higher strength, (iv) greater toughness, and so on. There are certain major factors that need to be considered before selecting the proper tool materials. They are (i) tool wear rate, (ii) weld quality, (iii) thermal conductivity, and (iv) coefficient of thermal expansion. The FSSW process involves the creation of plasticized material flow through heat generation in the tool. The tool should have a low wear rate and be able to withstand the heat generated during plunging and stirring. Based on this, the type of materials that could be used to manufacture the FSSW tools is briefly explained as follows:

17.3.1   Steel Tools (H13 Steel, AISI 4340, O-1 Tool Steel, and SKD61 Tool Steel)

Steel tools are one of the most common tools for producing spot welds via the FSSW process. This tool has higher toughness, greater abrasion and fracture resistance, better tensile strength, and excellent anti-wear characteristics [31,32]. The FSSW tool made from the steel holds good for producing spot welds in light alloys such as Al and Mg alloys as well as in aluminum matrix composites (AMC). In particular, the AMC consists of very hard abrasive phases. These phases constitute a greater wear rate in tools than the light alloys. While plunging in AMC, it is reported that the steel tools experience wear at the initial stage, and after this initial wear, the steel tool obtains a self-optimized shape. This optimized shape produces efficient spot welds in the AMCs with a lower tool wear rate [33,34]. Moreover, after prolonged usage, some researchers reported that threads can be introduced in the aged tools for improving the plasticized material flow. But the aged steel tools are able to produce high-quality spot welds even without threading them [35].

17.3.2   Polycrystalline Cubic Boron Nitride Tools

In general, the boron nitride is reported to have two crystal lattice structures, namely hexagonal closed packed (HCP) lattice and cubic lattice. The HCP has a layered structure and is not preferred for making tools. On the other hand, the cubic lattice in boron nitride was obtained by subjecting the HCP to elevated pressure and temperature, which is similar to producing diamond from graphite. This high temperature and pressure proliferate the mechanical characteristics of boron nitride, thereby making it a suitable material for producing the tools [36,37]. Moreover, cubic lattice boron nitride has outstanding characteristics such as greater hardness, excellent wear and abrasion resistance, high thermal conductivity, better thermal shock resistance, low thermal expansion, and so on. These characteristics paved the way for cubic boron nitride as a suitable material for producing spot welds in high-temperature applications. Furthermore, the cubic lattice boron nitride has a greater hardness; this material is mostly preferred to create joints in heavy weight alloys such as steel and titanium alloys [38–40]. It is also reported that grain growth and refinement in the weld bead zone increase while using the polycrystalline cubic boron nitride (pcBN) tools.

17.3.3 TUNGSTEN-BASED TOOLS (PURE W, W-3RE, W-25RE, MO-W ALLOY, DENSIMET, AND W-1LaO$_2$)

There are certain tool materials that could be adapted for producing spot joints in both light and heavy metallic alloys. One such tool material is tungsten. Pure tungsten can be used to produce spot welds in polymers and ceramic materials [41]. The hardness property of pure tungsten can be improved by adding or doping it with heavy metals such as rhenium. Rhenium-doped tungsten offers more hardness, which is nearly four times greater than the average hardness offered by pure tungsten. Moreover, this mixing (tungsten with rhenium) serves as the suitable material for creating spot welds in lightweight alloys such as aluminum, magnesium, and copper alloys and in dissimilar light alloys such as Al-Mg alloys and Al-Cu alloys [42,43].

On the other hand, this mixing is not suitable for heavy metallic alloys. For producing spot welds in steels, titanium alloys, and dissimilar heavy alloys such as Al-Fe alloys, WC composite is mostly preferred [44,45]. Moreover, the WC tool is good for producing spot welds in high-temperature applications. Some researchers reported that the hardness and wear resistance characteristics of the tungsten carbide could be improved by adding cobalt at a lower percentage. This combination (WC + cobalt) is preferred for fabricating spot joints in thick metallic alloys [46].

17.3.4 OTHER TOOL MATERIALS

Other than steel, pcBN, and W-based tools, silicon nitrides and nickel-based superalloys (Inconel 718, 738LC, Nimonic 90, Nimonic 105, and waspaloy) can be used to make the FSSW tools [47,48]. Particularly, nickel-based superalloys have excellent heat resistance, greater strength, and anti-wear behaviors. The frictional heat produced during the plunging and stirring process is reported to affect the FSSW tool to a greater extent. Also, this heat increases the wear rate of the tool, thereby reducing its life span. This could be overcome by fabricating the FSSW tools with nickel-based superalloys [49,50]. Table 17.1 represents the benefits of using various tool materials for the FSSW process in terms of their mechanical characteristics, whereas Table 17.2 lists the suitable tool materials based on the type of the base materials.

TABLE 17.1

Tool Materials and Benefits

Tool Materials	Benefits
Hot work tool steels	Most commonly used tool materials with ease of availability and machinability, resistance against thermal fatigue, high tool life.
Ni- and Co-based alloy tools	Excellent hardness, greater strength, high creep resistance, suitable for high-temperature working environments around 800°C.
Refractory metals (tungsten- and molybdenum-based tools)	Excellent mechanical strength, high creep and wear resistance, mostly suitable for high-temperature working environments around 1500°C.
Carbide particle-reinforced metal composites (WC, WC-Co, TiC)	Excellent hardness, greater wear resistance, superior fracture toughness and fatigue resistance.
Polycrystalline cubic boron nitride tools	Greater hardness, excellent wear and abrasion resistance, high thermal conductivity, better thermal shock resistance, low thermal expansion.

TABLE 17.2

Suitable Tool Materials Based on the Materials to Be Welded

Alloys to Be Welded	Tool Materials
Ti alloys	Tungsten alloys
Mg alloys	Steel tools (H13 steel), tungsten carbide composite
Al alloys	Steel tools (H13 steel), cobalt-based tungsten carbide composite
Cu alloys	Nickel alloys, pcBN, H13 steel
Steels	Tungsten alloys, pcBN, tungsten carbide composite

17.4 TOOL GEOMETRY AND ITS EFFECTS

Figure 17.2 represents the constituents of the FSSW tool. The pin's length and breadth determine the stir zone size, and the shoulder provides the necessary load or force in order to perform a spot weld. Except for the pinless FSSW tool, all other types of FSSW (refill and swing FSSW) tools contain the pin, which is attached to the shoulder. This section deals with the types of tool geometry (both shoulder and pin geometries) and its effects in relevant literature.

17.4.1 SHOULDER GEOMETRIES

In the FSSW process, the tool's shoulder performs three major functions:

    i. delivering the required load or force to perform the spot welding.
    ii. retaining the plasticized material flow within the plunge cavity.
    iii. ensuring the required amount of heat generation for localized melting at the faying site.

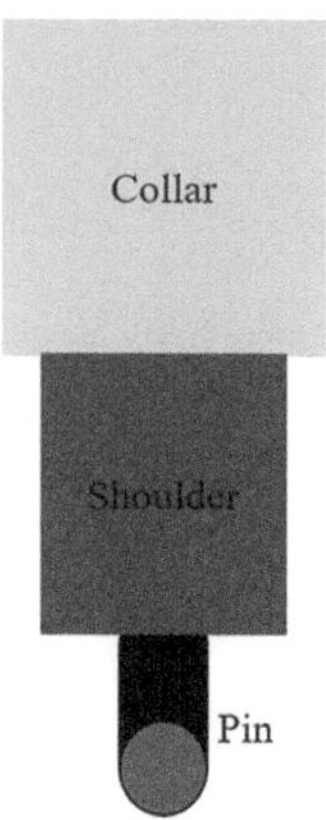

FIGURE 17.2   FSSW tool.

Moreover, in order to perform these functions, the geometry of the shoulder can be of any one of the three forms, namely flat (smooth), conical concave, and scrolled geometries.

I. Galvao et al. [51] analyzed the shoulder geometry effects in the weld bead formed on thin copper sheets. They compared the weld beads formed by the three types of shoulder geometries, such as flat, conical, concave, and scrolled shoulders, by varying the tool's speed (rpm). Their results reported that the weld bead formed by the flat (smooth) shoulder geometry consists of root defects, which make it unsuitable for producing the weld bead at the predefined rotational speed. The spot welds produced by the conical and scrolled geometries contain fewer defects at lower rpm. Particularly, the scrolled geometry is able to produce weld beads with better grain refinement and provides greater bonding strength. Casalino et al. [52] evaluated the effect of flat and conical small shoulders with cartilage-sized flat and conical shoulders. Their study revealed that both the conical small and cartilage-sized, large conical shoulders produced defect-free weld bead when compared with the small-sized flat and cartilage-sized flat shoulders. Moreover, conical shoulders result in high hardness when compared with flat shoulders. Aval et al. [53] examined the spot welds made in the AA5082-AA6061 alloy with three shoulder geometries (geometry 1: concave shoulder with a conical probe, geometry 2: flat shoulder with the threadless tool, and geometry 3: flat shoulder with the threaded tool). The concave shoulder with a conical probe (geometry) resulted in a greater amount of heat generation at the faying site. This high heat has resulted in homogenous stir zones at the faying site, followed by the fine grain orientation. Also, the weld bead formed by this geometry has greater hardness when compared to the weld bead formed with the aid of the other two geometries [54,55]. Mukuna et al. [56] examined the microstructural evolution that occurred in the spot weld beads obtained using the flat and concave shoulder via the FSSW process under various rpm and plunge depths. They reported that the micrographs captured at the weld bead formed using the concave shoulder have equiaxed grains. But the spot welds produced with the flat shoulder contain an uneven grain arrangement with intermediate precipitates. These precipitates will pave the way for a reduction in the mechanical characteristics offered by the weld bead.

17.4.2 Pin Geometries

The size and shape of the pin attached to the FSSW tool have a serious influence on determining the heat generation, plasticized material flow, hook formation, bonding strength, and other mechanical behaviors offered by the spot welds. Figures 17.3–17.7 represent a schematic illustration of the various pin geometries used in the FSSW process.

H. Badarinarayan et al. [57] analyzed the influence of tool pin geometries on the bonding strength and hook formation in the spot welds produced on AA 5754 sheets. Their study consists of 1.3 mm thick AA 5754 sheets, which are arranged in a lap joint configuration. For producing spot welds, five different tools with various pin geometries were used. The pin and shoulder geometries of these four tools are as follows: Tool 1: concave shoulder with a cylindrical pin, tool 2: flat shoulder with a cylindrical pin, tool 3: convex shoulder with a cylindrical pin, and tool 4: concave shoulder with a triangular pin. All the pins have the same pin length of 1.6 mm. On analyzing their results regarding the tool's shoulder, it is evident that among the three tools, the weld bead made from tool 1 shows higher static strength (~3550 N) than the other tools. Tool 2 and tool 3 have static strength values of 3400 and 3100 N, respectively. The higher static strength value obtained from tool 1 is due to the adequate

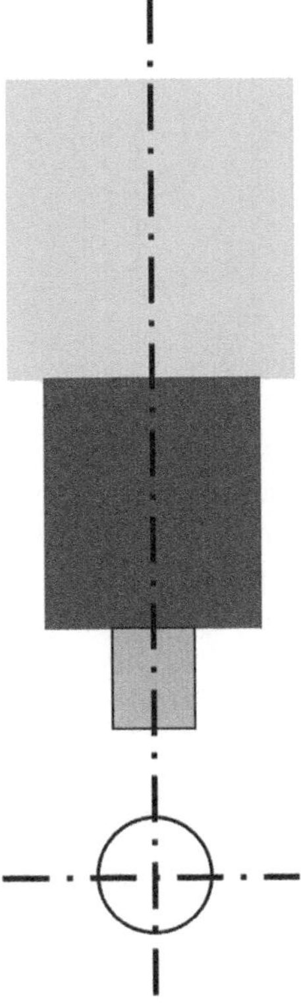

FIGURE 17.3 Cylindrical pin.

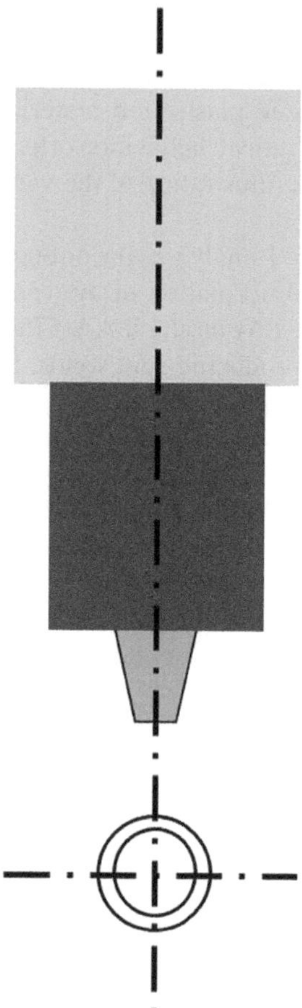

FIGURE 17.4   Tapered cylindrical pin.

amount of heat generation that happened in the faying interface. This sufficient heat results in better mixing of plasticized material, which results in fine grain growth at the weld bead. The hooks formed in the weld beads made from tool 1 and tool 4 were compared. The weld bead (spot weld) made from tool 1 experiences continuous hook formation. This hook started at the weld interface of the sheets and ended near the keyhole. On the other hand, a discontinuous hook was reported in the spot weld formed using the triangular pin. Moreover, the static strength obtained in the spot weld made using tool 4 has a higher static strength than tool 1 [58,59]. Hence, they concluded that for producing spot welds in the AA 5754 sheets, the FSSW tool with concave shoulder and triangular pin (tool 4) is more suitable than the other tools (tool 1, tool 2, and tool 3) by considering the static strength and the hook formed.

Yazunari et al. [60] examined the spot welds formed in the AA 6061-T4 alloy sheets with a thickness of 2 mm. Totally three tools with pin lengths of 2.4, 3.1, and 3.7 mm were used, and the process parameters such as the tool's rpm (2000, 2500, and 3000 rpm) and the dwell time (0.2, 1, and 3 s) were varied. On obtaining

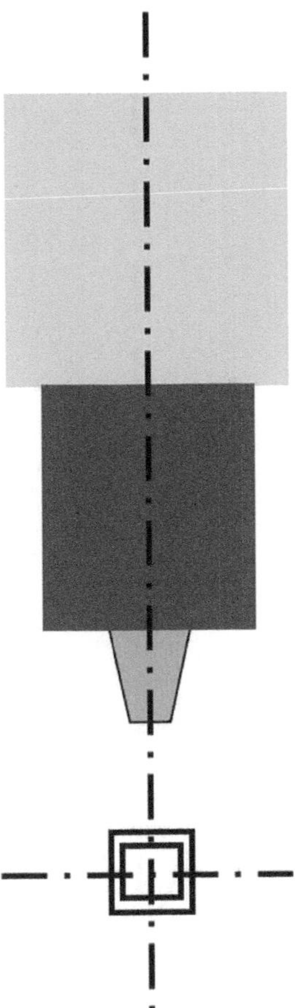

FIGURE 17.5 Trapezoidal pin.

the micrographs captured at various pin lengths at 2000 rpm at 0.2 s, the weld bead obtained at 3.7 mm pin length has a greater stir zone. This stir zone experiences greater plasticized material flow and reduces the upward movement of the plasticized materials. On the other hand, a higher tool's rpm (2500 and 3000 rpm) resulted in severe stirring and promoted the upward movement of the plasticized materials. Moreover, this upward movement, along with rigorous stirring, might affect the bonding strength. The static strength decreases as the pin length gets smaller. This is due to the fact that the amount of frictional heat generated tends to be less, whereas a higher pin length (3.7 mm) aids in generating a higher amount of frictional heat, which promotes the plasticized material flow, which will pave the way for enhancing the tensile shear strength at the spot weld zone. Furthermore, it is also reported that the cross-tension strength was decreased and the tensile shear strength tends to increase by increasing the tool holding time and rpm of the rotating tool.

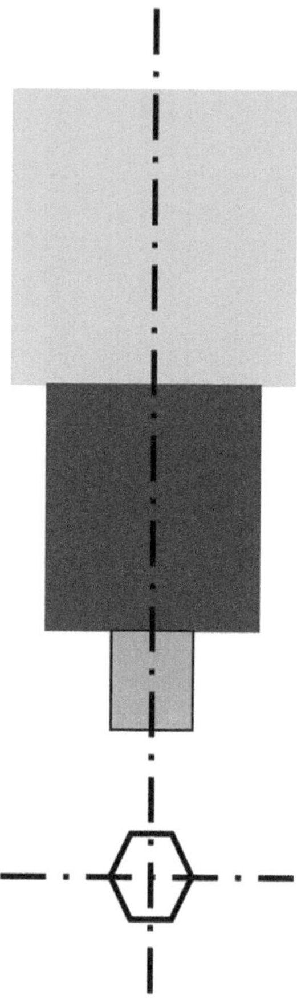

FIGURE 17.6 Hexagonal pin.

The characteristics of the thermal cycle need to be addressed since this cycle determines the cooling rate, which turns into the spot welds. Moreover, the mechanical characteristics offered by the spot welds are entirely dependent upon this thermal cycle [61]. Ilman et al. [62] investigated the thermal cycles obtained at various tool rpm with the cylindrical and stepped pin profiles, along with the evolution of microstructure that occurred in the spot welds. The spot welds were obtained from dissimilar metallic alloy sheets (AA 2024 and AA 6061-T6 alloys) with two tools. Tool 1 consists of a cylindrical pin with a diameter of 5 mm, whereas tool 2 has a stepped pin with three steps. The diameters of the first, second, and third steps are 8, 6, and 4 mm, respectively. At various rotational speeds (900, 1400, and 1800 rpm), the spot welds were made by using both tools, and the mechanical characteristics were investigated. On analyzing the thermal cycles obtained at various speeds via the two pin profiles, the cooling rate tends to be faster at lower rotational speeds. The frictional heat produced is also lower at lower rotational speeds. On the other hand,

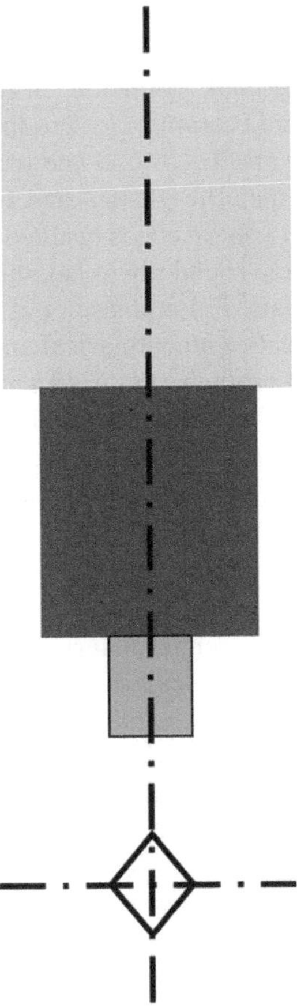

FIGURE 17.7 Squared pin.

at higher rpm (1800 rpm), more frictional heat is produced, and the cooling rate also becomes slow, which improves the mechanical characteristics of the spot welds. The stepped pin profile creates a large amount of frictional heat when compared with the cylindrical profile. The micrographs captured at a slow cooling rate have uniformly distributed grains with fine precipitates. Also, the grain coarsening was achieved at higher rpm with a slower cooling rate. The faster cooling rate affects grain growth and causes uneven grain distribution. Moreover, at 900 rpm, shear and nugget failure modes were observed. Thus, based on their research findings, higher rotational speeds are preferred for producing high-quality spot weld joints in terms of grain distribution, grain coarsening, precipitate hardening, and hardness characteristics.

Shigeki et al. [63] analyzed the temperature distribution obtained using the cylindrical pin and triangular pin profiles by means of the particle method. Their observations are as follows: The upward material flow is reported while using the cylindrical pin geometry, followed by a continuous hook formation near the keyhole. On the other hand, the triangular pin promotes the homogeneous mixing of materials during stirring, and this enhanced material flow controls the upward movement of materials, thereby minimizing the risk of hook formation. In recent years, the FSSW process has been adapted to polymers and ceramics for producing spot welds [64]. Shin et al. [65] examined the various pin profiles (round and triangular) used to make dissimilar spot welds between bulk metallic glasses and lightweight metallic alloys and reported that the triangular pin's sharp edges resulted in an increased amount of chip formation when compared to the round pin. Also, the round pin is good for making spot welds in the glass materials. F. Lambiase et al. [66] examined the spot welds made on the polycarbonate sheets with cylindrical and tapered pin geometries. They concluded that the tapered pin profile's spot weld shows better mechanical behavior than the spot weld obtained using the cylindrical pin. Thus, the thermal cycle of the spot welds depends on the type of parent materials (lightweight metallic alloys, ceramics, polymers), the tool's rpm, shoulder geometry, pin length, and so on.

17.5 CONCLUSION

In recent years, the usage of lightweight materials such as metallic alloys (Al, Cu, and Mg alloys), polymers, and ceramics has drastically increased due to their outstanding characteristics. Also, in the near future, it is expected that these lightweight materials could replace the heavy parts made from iron and steel. For producing weld joints in these materials, traditional techniques such as resistance spot welding and self-pierce riveting lack the ability to obtain high-quality welds. FSSW is reported as the most suitable method for producing spot welds in lightweight materials. This process produces high-quality weld joints with superior mechanical behavior. In this process, the FSSW tool is plunged to the required depth and stirred in order to produce heat via friction. This frictional heat creates a plasticized flow of materials, and after adequate cooling, the plasticized materials are turned into high-quality spot welds. The bonding strength and other mechanical characteristics offered by the spot weld are entirely dependent on the type of tool materials and the tool geometry. This chapter discussed the various types of tool materials and the tool's shoulder and pin geometries, along with their effects on the mechanical characteristics offered by the spot welds and relevant literature.

ACKNOWLEDGMENT

The authors would like to acknowledge the editors of this book for providing the opportunity to write this chapter.

REFERENCES

1. Verma RP, Lila MK. A short review on aluminium alloys and welding in structural applications. *Materials Today: Proceedings.* 2021;46:10687–91.
2. Jeyaprakash N, Duraiselvam M, Raju R. Modelling of Cr3C2-25% NiCr laser alloyed cast iron in high temperature sliding wear condition using response surface methodology. *Archives of Metallurgy and Materials.* 2018;63:1303–15.

3. Kataria R, Singh RP, Sharma P, Phanden RK. Welding of super alloys: a review. *Materials Today: Proceedings*. 2021;38:265–8.

4. Marques ES, Silva FJ, Pereira AB. Comparison of finite element methods in fusion welding processes-A review. *Metals*. 2020;10(1):75.

5. Tümer M, Schneider-Bröskamp C, Enzinger N. Fusion welding of ultra-high strength structural steels: A review. *Journal of Manufacturing Processes*. 2022;82:203–29.

6. Jeyaprakash N, Karuppasamy SS, Yang CH. Application of wear-resistant laser claddings. In: *Handbook of Laser-Based Sustainable Surface Modification and Manufacturing Techniques* edited by Hitesh Vasudev, Chander Prakash Jul 5, 2023 (pp. 1–26). CRC Press.

7. Karkhin VA, Plochikhine VV, Ilyin AS, Bergmann HW. Inverse modelling of fusion welding processes. *Welding in the World*. 2002;46:2–13.

8. Zinigrad M, Mazurovsky V, Borodianskiy K. Physico-chemical and mathematical modeling of phase interaction taking place during fusion welding processes. *Materialwissenschaft und Werkstofftechnik: Entwicklung, Fertigung, Prüfung, Eigenschaften und Anwendungen technischer Werkstoffe*. 2005;36(10):489–96.

9. Xue C, Blanc N, Soulié F, Bordreuil C, Deschaux-Beaume F, Guillemot G, Bellet M, Gandin CA. Structure and texture simulations in fusion welding processes-comparison with experimental data. *Materialia*. 2022;21:101305.

10. Jeyaprakash N, Yang CH, Tseng SP. Characterization and tribological evaluation of NiCrMoNb and NiCrBSiC laser cladding on near-α titanium alloy. *The International Journal of Advanced Manufacturing Technology*. 2020;106:2347–61.

11. Nassiri A, Abke T, Daehn G. Investigation of melting phenomena in solid-state welding processes. *Scripta Materialia*. 2019;168:61–6.

12. Bergmann JP, Petzoldt F, Schürer R, Schneider S. Solid-state welding of aluminum to copper-case studies. *Welding in the World*. 2013;57:541–50.

13. Cooper DR, Allwood JM. The influence of deformation conditions in solid-state aluminium welding processes on the resulting weld strength. *Journal of Materials Processing Technology*. 2014;214(11):2576–92.

14. Khan MI, Kuntz ML, Su P, Gerlich A, North T, Zhou Y. Resistance and friction stir spot welding of DP600: a comparative study. *Science and Technology of Welding and Joining*. 2007;12(2):175–82.

15. Karuppasamy SS, Jeyaprakash N, Yang CH. Application of corrosion-resistant laser claddings. *Handbook of Laser-Based Sustainable Surface Modification and Manufacturing Techniques*. 2023;51(51):27.

16. Yuan W, Mishra RS, Webb S, Chen YL, Carlson B, Herling DR, Grant GJ. Effect of tool design and process parameters on properties of Al alloy 6016 friction stir spot welds. *Journal of Materials Processing Technology*. 2011;211(6):972–7.

17. Ayaz A, Ülker A. Effects of process parameters on the lap joint strength and morphology in friction stir spot welding of ABS sheets. *Journal of Elastomers & Plastics*. 2021;53(6):612–31.

18. Jeyaprakash N, Yang CH, Ramkumar KR. Microstructure and wear resistance of laser cladded Inconel 625 and Colmonoy 6 depositions on Inconel 625 substrate. *Applied Physics A*. 2020;126:1–1.

19. Reilly A, Shercliff H, Chen Y, Prangnell P. Modelling and visualisation of material flow in friction stir spot welding. *Journal of Materials Processing Technology*. 2015;225:473–84.

20. Aota K, Ikeuchi K. Development of friction stir spot welding using rotating tool without probe and its application to low-carbon steel plates. *Welding International*. 2009;23(8):572–80.

21. Bilici MK, Yukler AI. Effects of welding parameters on friction stir spot welding of high density polyethylene sheets. *Materials & Design*. 2012;33:545–50.

22. Feng Z, Santella ML, David SA, Steel RJ, Packer SM, Pan T, Kuo M, Bhatnagar RS. Friction stir spot welding of advanced high-strength steels-a feasibility study. *SAE Transactions.* 2005:592–8.

23. Rashkovets M, Contuzzi N, Casalino G. Modeling of Probeless Friction Stir Spot Welding of AA2024/AISI304 Steel Lap Joint. *Materials.* 2022;15(22):8205.

24. Chen YC, Gholinia A, Prangnell PB. Interface structure and bonding in abrasion circle friction stir spot welding: a novel approach for rapid welding aluminium alloy to steel automotive sheet. *Materials Chemistry and Physics.* 2012;134(1):459–63.

25. Murugesan P, Satheeshkumar V, Jeyaprakash N, Yang CH, Karuppasamy SS. Effect of α-Al and Si precipitates on microstructural evaluation and corrosion behavior of laser powder bed fusion printed AlSi10Mg plates in seawater environment. *Metals and Materials International.* 2023:1–8.

26. Gerlich A, Avramovic-Cingara G, North TH. Stir zone microstructure and strain rate during Al 7075-T6 friction stir spot welding. *Metallurgical and Materials Transactions A.* 2006;37:2773–86.

27. Farmanbar N, Mousavizade SM, Ezatpour HR. Achieving special mechanical properties with considering dwell time of AA5052 sheets welded by a simple novel friction stir spot welding. *Marine Structures.* 2019;65:197–214.

28. Jeyaprakash N, Sivasankaran S, Prabu G, Yang CH, Alaboodi AS. Enhancing the tribological properties of nodular cast iron using multi wall carbon nano-tubes (MWCNTs) as lubricant additives. *Materials Research Express.* 2019;6(4):045038.

29. Chu Q, Li WY, Yang XW, Shen JJ, Vairis A, Feng WY, Wang WB. Microstructure and mechanical optimization of probeless friction stir spot welded joint of an Al-Li alloy. *Journal of Materials Science & Technology.* 2018;34(10):1739–46.

30. Karthikeyan R. Establishing relationship between optimised friction stir spot welding process parameters and strength of aluminium alloys. *Advances in Materials and Processing Technologies.* 2022;8(1):1173–95.

31. Jeyaprakash N, Yang CH, Karuppasamy SS, Duraiselvam M. Stellite 6 cladding on AISI Type 316L stainless steel: Microstructure, nanohardness and corrosion resistance. *Transactions of the Indian Institute of Metals.* 2023;76(2):491–503..

32. Thomas WM, Nicholas ED, Watts ER, Staines DG. Friction based welding technology for aluminium. *Materials Science Forum.* 2002;396:1543–1548.

33. Fernandez GJ, Murr LE. Characterization of tool wear and weld optimization in the friction-stir welding of cast aluminum 359+ 20% SiC metal-matrix composite. *Materials Characterization.* 2004;52(1):65–75.

34. Nami H, Adgi H, Sharifitabar M, Shamabadi H. Microstructure and mechanical properties of friction stir welded Al/Mg2Si metal matrix cast composite. *Materials & Design.* 2011;32(2):976–83.

35. Prado RA, Murr LE, Soto KF, McClure JC. Self-optimization in tool wear for friction-stir welding of Al 6061+ 20% Al_2O_3 MMC. *Materials science and engineering: A.* 2003;349(1–2):156–65.

36. Yeh CH, Jeyaprakash N, Yang CH. Nondestructive characterization of elastic modulus of APS Ni-5Al/10hBN coating on stainless steel 304 under high temperature. *Archives of Civil and Mechanical Engineering.* 2020;20:1–5.

37. Vicharapu B, Lemos GV, Bergmann L, dos Santos JF, De A, Clarke T. Probing underlying mechanisms for PCBN tool decay during friction stir welding of nickel-based alloys. *Tecnologia em Metalurgia, Materiais e Mineração.* 2021;18.**

38. Zhang Y, Sato YS, Kokawa H, Park SH, Hirano S. Stir zone microstructure of commercial purity titanium friction stir welded using pcBN tool. *Materials Science and Engineering: A.* 2008;488(1–2):25–30.

39. Sato YS, Yamanoi H, Kokawa H, Furuhara T. Microstructural evolution of ultrahigh carbon steel during friction stir welding. *Scripta materialia.* 2007;57(6):557–60.

40. Karuppasamy SS, Jeyaprakash N, Yang CH. Microstructure, nanoindentation and corrosion behavior of colmonoy-5 deposition on SS410 substrate using laser cladding process. *Arabian Journal for Science and Engineering.* 2022:1–7.

41. Mashinini PM, Dinaharan I, Selvam JD, Hattingh DG. Microstructure evolution and mechanical characterization of friction stir welded titanium alloy Ti-6Al-4V using lanthanated tungsten tool. *Materials Characterization.* 2018;139:328–36.

42. Schneider J, Terrell J, Farris L, Tucker D, Leonhardt T, Goldbeck H. Low-cost fabrication of tungsten-rhenium alloys for friction stir welding applications. *Metallurgical and Materials Transactions B.* 2020;51:35–44.

43. Leonhardt T. Properties of tungsten-rhenium and tungsten-rhenium with hafnium carbide. *JOM.* 2009;61(7):68–71.

44. Siddiquee AN, Pandey S. Experimental investigation on deformation and wear of WC tool during friction stir welding (FSW) of stainless steel. *The International Journal of Advanced Manufacturing Technology.* 2014;73:479–86.

45. Raj S, Pankaj P, Biswas P. Friction stir welding of Inconel-718 alloy using a tungsten carbide tool. *Journal of Materials Engineering and Performance.* 2021:1–6.

46. Jeyaprakash N, Yang CH, Duraiselvam M, Prabu G. Microstructure and tribological evolution during laser alloying WC-12% Co and Cr3C2− 25% NiCr powders on nodular iron surface. *Results in Physics.* 2019;12:1610–20.

47. Funaki K, Morisada Y, Fukasawa T, Abe Y, Fujii H. Effect of silicon nitride microstructure on characteristics of FSW tool for steel and tool life. *Welding International.* 2023:1–8.

48. Ahn BW, Choi DH, Kim DJ, Jung SB. Microstructures and properties of friction stir welded 409L stainless steel using a Si3N4 tool. *Materials Science and Engineering: A.* 2012;532:476–9.

49. Nakazawa T, Sato YS, Kokawa H, Ishida K, Omori T, Tanaka K, Sakairi K. Friction stir welding of steels using a tool made of iridium-containing nickel base superalloy. *Friction Stir Welding and Processing.* VIII. 2016:77–82.

50. Jeyaprakash N, Yang CH, Sivasankaran S. Laser cladding process of cobalt and nickel based hard-micron-layers on 316L-stainless-steel-substrate. *Materials and Manufacturing Processes.* 2020;35(2):142–51.

51. Galvão I, Leal RM, Rodrigues DM, Loureiro A. Influence of tool shoulder geometry on properties of friction stir welds in thin copper sheets. *Journal of Materials Processing Technology.* 2013;213(2):129–35.

52. Casalino G, Campanelli S, Mortello M. Influence of shoulder geometry and coating of the tool on the friction stir welding of aluminium alloy plates. *Procedia Engineering.* 2014;69:1541–8.

53. Jamshidi Aval H, Serajzadeh S, Kokabi AH, Loureiro A. Effect of tool geometry on mechanical and microstructural behaviours in dissimilar friction stir welding of AA 5086-AA 6061. *Science and Technology of Welding and Joining.* 2011;16(7): 597–604.

54. Jeyaprakash N, Yang CH, Karuppasamy SS, Dhineshkumar SR. Evaluation of microstructure, nanoindentation and corrosion behavior of laser cladded Stellite-6 alloy on Inconel-625 substrate. *Materials Today Communications.* 2022;31:103370.

55. Kunnathur Periyasamy Y, Perumal AV, Kunnathur Periyasamy B. Influence of tool shoulder concave angle and pin profile on mechanical properties and microstructural behaviour of friction stir welded AA7075-T651 and AA6061 dissimilar joint. *Transactions of the Indian Institute of Metals.* 2019;72:1087–109.

56. Mubiayi MP, Akinlabi ET. Friction stir spot welding between copper and aluminium: microstructural evolution. In: *Proceedings of the International MultiConference of Engineers and Computer Scientists,* March 18, 2015 IMECS 2015, March 18 - 20, 2015, Hong Kong (Vol. 2, pp. 18–20).

57. Badarinarayan H, Shi Y, Li X, Okamoto K. Effect of tool geometry on hook formation and static strength of friction stir spot welded aluminum 5754-O sheets. *International Journal of Machine Tools and Manufacture*. 2009;49(11):814–23.

58. Badarinarayan H, Yang Q, Zhu S. Effect of tool geometry on static strength of friction stir spot-welded aluminum alloy. *International Journal of Machine Tools and Manufacture*. 2009;49(2):142–8.

59. Jeyaprakash N, Prabu G, Yang CH. Inclusion of nano silicon particle in SS316L through LPBF and its responses on corrosion behaviour. *Ceramics International*. 202319.

60. Tozaki Y, Uematsu Y, Tokaji K. Effect of tool geometry on microstructure and static strength in friction stir spot welded aluminium alloys. *International Journal of Machine Tools and Manufacture*. 2007;47(15):2230–6.

61. Jegadheesan C, Somasundaram P, Kumar P, Singh AP, Jeyaprakash N. State of art: Review on laser surface hardening of alloy metals. *Materials Today: Proceedings*. 2023.

62. Ilman MN. Microstructure and mechanical performance of dissimilar friction stir spot welded AA2024-O/AA6061-T6 sheets: Effects of tool rotation speed and pin geometry. *International Journal of Lightweight Materials and Manufacture*. 2023;6(1):1–4.

63. Hirasawa S, Badarinarayan H, Okamoto K, Tomimura T, Kawanami T. Analysis of effect of tool geometry on plastic flow during friction stir spot welding using particle method. *Journal of materials processing technology*. 2010;210(11):1455–63.

64. Jeyaprakash N, Yang CH, Karuppasamy SS, Rajendran DK. Correlation of microstructural with corrosion behaviour of Ti-6Al-4V specimens developed through selective laser melting technique. *Proceedings of the Institution of Mechanical Engineers, Part E: Journal of Process Mechanical Engineering*. 2022;236(5):2240–51.

65. Shin HS. Tool geometry effect on the characteristics of dissimilar friction stir spot welded bulk metallic glass to lightweight alloys. *Journal of Alloys and Compounds*. 2014;586:S50–5.

66. Lambiase F, Paoletti A, Di Ilio A. Effect of tool geometry on mechanical behavior of friction stir spot welds of polycarbonate sheets. *The International Journal of Advanced Manufacturing Technology*. 2017;88:3005–16.

18 Wettability and Corrosion Resistance on Hydrophobic Steel Surfaces

A. Vivek Anand, R. Arvind Singh,
S. Jayalakshmi, M. Satyanarayana Gupta,
Jaligama Yogesh, and Balakrishnan Deepanraj

18.1 THEORY OF NON-WETTING SURFACES

Wettability is the ability of a liquid to spread or cohere with solid surfaces when exposed to an environmental condition. The wettability depends on the attraction between the molecules residing on the surface of the solid or liquid, i.e., the cohesive force, and also on the attraction of molecules in between the solid and liquid, i.e., the adhesive force. The nature of wettability relies on the equilibrium between the cohesive and adhesive forces and whether the surface of the solid acts as a hydrophobic surface (non-sticky surface) or a hydrophilic surface (sticky surface) [1].

Fluid and surface interaction is observed in all engineering systems and components, implying that wetting is an unavoidable process. The surface area covered by a water droplet is studied by measuring the contact angle. The contact angle formed due to the interaction of two media is influenced by the following factors: the nature of surface roughness, cleanliness, and energy [2]. The angle of the liquid droplet with respect to the material surface is measured using a goniometer. The contact angle determines whether the surface is hydrophobic or hydrophilic. The correlation between the shape of the droplet and the static contact angle is provided using the Young-Laplace equation.

$$\cos \theta = \frac{\gamma_{SV} - \gamma_{SL}}{\gamma_{LV}}$$

In the above equation, θ signifies contact angle, and γ_{SV}, γ_{SL}, and γ_{LV} indicate solid-vapor, solid-liquid, and liquid-vapor interface, respectively. The surface with a contact angle of $0°$–$90°$ is said to be hydrophilic in nature. In the hydrophilic surface, the water droplet will completely spread over its area, whereas on a hydrophobic surface, the contact angle will be between $90°$ and $150°$, where the liquid droplet will not wet the surface. The superhydrophobic surface will exhibit a contact angle higher than $150°$, and contact angle hysteresis will be very low [3]. The composite interface

DOI: 10.1201/9781003432289-18

of solid-liquid-air will be in thermodynamic equilibrium. In hydrophobic surfaces, when the contact angle value is further increased between 150° and 180°, the surface becomes superhydrophobic [3], where the droplet rolls over the surface and the material is said to be self-cleaning [4,5].

Wetting can broadly be classified into two categories: homogeneous wetting and heterogeneous wetting. Homogeneous wetting follows the Wenzel model, and heterogeneous wetting follows the Casie Baxter model [6]. The morphology of the surface and the roughness asperities determine the nature of the wetting [7,8]. To maintain the surface's hydrophobic nature, it has to follow the Casie Baxter model. Superhydrophobic surfaces are widely available in nature and are observed in most plants, birds, and insects. For the past few decades, researchers have been trying to mimic this kind of surface on man-made materials.

18.2 MICROPATTERNED SURFACE

The property of non-wetting nature can be introduced on the surface by means of texturing, i.e., micropatterns. The surface morphology can be formed/varied by chemical etching, Electrical Discharge Machining (EDM), laser cladding, etc. [9–13]. The problem identified was that all the textured surfaces would not act as hydrophobic surfaces; some of them behaved like hydrophilic surfaces, which is quite the opposite of hydrophobic surfaces; the liquid would stick to the surface. The nature of the surface was determined by two parameters, namely surface roughness and a solid fraction [14]. The contact angle leads to an increase as the pinning of a three-phase contact line happens within the textures of the surface. For hydrophobic surfaces, contact line pin is at the edges due to the sharp corners. At the same time, if it is a superhydrophobic surface, pinning will not occur, and the liquid droplet will simply roll off. This property triggers a self-cleaning effect under external disturbances such as pressure, impact, wind flow, or vibration [15,16]. The origin of the self-cleaning effect was noted in nature in lotus leaves. The micro- and nano-hair-like structures on the lotus leaf were responsible for this self-cleaning effect, which kept the surface in better wetting resistance with a higher contact angle of more than 150° [17].

The wetting behavior of the micropillar surface, i.e., isotropic surface – the surface will be average in all directions and in different shapes with different radii of droplet size ranging from 300 to 700 microns during evaporation and condensation [18]. A strong correlation is observed between the surface morphology and size of the droplet with respect to the increase in the radii of the droplet. As the contact angle decreases, the droplet can easily sink between the pillars. The above problem can be overcome by decreasing the distance between the pillars, which may increase the contact area but is still effective in avoiding the sinking of droplets between the pillars. If there is any external force acting, the above method will not work, and wetting shifting will occur [19,20].

The distance between the pillars alone was not affecting the contact angle of the droplet; the height of the pillar also played a major role [21]. The wetting studies are conducted on micropillar surfaces with different pillar diameters and heights. The result observed was the difference between advancing and receding contact angles, i.e., contact angle hysteresis leads to an increase or decrease depending upon the variation between the pillar pitch and height. The morphology of the substrate and the geometry of the roughness dictate the wetting behavior of the micropatterned surface [22].

Most of the researchers focused on exploring the wetting nature of the flat, textured surface. In practical cases, all the surfaces are not flat. The whole phenomenon will tend to vary with any curvature on the surface. In curved micro-textured surfaces for both concave and convex cases, the sliding angle of both static and dynamic contact angles varies with respect to the change in nature of the curvature [23]. The sliding angle of a liquid droplet leads to an increase concerning the decrease in substrate curvature.

18.3 CORROSION PREVENTION TECHNIQUES

Conventional coatings, even magnetite coatings, show high instability in corrosion after a certain time period [24]. Therefore, there is a need to look for other ways of reducing corrosion, like the creation of hydrophobic surfaces with a high water contact angle and a low sliding angle. Hydrophobic surfaces have better non-wetting characteristics, which make them suitable for a wide range of applications such as self-cleaning, corrosion resistance, anti-icing, anti-biofouling, drag reduction, structural color, and self-cleaning [25].

Superhydrophobic surfaces show better corrosion resistance by decreasing wetting, i.e., reducing the contact area between the substrate and the corrosive medium due to the formation of an air cushion between them [26,27]. The anticorrosive coating on steel mimics the *Xanthosoma sagittifolium* leaf-like structure using electroactive epoxy coating [28]. The epoxy coating forms an anode layer, which reduces the corrosion of the substrate significantly, and it is supported by an air barrier formed between the coating and the environment.

As per the earlier researchers' suggestions, coatings with hydrophobic characteristics may reduce corrosion, which triggered a study of the corrosion resistance property of superhydrophobic coatings. The fabricated superhydrophobic surface is fabricated on titanium by the coupled action of nanocomposite coating with plasma electrolytic oxidation, and the wetting and corrosion behavior of both hydrophobic and superhydrophobic surfaces are compared [29]. The result shows superhydrophobic were better than hydrophobic surfaces in both wetting and corrosion perspectives.

Composite coating has the advantage of dual layer roughness, and its stability is higher when compared to conventional coatings, which trap the air between the roughness asperities, analogous to what happens in valleys [30]. Simultaneously, it will decrease the surface energy of the material, which accelerates the mobility of the liquid droplet sitting on the surface as well as its contact area. Composite coating increases the thickness of the coating when compared to other conventional coatings such as single-layer and bilayer coatings, which increases its durability for prolonged time period [31,32]. Later, the surface was slightly modified through grafting by polymerization instead of composite nanocoatings [33]. Significant improvements in wetting and corrosion resistance properties were noted. The water contact angle achieved on the surface fabricated by the above-mentioned method was greater than 150° in copper. The morphology of the surface and its roughness play a vital role in determining the wetting and corrosion characteristics of the material. Apart from surface roughness, the distance between the roughness asperities also has equal importance in maintaining the corrosion resistance behavior of the material. If the distance between the roughness asperities is minimum, the corrosive rate will be maintained constant at a minimum level.

The effect of roughness with coatings was combined by forming the hydrophobic surface on titanium and steel using etching, followed by anodization and myristic acid dip coating [34]. The contact angle achieved on titanium is 150°, which is similar to composite nanocoatings, but in steel, the observed contact angle was only 109°. The wetting resistance, biofouling, and corrosion resistance of both materials were tremendously increased by implementing hydrophobic surfaces [35].

18.4 SURFACE STRUCTURE PREPARATION

Stainless steel S304 with dimensions of 70 mm×10 mm×0.5mm and 5 mm ×5 mm×0.5mm were used as specimens in the current study. The surface of the specimen was polished by using different grades of emery sheets to remove roughness elements formed on the surface of the specimen during the fabrication process. A high-precision diamond saw cutter was used to cut the specimen for a better finish. Later, the specimens were thoroughly washed with acetone and distilled water to remove the dust particles that had settled on the surface of the specimen.

Chemical etching is the widely preferred top-down fabrication technique for surface modification in small and flat specimens. To make the surface hydrophobic, micropatterns have to be formed on the surface at the micron level, and the surface should be free from any forms of debris that will directly affect the pinning of the three-phase contact line. Hence, extreme care is needed for a better surface finish. Chemical etching is well known for its precision and durability. In chemical etching, there is no need for coolants like oil during fabrication process, which makes the substrate free from contaminants. Moreover, chemical etching is possible even if the thickness of the specimen is in the range of 0.5 mm. A chemical etching process was done on the specimen to create micropatterns on the surface. The steps involved in chemical etching are shown in Figure 18.1.

If the surface of the specimens are thoroughly cleaned with different grades of emery sheets to make sure they are free from debris or any macro-impurities, then only the surface will have a good grip with photoresist. The surfaces are then cleaned with acetone and de-ionized water for 15 minutes to dispose of any natural debasement residue at first glance. The samples are then set in the oven at 150°C for 30 minutes to expel the dampness from the surface.

Photoresist was framed on the steel substrate by methods for spin coating. The essential turning was done at 250°C with a stay time of 30 seconds at a turning velocity of 500 rpm. The optional turning was done at the same temperature with double the abide time and rpm. The holding between the film and the example surface will be great if the sample surface is free from dampness. Subsequently, the sample was

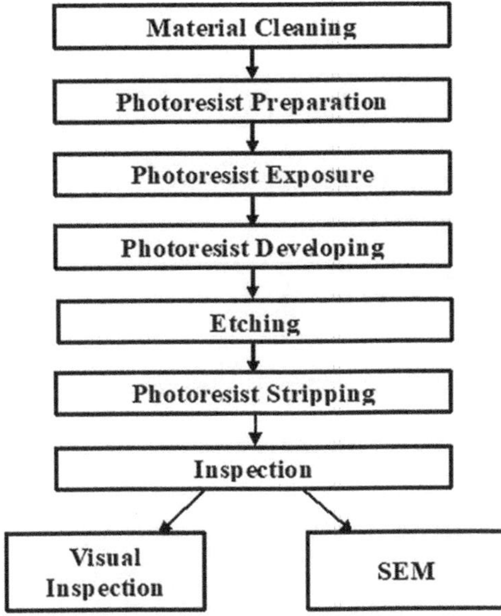

FIGURE 18.1 Schematic sketch of chemical etching process and high velocity of micron-sized.

dried by putting it in the oven for 20 minutes at 900°C. Exposure: The micropatterns were engraved on the photoresist for the desired dimensions using a laser. To eradicate the waviness on the photoresist, the sample was set in the oven for 10 minutes at 100°C.

The excess photoresist on the sample was removed and dipped in warm, refined water until the film solidified. Then the samples were dried in the oven for 1 hour at 120°C. Etching solution, water, and ferric chloride are mixed in equal proportion and kept idle for 2 hours to avoid heating while adding hydrofluoric acid in the further process, which also helps to stabilize the chemical proportion. Hydrofluoric acid is added to the prepared solution in a ratio of 1:10 and kept idle for 5 minutes. SS304 steel with a mask on the surface was immersed in the etching solution. The surface area without a mask undergoes etching and forms a channel on the surface. The remaining area covered with the mask was not exposed to the etching solution, which will act as ridges.

The acetic acid blended with refined water in the proportion of 1:10 is preferred to clean the micropatterned SS304 steel samples to expel the photoresist stick on the surface of ridges and for a better surface finish. The micropatterned surface was outwardly reviewed to ensure it was free from debris. The last review was done by taking the substrate picture using a scanning electron microscope to check the nearness of any imperfections on the micropatterns. Figure 18.2 shows the sample image of the substrate taken using a scanning electron microscope at lower and higher magnification.

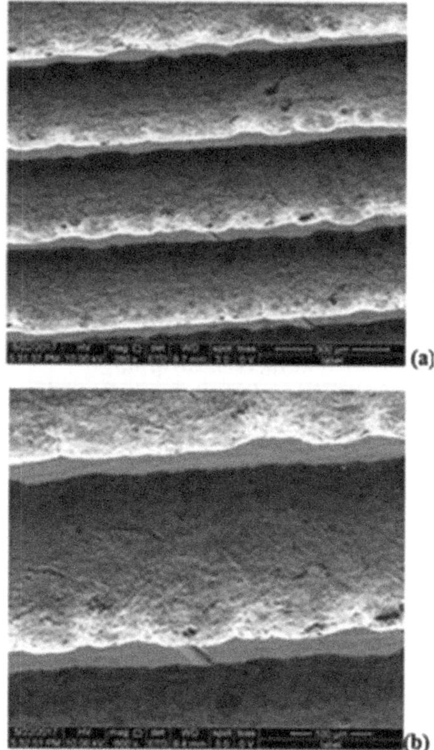

FIGURE 18.2 SEM image of microgroove-patterned surface fabricated on SS304 steel surfaces using chemical etching: (a) low magnification and (b) high magnification.

TABLE 18.1

Geometrical Dimensions of Micropatterns SS304 Steel Samples

S. No.	Sample	Ridge Width (μm)	Channel Width (μm)	Ridge Width/Channel Width
1	MP 1	48.56	245.43	0.19
2	MP 2	53.9	129.45	0.41
3	MP 3	104.27	136.57	0.76
4	MP 4	115.39	139.18	0.82

The geometry of the microgroove patterns measured using a 3D optical profilometer is listed in Table 18.1. The height of the grooves and the depth of the channels were maintained in the range between 41.03 and 41.25 μm. The variation between channel depths is very minimal for all cases, hence the average value of 41.07 μm, which is considered the channel depth and ridge height for all the cases.

18.5   WETTING CHARACTERISTICS EXAMINATION

The wetting conduct of the loaded substrate is determined by the static contact point of the fluid drop. A schematic of the test setup used for static contact edges is given in Figure 18.3. A fluid volume of 10 µl was administered from a 20 ml syringe through a 24-check hypodermic needle to shape a sessile drop of distance across 2.67 mm on the substrate, and the miniaturized scale movement of the syringe cylinder was controlled physically using the droplet generator. At the beginning of every test, the surface would be cleaned first with acetone and then with distilled water. The static contact angle was estimated from the fluid drop picture shaped on the substrate using a Keygence video microscope with a resolution of 768×576 pixels through the backdrop illumination enlightenment of the fluid drop by strobe light (streak span 10–15 µs).

The static contact angle and contact diameter across were estimated using ImageJ software for describing the wetting conduct of the micropatterned surface [36] and recorded in Table 18.2. The contact angle and also the contact distance across were estimated for each surface and, furthermore, for each case through 10–15 isolate

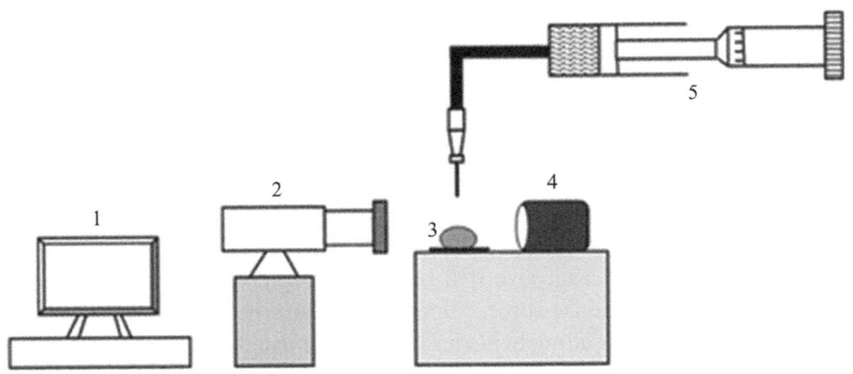

1. Computer, 2. High speed camera, 3. Droplet, 4. LED lamp and 5. Droplet generator

FIGURE 18.3   Schematic sketch of surface wettability measurement setup.

TABLE 18.2
Wetting Parameters of Micropatterns SS304 Steel Samples

S.No.	Sample	Contact Angle θ_{per}	Contact Angle θ_{Par}	Contact Diameter D_{per} (mm)	Contact Diameter D_{Par} (mm)
1	Flat	79		4	
2	MP 1	129	103	1.6	2.7
3	MP 2	143	109	1.2	2.1
4	MP 3	135	108	1.4	2.2
5	MP 4	131	106	1.5	2.4

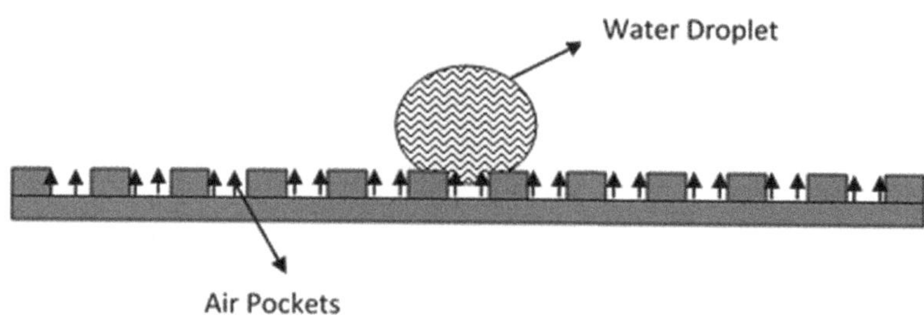

FIGURE 18.4 Schematic sketch illustrates wetting on micropatterned surface.

tests to minimize the standard deviation. The fluid drop shows the distinction in contact angle and, additionally, contact width between the furrow grooves opposite and parallel to the channel.

Adhesion and cohesion manage the marvel of capillarity impact. At the point when a fluid bead is set on the micropatterned surface, the power of the bond between the particles of the micropatterned surface and the fluid bead is lower than the cohesion force between the atoms of the fluid bead. The static contact angle will prompt an increment in the opposite direction of the contact diameter, which prompts a diminishment. The schematic outline in Figure 18.4 clarifies the wetting on the micropatterned surface. The air pockets are framed between the channels of the furrow finished surface, as depicted by Casie Baxter, which controls the sticking conduct of a three-stage contact line between solid, fluid, and air. These air molecules control the wetting nature of the micropatterned surface and make it carry on like a hydrophobic surface.

The wetting on the microgroove will be non-isotropic, while the wetting on the micropillar surface will be isotropic. Microgroove-patterned surfaces follow anisotropic wetting, i.e., the spreading behavior of the water droplet varies from one direction to another. The morphology of the fluid bead changes from perpendicular to the ridges to parallel to the ridges. The schematic drawing of the morphology of water bead variation is shown in Figure 4.9. Perpendicular to the edges, the water bead will not wet the surface due to the ridges in the direction perpendicular to the spreading of water droplets. The sticking of the three-phase contact line is stronger close to the edges. Hence, parallel to the edges, the water bead can easily spread along the channels due to the absence of ridges. Subsequently, the estimated contact angle and contact diameter, both perpendicular and parallel to the ridges, will differ because of variations in bead morphology. The perpendicular to the ridge case will indicate the least wetting when contrasted with the parallel to the ridge for every single geometrical measurement of the ridge and channels because of the sticking of the contact line at the corners of the ridge. The state of the bead spreading accomplished on the microgroove surface additionally looked spherical.

The micropatterned surface takes after Wenzel or Casie Baxter wetting, which will be assessed in the further sections. The wetting resistance of the microgroove-patterned surface was high as long as the surface followed the Casie Baxter model, which implies the water bead wets just the ridges, i.e., the projected region. The surface roughness plays a significant role in determining the wetting properties of the substrate. Therefore, it is necessary to study the wetting properties as a function of roughness. The roughness factor was estimated for both perpendicular and parallel to the groove. For a perpendicular configuration, the overall effective roughness was higher than that in the case of a groove parallel to the groove because of the nearness of the groove as a valley. For the parallel case, roughness was estimated either at the top of the ridges or the top of the channels. The effect of nano-level roughness on a simple flat surface is insignificant, but the same nano-level roughness over a micropatterned surface causes a significant effect on wetting parameters. A similar phenomenon is observed in lotus leaves.

18.6 CORROSION CHARACTERISTICS EXAMINATION

Electrochemical corrosion experiments were conducted for the test samples in a saline water environment with a pH value of 8. An electrochemical workstation (K-Lyte1.2) was used to conduct potentiodynamic polarization experiments. The intensity of corrosion on the samples can be resolved using the TAFEL plot strategy or the polarization resistance technique. The polarization strategy measures the weight reduction in the sample because of corrosion. Henceforth, TAFEL plot strategy was preferred in the current research work. TAFEL plots were generated using the customized software, and the fitting parameters were acquired.

The experimental setup consists of three electrodes, namely, the working electrode, the reference electrode, and the counter electrode. The working electrode is the test sample; the reference electrode is silver filled with saturated KCl solution; and the counter electrode is platinum. The exposed surface area of the test samples was $1\,cm^2$. The test samples are immersed in the saline water for 1 hour before starting the experiment to eradicate the variation in open corrosion potential (E_{corr}). The potential between the reference electrode and the working electrode was fixed at $-0.8\,V$. The potentiodynamic polarization curves were registered at a scan rate of $1\,mV/s$. The polarization resistance (R_p) was calculated from the current vs. voltage curve by taking the slope near the corrosion potential. The corrosion current (I_{corr}) was determined using the Stern-Geary relationship, and the curve fitting parameters were interpreted from TAFEL plots of the test samples. The microgroove surface was considered for electrochemical examinations because of its better protection from wetting. The corrosion rate extracted from the TAFEL plot for the microgroove-patterned surface is listed in Table 18.3.

The widefield WF-10X was used as an eyepiece, with an amplification scope of 100× to 1000×. The measurement of the sample bed was 200 mm × 152 mm, and the C mount was used to set the focal point in the camera. The halogen light (6 V, 20 W) was used as an illuminator, and the diaphragm had three channels. The metallographic image analysis software was used to decipher the picture of the SS304 sample. The sample was thoroughly cleaned with cotton material before beginning the trial to ensure it was free from contaminants other than corrosion. Ethanol or distilled water was not used for cleaning the surface because of the likelihood of removing the

TABLE 18.3
Wetting Parameters of Microgroove
Patterns on SS304 Steel

S. No.	Sample	Corrosion rate (mm/year)
1	Flat Surface	1.39E−2
2	MP 1	10.63E−4
3	MP 2	4.24E−4
4	MP 3	7.38E−4
5	MP 4	9.00E−4

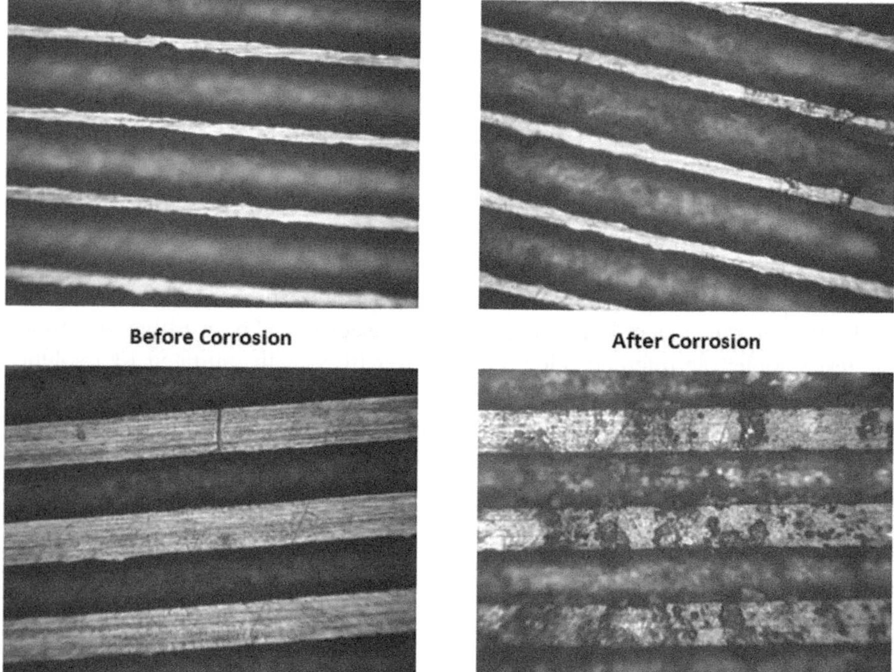

FIGURE 18.5   Micropatterned surface image taken before and after corrosion using optical microscope.

impurities caused by corrosion. Corrosion cannot be quantified through an optical magnifying lens because both the magnification level and scale factor are unknown. However, corrosion can be roughly estimated using ImageJ software by contrasting the obscure pixel length with the known pixel length from the picture of the ruler taken using the same optical magnifying lens. The sample picture of the micropatterned surface taken using an optical microscope is shown in Figure 18.5, which shows the extent of corrosion caused on the surface of the SS304 sample. The dark

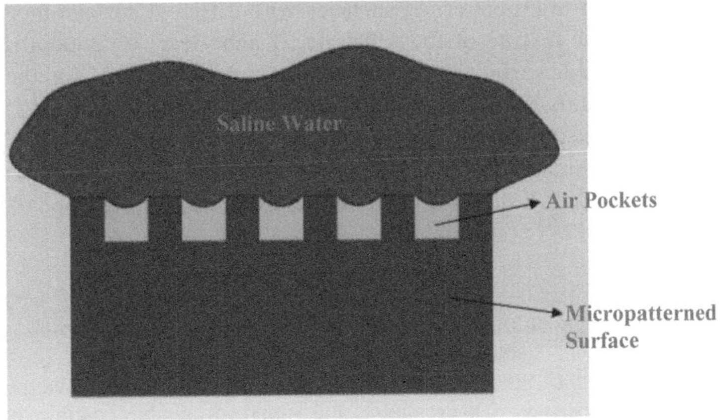

FIGURE 18.6   Corrosion mechanism of microgroove surface.

patches and the unevenness on the ridges were observed in the image taken on the surface of the specimen, which are all defects caused by corrosion.

The micropatterned SS304 steel surfaces diminish corrosion inferable from the decrease in the region of contact with the corrosion fluid media, i.e., lessening in the region of the solid-liquid interface, and the protection offered by air caught in the samples to the passage of corrosive species (Cl⁻), rendering the surfaces impervious to corrosion in NaCl condition. The schematic of the above explanation is shown in Figure 18.6.

18.7   CONCLUSION AND DIRECTION FOR FUTURE RESEARCH

The micropatterned SS304 steel surfaces control corrosion by decreasing the contact region with the corrosion fluid medium, i.e., decreasing the solid-liquid interface. The protection offered is due to the entrapment of air on the surfaces of the samples, which opposes the passage of corrosive elements (Cl⁻) when the surfaces are exposed to corrosion in a saline condition. SS304 steel experiences corrosion within sight of Cl⁻ ions, which confines its applications in saline water conditions. By making its design hydrophobic, corrosion on the steel surfaces can be reduced significantly. The striking highlights of micropatterned surfaces are their ability to decrease corrosion through their inhibition of wetting properties. Micropatterns can be formed on the structures/components during manufacturing itself by methods such as laser, electrochemical processes, machining (e.g., granulating), micromachining (e.g., wire-cut electrical release machining), ultrasonic circular vibration cutting, and so on. These structures/components can withstand stresses induced during their service lives because of their lower corrosion.

Hydrophobic surfaces have potential uses other than corrosion inhibition, such as drag reduction, self-cleaning, auxiliary shading, biofouling, and anti-icing. The drag reduction and self-cleaning surface will fall under the category of dynamic wetting studies such as rolling and sprinkling of the water bead. Dynamic wetting studies can be conducted on micropatterned surfaces with different surface morphologies, even by changing the texture of the substrate. If nano-level roughness is superimposed on a hydrophobic surface, it will become superhydrophobic, and the corrosion rate will be significantly less. However, the durability or shelf life of such surfaces is severely limited. The studies may improve the methods for manufacturing hydrophobic surfaces in bulk and can be economical for creating a double layer of roughness on the metal substrate. All these can be considered for the future scope of research.

REFERENCES

1. T. Young, An essay on the cohesion of fluids, *Philosophical Transaction of the Royal Society of London.* 95 (1805) 65–87.
2. S.V Gnedenkov, S.L. Sinebryukhov, V.S Egorkin, D.V. Mashtalyar, Wetting and electrochemical properties of hydrophobic and superhydrophobic coatings on titanium, *Colloids and Surfaces A: Physico Chemical and Engineering Aspects.* 383 1–3 (2011) 61–66.
3. B. Bhushan, Y.C. Jung, K.K. Micro, nano- and hierarchical structures for superhydrophobicity, self-cleaning and low adhesion, *Philosophical Transactions of Royal Society A Mathematical, Physical and Engineering Services.* 367 (2009),1631–1672.
4. P. Forbes, Self-cleaning materials, *Scientific American.* 299 (2008) 68–75.
5. P. Ivan, R.G. Pagrave, Self-cleaning coatings, *Journal of Material Chemistry.* 15 (2005) 1689–1695.
6. A. Marmur, Wetting on hydrophobic surfaces: to be heterogeneous or no to be?, *Langmuir.* 19 (2003) 8343–8348.
7. C. Huh, S.G. Mason, Effects of surface roughness on wetting, *Journal of Colloid and Interface Science.* 601 (1977) 11–38.
8. K. Narayan Prabhu, P. Fernades & G. Kumar, Effect of substrate roughness on wetting behavior of vegetable oils, *Materials & Design.* 30 2 (2009) 297–305.
9. N. Jeyaprakash, C.-H. Yang, S. Sivasankaran, Laser cladding process of Cobalt and Nickel based hard-micron-layers on 316L-stainless-steel-substrate, *Materials and Manufacturing Processes,* 35 2 (2020) 142–151.
10. N. Jeyaprakash, C.-H. Yang, M. Duraiselvam, G. Prabu, Microstructure and tribological evolution during laser alloying WC-12% Co and Cr_3C_2-25%NiCr powders on nodular iron surface, *Results in Physics.* 12 (2019) 1610–1620.
11. N. Jeyaprakash, M. Duraiselvam, S.V. Aditya, Numerical modelling of WC-12% Co laser alloyed cast iron in high temperature sliding wear condition using response surface methodology, *Surface Review and Letters.* 25 7 (2018) 1950009.
12. N. Jeyaprakash, C.-H. Yang, S.-P. Tseng, Wear tribo-performances of laser cladding colmonoy-6 and stellite-6 micron layers on stainless steel 304 using Yb: YAG disk laser, *Metals and Materials International,* 27 (2021) 1540–1553.
13. N. Jeyaprakash, C.-H. Yang, Comparitive study of NiCrFeMoNb/ FeCrMoVC laser cladding process on nickel-based superalloy, *Materials and Manufacturing Processes,* 35 12 (2020) 1383–1391.
14. J. Bico, U. Thiele, D. Quére, Wetting of textured surfaces, *Colloids and Surfaces A: Physico Chemical and Engineering Aspects.* 2061-3 (2002) 41–46.
15. E. Bormashenko, Progress in understanding wetting transitions on rough surfaces, *Advances in Colloid and Interface Science.* 222 (2015)92-103.
16. S. Manakasettharn, T.-H. Hsu, G. Myhre, S. Pau, A. Taylor, T. Krupenkin, Transparent and superhydrophobic Ta_2O_5 nanostructured thin films, *Optical Materials Express.* 22 (2012) 214–221.

17. Y.T. Cheng, D.E. Rodak, C.A. Wong, C.A. Hayden, Effects of micro- and nano-structures on the self-cleaning behaviour of lotus leaves, *Nanotechnology.* 17 (2006) 1359–1362.
18. B. Bhushan, Y.C. Jung, K. Koch, Micro-, nano- and hierarchical structures for superhydrophobicity, self-cleaning and low adhesion, *Philosophical Transactions of Royal Society A Mathematical, Physical and Engineering Services.* 367 (2009) 1631–1672.
19. M. Callies, D. Quere, On water repellence, *Soft Material.* 1 (2005) 55–61.
20. M. Reyssat, A. Pepin, F. Marty, Y. Chen, D. Quere, Bouncing transitions on microtextured materials, *Europhysics Letters.* 74(2006) 306–312.
21. N. Moronuki, K. Ryu, A. Kaneko, Design of surface for the control of wettability transaction, *Japan Society of Mechanical Engineers Series B.* 70(693) (2004) 1244–1249.
22. N. Izumi, T. Minami, H. Mayama, A. Takata, S. Nakamura, S. Yokojima, K. Tsujii, K. Uchida, Super water-repellent fractal surfaces of a photochromic diarylethene induced by UV light, *Japanese Journal of Applied Physics.* 47(9) (2008) 7298–7302.
23. A.H. Cannon, W.P. King, Hydrophobicity of curved microstructured surfaces, *Journal of Micromechanics and Microengineering.* 201 (2010) 025018.
24. I. Kuznetsov Yu, D.B. Vershok, S.F. Timashev, A.B. Soloveva, P.L. Misurkin, V.A. Timofeeva, S.G. Lakeev, Features of formation of magnetite coatings on low carbon steel in hot nitrate solutions, *Russian Journal of Electrochemistry.* 4610 (2010) 1155–1166.
25. A.M.A. Mohamed, A.M. Abdullah, N.A. Younan,Corrosion behavior of superhydrophobic surfaces: A review, *Arabian Journal of Chemistry.* 8 (2015) 749–765.
26. D. Zhang, L. Wang, H, Qian, X. Li, Superhydrophobic surfaces for corrosion protection: a review of recent progresses and future directions, *Journal of Coatings Technology and Research.* 13 1 (2016) 11–29.
27. S.S. Latthe, C. Terashima, K. Nakata, A. Fujishima, Superhydrophobic surfaces developed by mimicking hierarchical surface morphology of lotus leaf, *Molecules.* 19 (2014) 4256–4283.
28. C.-J. Weng, C.-H. Chang, C.-W. Peng, S.-W. Chen, J.-M. Yeh, C.-L. Hsu, Y. Wei, Advanced anticorrosive coatings prepared from the mimicked Xanthosoma Sagittifolium-leaf-like electroactive epoxy with synergistic effects of superhydrophobicity and redox catalytic capability, *Chemistry of Materials.* 23 (2011) 2075–2083.
29. S.V. Gnedenkov, S.L. Sinebryukhov, V.S. Egorkin, D.V. Mashtalyar, D.A. Alpysbaeva, L.V. Boinovich, Wetting and electrochemical properties of hydrophobic and superhydrophobic coatings on titanium, *Colloids and Surfaces A: Physiochemical and Engineering Aspects.* 383 1 (2011) 61–66.
30. X.H. Xu, Z.Z. Zhang, F. Guo, J. Yang, X.T. Zhu, Fabrication of superhydrophobic binary nanoparticles/PMMA composite coating with reversible switching of adhesion and anticorrosive property, *Applied Surface Science.* 257 (2011) 7054–7060.
31. D. Yu, J. Tian, J. Dai, X. Wang, Corrosion resistance of three-layer superhydrophobic composite coating on carbon steel in sea water, *Electrochemica Acta.* 97 (2013) 409–419.
32. B.D. Mert, Corrosion protection of aluminum by electrochemically synthesized composite organic coating, *Corrosion Science.* 103 (2016) 88–94.
33. S. Yuan, S.O. Pehkonen, B. Liang, Y.P. Ting, K.G. Neoh, E.T. Kang, Superhydrophobic fluoropolymer-modified copper surface via surface graft polymerization for corrosion protection, *Corrosion Science.* 53 (2011) 2738–2747.
34. P.V. Mahalakshmi, S.C. Vanithakumari, J. Gopal, U. Kamachi Mudali, B. Raj, Enhancing corrosion and biofouling resistance through superhydrophobic surface modification, *Current Science.* 10110 (2011) 1328–1336.

35. A. Abdolahi, E. Hamzah, Z. Ibrahim, S. Hashim, Application of environmentally-friendly coatings towards inhibiting the microbially influenced corrosion (MIC) of steel: a review, *Polymer Reviews*. 54 (2014) 702–745.
36. A. Vivek Anand, V. Arumugam, S. Jayalakshmi, R. Arvind Singh, Innovative approach for suppressing corrosion of SS304 steel in saline water environment, *Anti- Corrosion Methods and Materials*. 65(5) (2018) 484–491.

Index